Praise for *The Masters of Medicine*

"This book is magnificent! Dr. Andrew Lam guides us through the audacious history of medical innovation in all its genius, serendipity, hubris, and even psychopathy with the warm-hearted and clear-eyed touch of a true writer. I enjoyed every twist through time of this beguiling reminder of how far we've come in medicine—and how far we still have to go."

—Susannah Cahalan, #1 *New York Times* bestselling author of *Brain on Fire*

"This is a book filled with extraordinary tales of extraordinary personalities who made history—a history that has created the world we live in. It's accurate, clear sighted, well written, and where necessary it cuts through myth to get to what really happened. In truth, no myths are necessary—the reality is fascinating enough."

—John Barry, #1 *New York Times* bestselling author of *The Great Influenza*

"Andrew Lam's *The Masters of Medicine* is a fascinating book that describes some of medicine's greatest discoveries. Lam himself is a master surgeon, storyteller, and writer, and this combination produces a book that is both a pleasure to read and rich with information. *The Masters of Medicine* is a must-read, and Lam is a stellar guide to the breakthroughs that have defined and continue to shape modern medicine."

—Adam Alter, Ph.D., professor of marketing and psychology, New York University Stern School of Business, and *New York Times* bestselling author of *Drunk Tank Pink* and *Irresistible*

"*The Masters of Medicine* is a skillful and engaging portrayal of the journeys through some of the pivotal accomplishments in medicine. Lam vividly describes the very human exploits, failures, and tenacity of clinicians who found solutions to some of nature's most perplexing problems. Andrew Lam has furnished a major contribution to the history of medicine."

—Thomas Helling, M.D., professor of surgery, University of Mississippi School of Medicine, and author of *The Great War and the Birth of Modern Medicine*

"This wonderfully absorbing book describes the efforts of individuals that led to key medical advances, facilitated by their skill, knowledge, determination, courage and, quite often, sheer self-confidence coupled to luck. All of which only goes to show, if we ever doubted it, that scientists and clinicians are human. The reader feels as though he is in the same room as these diligent trailblazers. You are almost looking over Dwight Harken's shoulder as he plugs shrapnel-damaged hearts with his finger while stitching up the holes. It's a thoroughly ripping yarn that should be told to all. Remember: the next patient could be you."

—Robin Hesketh, Ph.D., Department of Biochemistry, University of Cambridge, and author of *Betrayed by Nature: The War on Cancer* and *Understanding Cancer*

"An assemblage of many great triumphs in the human effort to achieve better health, including stunning narratives of pain, disappointment, luck, and—above all—the hard work that it took to succeed."

—Robert Bazell, Ph.D., past chief science and health correspondent for NBC News, and author of *Her-2: The Making of Herceptin, a Revolutionary Treatment for Breast Cancer*

"A fascinating book! Andrew Lam's narrative will pull you in. *The Masters of Medicine* is a character-driven book that takes you inside the minds of some of the most innovative thinkers on the planet, and along the way you learn a great deal more about your own body and how it functions!"

—Michael Tougias, *New York Times* bestselling coauthor of *The Finest Hours* and author of *Extreme Survival*

"Dr. Andrew Lam skillfully dissects some of the most seminal discoveries in the fight to combat humanity's deadliest diseases. I highly recommend this fascinating book to anyone interested in the history of medicine."

—Sandeep Jauhar, M.D., Ph.D., *New York Times* bestselling author of *Intern*, *Heart: A History*, and *My Father's Brain: Life in the Shadow of Alzheimer's*

"This is a gripping and fascinating book by someone who knows what they are writing about, and writes with flair. Dr. Lam takes you through the strokes of genius—and strokes of absolute luck—that brought us some of the world's most important lifesaving treatments. A delight."

—Charles Kenny, author of *The Plague Cycle: The Unending War Between Humanity and Infectious Disease*

"*The Masters of Medicine* recounts the struggles, successes, disasters, and staggering nonlinear path of medical discovery. With an engaging narrative style, Andrew Lam presents unforgettable tales of life-changing medical breakthroughs and the courageous physicians who devoted their lives to saving mankind from disease. Readers will find *The Masters of Medicine* riveting, and it will likely excite young people to consider a medical research career just as Paul de Kruif's *Microbe Hunters* did decades ago."

—Charlotte Jacobs, M.D., professor of medicine (emerita),
Stanford University, and author of *Jonas Salk: A Life*

"A comprehensive and fun overview of key discoveries that have improved public health. Andrew Lam conveys the personalities of the key players, both famous and overlooked, who contributed to the medical breakthroughs that benefit us all. Rather than placing famous scientists on a pedestal, Dr. Lam shows that not all these characters were necessarily likable and that some had foibles that both helped and hindered their work."

—Michael Kinch, Ph.D., author of *The End of the Beginning: Cancer, Immunity, and the Future of a Cure*

"Andrew Lam is the perfect combination of physician, historian, and storyteller. *The Masters of Medicine* vividly captures the mix of well-planned experiments, serendipitous findings, and eureka moments leading to some of the most important breakthroughs of modern medicine."

—Gregg Semenza, M.D., Ph.D., professor of genetic medicine,
Johns Hopkins University School of Medicine,
and winner of the Nobel Prize in Physiology or Medicine

The
MASTERS
of
MEDICINE

Also by Andrew Lam, M.D.

Nonfiction
*Saving Sight: An Eye Surgeon's Look at Life Behind the Mask
and the Heroes Who Changed the Way We See*

Fiction
Two Sons of China
Repentance

The MASTERS of MEDICINE

*Our Greatest Triumphs in the Race
to Cure Humanity's Deadliest Diseases*

ANDREW LAM, M.D.

BenBella Books, Inc.
Dallas, TX

The Masters of Medicine copyright © 2023 by Andrew Lam

All rights reserved. No part of this book may be used or reproduced in any manner whatsoever without written permission of the publisher, except in the case of brief quotations embodied in critical articles or reviews.

BenBella Books, Inc.
10440 N. Central Expressway
Suite 800
Dallas, TX 75231
benbellabooks.com
Send feedback to feedback@benbellabooks.com

BenBella is a federally registered trademark.

Printed in the United States of America
10 9 8 7 6 5 4 3 2 1

Library of Congress Control Number: 2022040441
ISBN 9781637742631 (hardcover)
ISBN 9781637742648 (electronic)

Editing by Vy Tran
Copyediting by Leah Baxter
Proofreading by Isabelle Rubio and Lisa Story
Indexing by Debra Bowman
Text design and composition by Jordan Koluch
Cover design by Kara Klontz
Cover image © Shutterstock / Morphart Creation (lungs), Yevheniia Lytvynovych (virus), and vectortatu (heart); icons made by Freepik (Facebook), vidyavidz (Twitter), and Laisa Islam Ani (Instagram) from www.flaticon.com
Printed by Lake Book Manufacturing

Special discounts for bulk sales are available.
Please contact bulkorders@benbellabooks.com.

For Bernard and Alice Chang

Contents

INTRODUCTION: *To Wrest from Nature*	1
1. HEART DISEASE: *The Mavericks*	6
2. DIABETES: *The Pissing Evil*	46
3. BACTERIAL INFECTION: *The Magic Bullet*	75
4. VIRAL INFECTION: *Pandemic*	126
5. CANCER: *A Bewilderingly Complex Array*	169
6. TRAUMA: *The Only Winner in War Is Medicine*	214
7. CHILDBIRTH: *The Mysterious Killer*	254
CONCLUSION: *The Masters of Medicine*	299
Acknowledgments	309
Notes	312
Selected Bibliography	349
Image Credits	354
Index	356

INTRODUCTION

To Wrest from Nature

> To wrest from nature the secrets which have perplexed philosophers of all ages, to track to their sources the causes of disease, to correlate the vast stores of knowledge, that they may be quickly available for the prevention and cure of disease—*these are our ambitions.*
>
> —William Osler, M.D.

By the end of this century we will have found a cure for cancer. We will use stem cells to regrow nervous tissue and repair damaged spinal cords. Our knowledge of each individual's genetic makeup will allow us to diagnose, and preemptively treat, many diseases before they become manifest. The average life span will exceed ninety years, and reaching centenarian status will be commonplace.

Most, if not all of us, believe that attainment of these ambitious goals is not only possible, but probable. We should be forgiven our hubris, for we have been conditioned to believe in mankind's almost limitless potential. Just think of what we have had the good fortune to witness in our lifetimes—an extraordinary, exponential increase in scientific knowledge that dwarfed the modest achievements of prior centuries. Consider the life of just one person, my great-grandfather who was born in 1893 and died in 1977. In his youth, horses remained a primary mode of transportation. The airplane had not been invented. No one in his family had ever used a telephone or seen an electric light bulb. But by the time he died, man had landed on the moon. Since then, our world has continued to evolve in innumerable ways, benefiting from such varied advances as the Internet and the deciphering of the human genome. We have also beheld a quantum leap in medical knowledge and treatment that would be unfathomable to our ancestors.

At the turn of the twentieth century, hundreds of thousands died annually from infectious diseases like smallpox, tuberculosis, and cholera. The germ theory of disease was just beginning to gain widespread acceptance over prevailing nineteenth-century theories that blamed "bad air" or "miasmas." Legitimate medical doctors believed harmful treatments like radium water could cure a plethora of ailments including arthritis, impotence, and anemia. Others used mercury to treat syphilis. Quackery and snake oil salesmen abounded. The average life expectancy, worldwide, was only thirty-three years (forty-seven in developed countries).

Then several changes occurred. Medical education became standardized. Improvements in sanitation, the introduction of vaccines and antibiotics, and advances in surgery and anesthesia had dramatic, beneficial effects. To a physician of the late nineteenth century, the idea that man might one day tame fatal infectious diseases, use a patient's own immune system to cure cancer, and operate on the human heart would be just as unbelievable as the idea of landing on the moon. If life expectancy can double, so that today average worldwide longevity has reached 73.4 years (77.8 in the United States), is it so far-fetched to envision a future in which ordinary people will live to 100? Or even 110?

Historian and philosopher Yuval Noah Harari, author of the international bestselling books *Sapiens* and *Homo Deus*, goes even further, contemplating a day when mankind may even conquer death due to disease entirely. Pointing to advances in genetic engineering, regenerative medicine, and nan-

INTRODUCTION: *To Wrest from Nature*

otechnology, he notes that some futurists anticipate the achievement of a form of human "immortality" by 2100 or 2200, wherein diseases have been conquered and aging tissues will be regenerated, upgraded, or transplanted. This view of the future sees no limit to science's ability to prolong life and postpone death from physiological causes; in such a world, death might only occur due to war, accidents, and homicide.

The future is always uncertain, but it is clear that we are living in the midst of a remarkable continuum of technological progress. It's a journey in which our recent past informs our view of a potentially limitless future. And yet, despite our penchant for optimism, a close study of history yields this crucial lesson: progress is not inevitable, nor inexorable.

In 1961, my father was a teenager who watched the first American, Alan Shepard, launched into space on a trip that lasted only fifteen minutes and twenty-eight seconds. Eight short years later, Neil Armstrong and Buzz Aldrin were walking on the lunar surface. If you asked my father in 1969 what he thought the future held for space exploration, these astonishingly rapid scientific advances might have prompted him and many others to enthusiastically predict manned flights to Mars within a couple more decades, perhaps by the year 2000 at the latest.

What we assume or expect to happen will not always come to pass. What seems assured or even everlasting is not. Nothing is certain. The Western Roman Empire encompassed five centuries of remarkable technological progress in Europe, yet receded from world history and was followed by the Dark Ages—a greater number of centuries marked by regression and ignorance.

Perpetual medical progress is no more assured than any other aspect of human history. Our greatest medical accomplishments—the discoveries that saved millions of lives and alleviated untold suffering—were not preordained, nor inevitable. Far from it. More often than not, these crucial advances depended on just one, or a handful, of individuals—fallible people who, through perseverance, skill, and sometimes dumb luck, took risks or made key observations that no one else had previously been willing or able to. Such physicians and scientists were often ridiculed and even ostracized by their peers, especially if their newfangled ideas threatened medical elites of their day. They dared to confront problems others assumed were unsolvable, and sought answers to questions no one else in history had thought to ask.

And yet, none of these innovators were martyrs or saints. Their brilliance was often marred by jealousy, pettiness, and astounding arrogance. Their remarkable stories illustrate that medical progress is not linear. It advances in fits and starts, and sometimes even moves backward. An examination of medical history reveals that the most crucial discoveries—ones that have proved more consequential than any atomic bomb or world war—hinged upon fleeting moments of breathtaking risk, mundane observation, or serendipitous error.

This is a book about those moments. This is a book about those people—heroes, most of them unsung, whose discoveries were milestones in our perpetual quest to heal the sick, mitigate suffering, and delay death. It is impossible to understate the impact of these singular moments, for they relate to diseases that will afflict, and kill, us all. As such, this narrative of mankind's greatest modern medical achievements is organized by disease—the world's most common afflictions, and most important killers.

Heart disease, the world's number one predator, rightly sits atop any list of most crucial disorders for laypeople to understand.

Diabetes, an epidemic instigated by the tragic irony that for the first time in human history, people are dying from eating too much instead of too little.

Infectious disease, that tireless enemy against which we have celebrated many successes, but that continues to threaten global health like no other condition can.

Cancer, our loathsome nemesis that has touched every family and strikes down young as well as old.

Trauma, the ubiquitous infirmity that will surely persist even after all other diseases are someday defeated.

And the "affliction" of childbirth, a top killer of mothers and children—until doctors uncovered the secrets that would finally bring the practice of

INTRODUCTION: *To Wrest from Nature*

obstetrics kicking and screaming into a new age of improved health and better outcomes.

This is not a comprehensive history of medicine. Instead, it is a tale of human audacity and courage that cannot fail to leave one with the astonishing impression that, save for one person, fortuitous observation, or sometimes, mistake, life for all of us could be drastically different today.

In the process, we will investigate the pathophysiology of diseases that have touched all our lives, and the lives of our loved ones. An understanding of these conditions should not and cannot be reserved for the medical school graduate. Just as any parent would encourage a son or daughter to understand how a car runs, or the basics of the Internet, we should all know how our most common maladies affect us. To understand the human body and what can malfunction within it is not mere intellectual exercise; it is practical knowledge that will help us each recognize dangerous symptoms of disease, live healthier lives, and better understand our own conditions when we are the ones sitting across from the doctor. Each of us will one day be that person, and most likely, the news will not be pleasant.

Yet, now more than ever we are in position to wrest from nature its secrets and use this knowledge to combat disease in ways inconceivable to our predecessors. In this journey we will also look to the future and, with luck, inspire the young to take up the mantle of discovery so that the next generation will succeed in achieving breakthroughs that transform our expectations of health and longevity. Most people born today will enjoy lifetimes that reach into the twenty-second century. We hope that they will look back with astonishment at how much has been accomplished between now and then. We pray that the scourges that kill and perplex us today will become mere footnotes in tomorrow's books of history and science, and that our legacy of progress continues unabated.

These are our ambitions.

1

HEART DISEASE

The Mavericks

As a young adult, Richard was a three-pack-a-day smoker with a stressful job. One night during the summer of 1978, when he was just thirty-seven years old, Richard awoke with a tingling sensation in two fingers of his left hand. He decided to go to the emergency room, where he passed out shortly after arriving. When he woke up, a doctor informed Richard that he'd had a heart attack.

This was a shock. Richard had always considered himself athletic and very healthy. His physician prescribed bed rest, and Richard remained in the hospital for eleven days.

After this alarming incident, Richard stopped smoking, exercised more, and lost weight. He underwent a relatively new procedure called cardiac catheterization, which permitted imaging of his coronary arteries. This showed a 50 percent narrowing of the right coronary artery (RCA) and a 75 percent blockage of the circumflex branch of the left anterior descending artery (LAD). No further treatment was recommended.

Richard went on with his life, working in a career that he loved but that seemed to grow increasingly stressful with each passing season. Even so, he thrived under the pressure, and his mentality reflected that of the young—he

HEART DISEASE: *The Mavericks*

was invincible, too youthful to have a serious problem or disease that should limit his activities, career, or life. He began to view his heart attack as a one-off—a problem he'd solved through diet and exercise.

Then, six years later, Richard felt a mild discomfort in his chest and throat. He went to the hospital for an evaluation; his electrocardiogram (EKG) was normal, but cardiac enzyme tests showed he'd had a second, mild heart attack. Four years after that, he awoke early one morning with chest pain. This was yet another heart attack. Catheterization revealed a clot in his RCA, and an echocardiogram showed he had lost 30 percent of his normal heart function.

By now, Richard understood that he had a very serious disease that was likely to kill him one day, possibly much sooner than he'd ever imagined. He was only forty-seven, with a wife and two wonderful daughters, and he couldn't imagine not being there to take care of them. Richard also didn't want to slow down his career, and he hated living in constant fear of another heart attack. So, when his cardiologist suggested he undergo quadruple coronary artery bypass surgery to improve his exercise tolerance and hopefully prolong his life, Richard agreed. Thankfully, the surgery went well and Richard was discharged from the hospital seven days after the operation.

Shortly thereafter, Richard's medical condition began to draw a lot of attention from people outside of his immediate family and even from the press. This was because President George H. W. Bush nominated him for a new job: United States Secretary of Defense.

Richard "Dick" Cheney had previously served as Gerald Ford's chief of staff, at the age of thirty-four. His first heart attack had occurred during his first congressional race in Wyoming, which he won. At his Senate confirmation hearing for Secretary of Defense in 1989, his cardiologist, Dr. Allan Ross, testified that Cheney had "no functional limitations whatsoever, and a prognosis not substantially different from men of the same age group without such a previous cardiologic history . . . I see no medical reason for him not to perform well in the highest and most sensitive of public offices."

But Dick Cheney's medical odyssey was only just beginning, and his was destined to be the story of a man who fortuitously benefited from almost every major cardiac advance of the last half century—advances that sometimes seemed to hit the scene just in time to save his damaged heart.

After Cheney's bypass surgery, he successively served as Secretary of Defense, CEO of the Halliburton Company, and George W. Bush's vice-presidential running mate. On November 22, 2000, in the midst of the post-election legal battle to determine the winner of the Bush/Gore election, Cheney again experienced nighttime chest pain—his fourth heart attack. Catheterization revealed a new, severe narrowing of a branch of the LAD called the diagonal coronary artery. This was causing significant ischemia, an inadequate supply of blood to the heart. To treat it, Cheney's cardiologist performed an angioplasty procedure using a catheter to place a stent and reestablish normal blood flow.

Four months later, Cheney had chest pain again. Though this was not determined to be a heart attack, he underwent another catheterization and a balloon angioplasty procedure to open up a narrowing that had formed in his stent. Upon further evaluation with a Holter monitor, which continuously recorded his heartbeat over a couple of days, Cheney was found to have episodes of very fast heart rate—termed *ventricular tachycardia*. Ventricular tachycardia puts patients at risk for more severe arrhythmias like *ventricular fibrillation*, so to reduce this risk, Cheney had an implantable cardioverter-defibrillator (ICD) inserted in his body. This electronic device is about the size of a small pager and resides under the skin of the upper chest. It continually monitors a patient's heartbeat and can overcome abnormal heart rhythms by acting as a pacemaker or by delivering an electric shock.

On December 9, 2009, after Cheney's last term as vice president ended and he had returned to civilian life in Wyoming, he was backing his car out of his garage when he blacked out. His car crashed into a large boulder in his front yard. Secret Service agents rushed to the vehicle and pounded on the windows because the car doors had automatically locked. Cheney regained consciousness, unaware of what had happened but with a bruised forehead from where it had hit the steering wheel.

At the hospital, data from his ICD revealed he had gone into spontaneous ventricular fibrillation with a heart rate of 222 beats per minute at 3:11 p.m. Upon detecting this irregularity, the ICD charged its capacitor and discharged 34.5 joules of energy to his heart to restore normal heart rhythm. The episode had lasted sixteen seconds. The ICD saved Cheney's life.

But by this point, Cheney's heart had suffered so much damage that he began to experience classic symptoms of end-stage heart failure. His heart's inability to pump enough blood through his body led to shortness of breath due to fluid buildup in his lungs. He experienced fatigue and leg swelling. He could not climb stairs or walk down the driveway to retrieve the morning newspaper. He felt so exhausted that he could barely get out of bed. Because he used blood thinners to prevent clot formation in his heart and legs, he was prone to severe nosebleeds. Sometimes these nosebleeds were so bad that he had to be hospitalized and receive blood transfusions. Meanwhile, his liver and kidneys began to fail from inadequate blood perfusion.

Dick Cheney was dying.

He had lived a remarkable life. He was proud of what he'd accomplished and felt at peace with the reality of dying. But he and his family were not yet ready to give up. To prolong his life and hopefully keep him alive long enough to receive a heart transplant, Cheney underwent surgery to implant a left ventricular assist device (LVAD). This remarkable piece of machinery is attached to the left ventricle at one end and the aorta at the other. Inside, a pump speeds at 9,000 revolutions per minute to propel blood from the heart to the aorta and the rest of the body, effectively replacing the function of the left ventricle. The pump is connected to a driveline that goes through the patient's chest wall and is attached to batteries that the patient wears in a vest. The LVAD delivers a smooth, continuous flow of blood to the body that is not dependent on the heart's pumping ability. Because of this, if one were to check an LVAD patient's wrist, no pulse would be felt.

Now the most important objects in Cheney's world became his batteries. His life was dependent on them, and he went everywhere with a bag or rolling suitcase full of extra batteries and spare components. Or he tried to. Once, while running errands in town, his batteries started beeping, indicating they had only ten minutes of power remaining. To Cheney's dismay, he realized he had forgotten his spare batteries at home, some twenty minutes away. He immediately called his daughter Mary, who happened to be visiting his home, and told her to grab the batteries and meet him at a halfway point between them. They made it just in time.

Dick Cheney did not receive any special consideration on the heart transplant waitlist. He spent twenty months waiting for a new heart (almost double the average wait time), steadily moving up a list that is dependent not only on one's place in line, but also blood type, body size, and degree of medical need—the sickest patients move to the top of the list.

On March 23, 2012, Cheney got a nighttime call to tell him a heart was available. The surgery was successfully performed at Inova Fairfax Hospital in Virginia, giving Cheney a new chance at many more years of life with his family. His unique medical journey had progressed from his first heart attack in 1978, when his primary treatment had been bed rest, through the development of numerous medical advances that benefited him—including cardiac catheterization, angioplasty, coronary artery bypass surgery, and heart transplant. In the span of just over thirty years, treatment for heart attack victims had dramatically improved. These advances were propelled by the courageous work of medical mavericks: physicians and surgeons who aspired to solve the mystery of the world's number one killer—heart disease.

A SILENT, MYSTERIOUS ADVERSARY

At the beginning of the last century, it was not uncommon for middle-aged people to just drop dead. They would look fine one second, then suddenly collapse and expire within a few minutes. Physicians had no grasp of what was happening inside these unlucky individuals. Astute practitioners correctly surmised the problem somehow lay with a failure of the heart—especially since victims often suffered crushing chest pain and grew short of breath in the agony of their final moments—but doctors were at a loss to explain the pathophysiology of the condition.

For millennia, the workings of the heart itself had likewise been shrouded in mystery. The organ was inaccessible, hidden within the center of the chest, guarded by a wall of ribs, and impossible to examine in life. Truthfully, uncovering the heart's mysteries was not a pressing concern to practitioners of centuries past. Infectious diseases were far more deadly, and most people simply did not live long enough to develop heart disease—in the United States, average life expectancy in 1900 was only forty-seven years. By 1930, however, average American longevity had increased to sixty years, and heart disease became the leading cause of death. Today, heart disease is responsible for about 25 percent of all deaths in the U.S., approximately 659,000 per year—greater than mortality due to all types of cancer combined.

The heart is an incredibly effective pump that pushes 6,000 liters of blood through the body every day. In doing so, it fulfills the essential job of delivering oxygen and nutrients to all the body's tissues. Under normal conditions, deoxygenated blood returning though the venous system to the four-chambered heart enters the right atrium and passes through the tricuspid valve into the right ventricle. From there, blood temporarily exits the heart via the pulmonary artery, which bifurcates, sending a branch into each lung. These branches arborize into intricate webs of increasingly smaller vessels, progressing from arteries to arterioles to capillaries so small that individual red blood cells must pass through them single file. These capillary walls surround and fuse with the lungs' alveoli—equally tiny, grape-like clusters at the termini of the bronchial tree. In millions of alveoli, during every breath of life, inhaled air undergoes gas exchange—oxygen binds to the hemoglobin protein residing inside each red blood cell, which simultaneously unloads carbon dioxide that is then exhaled.

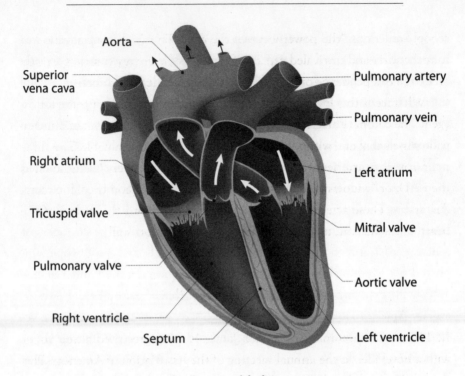

Anatomy of the heart

Like the myriad tributaries of a mighty river, tiny capillaries grow in diameter, join with other vessels, and ultimately spill into the sizable pulmonary veins that return newly oxygenated blood to the left atrium of the heart. From the left atrium, blood passes through the mitral valve into the left ventricle, a larger cavity with muscular walls well suited to propel blood up into the aorta and out to the farthest reaches of the body and its extremities.

It may be more accurate to call this remarkable pump an extremely reliable muscle, one run by an enduring battery capable of lasting an entire lifetime. This battery is the sinoatrial node, a small nexus of specialized cells in the wall of the right atrium that spontaneously generates electrical pulses to act as the heart's pacemaker. Cardiac cells that respond to the demands of the sinoatrial node are capable of normal function at 60 to 100 contractions per minute and much more as the need arises. Their key distinguishing feature is a remarkable abundance of mitochondria—known to every high school student as "the powerhouse of the cell." This high concentration of mitochondria and unrivaled capacity for contractile energy comes at a price: an extreme dependence on *aerobic*, or oxygen-dependent, metabolism.

What feeds this life-giving muscle-pump with a voracious appetite for oxygen? The coronary arteries: three five-to-ten-centimeter-long, four-millimeter-wide vessels that are, without a doubt, humankind's most valuable. The three main coronary arteries are the left anterior descending artery (also known as the LAD, or "widow-maker"), the right coronary artery, and the left circumflex artery. These arteries stem from the aorta, run along the surface of the heart, and send tiny, nourishing branches into the heart wall.

THE MAVERICK

In 1912, a Chicago internist named James Herrick presented a new paper with a novel idea at the annual meeting of the Association of American Physicians. He postulated that sudden cardiac death occurred when a thrombus, or clot, blocked a coronary artery and prevented oxygen-rich blood from reaching cardiac muscle, resulting in tissue death. To Herrick's chagrin, his presentation elicited little reaction from the audience.

"It fell like a dud," he later wrote. No one recognized it as the groundbreaking (and correct) theory that it was. It was not until 1918—when Herrick reintroduced his idea with results from experiments performed on the coronary arteries of dogs—that his peers began to take notice. The perplexing condition that could fell young men and women in the prime of their lives was termed a *myocardial infarction*. For the first time, doctors began to understand the importance of the coronary arteries and their complicity in ischemic heart disease. But by what means could these tiny, constantly moving vessels be studied and understood?

In 1929, a twenty-four-year-old German medical intern named Werner Forssmann theorized that a catheter could be threaded up into the heart though a vein in the arm. If successful, he thought this might be an effective way of injecting drugs directly into the right atrium. When he proposed this radical idea to the chief of his department, Dr. Richard Schneider, Schneider immediately replied, "I cannot possibly let you carry out such an experiment on a patient."

To intentionally introduce a foreign object like a catheter into the heart was mad—conventional wisdom held that doing so would precipitate an arrhythmia or air embolism that could kill the patient.

Undeterred, the impetuous Forssmann then suggested, "Well, in that case there's another way to prove it's not dangerous. I'll experiment on myself."

Shocked, Schneider insisted, "My no is final and absolute." The mere suggestion of such a dangerous and irresponsible idea convinced him that the impulsive young man standing before him lacked good sense.

Forssmann pretended to accept his boss's judgment, but in secret, he hatched a plot. "I decided to override Schneider's prohibition and go ahead with the experiment on my own heart, secretly and quickly," Forssmann admitted in his 1972 autobiography.

He first needed to obtain a catheter and the sterile instruments he'd use to perform the procedure, but they were kept in a locked storage area. The custodian of the keys was a nurse named Gerda Ditzen. How to gain her cooperation? Forssmann, the handsome, rakish doctor-in-training, decided his best option was to beguile the young woman.

"I let a few days go by and then started to prowl around Nurse Gerda Ditzen like a sweet-toothed cat around the cream jug," wrote Forssmann. "It was easy to find something to gossip about; and she'd invite me back to her little office . . . So, little by little, I won over my essential disciple." Within a couple weeks Ditzen had grown so enamored with Forssmann that he decided the time was right to put his plan into action. "The following afternoon the good lady was sitting in her cubicle when I breezed in, whistling cheerfully," Forssmann recalled.

"Nurse Gerda," he said, "I want you to give me a set of instruments for a venesection under local anesthesia, and a ureteral catheter."

The nurse was suspicious. "But no one in the ward's scheduled for a venesection. You're not planning to do that experiment of yours against boss's orders, are you?" she asked.

Forssmann had been found out, even before he'd begun. But instead of giving up, he doubled down.

"Nurse Gerda, you need know nothing about what I am going to do. But supposing I *were* to do the experiment—it'd be quite safe."

"Are you absolutely sure there's no danger?" the nurse said.

"Absolutely," Forssmann answered.

Ditzen considered this and eyed Forssmann closely. "All right then, do it to me," she said. "I put myself in your hands."

This turn of events must have taken Forssmann by surprise. In Ditzen he had cultivated a willing and devoted accomplice who was now prepared to risk all for him and to advance medical science. Forssmann accepted her offer. Ditzen took him to a locked cabinet and removed a scalpel, sterile drapes, anesthetic, and a long, thin urinary catheter intended for use in the bladder. Ditzen then bravely, and perhaps a little dramatically, lay herself down on the operating table, placing herself at Forssmann's mercy. Forssmann began to work quickly, strapping down her arms and legs, which he explained as precautions against her falling off the table.

But Forssmann did not actually intend to exploit the young nurse. It was all a charade, necessary because the nurse may not have consented to help Forssmann go through with the experiment on himself. According to Forssmann:

> I'd pushed the instrument tray behind her head so she couldn't see what I was doing. In the twinkling of an eye I had anesthetized my left elbow . . . I quickly made an incision in my skin, inserted a Dechamps aneurism [sic] needle under the vein, opened it and pushed the catheter about a foot inside. I packed it with gauze and laid a sterile split over it. Then I released Nurse Gerda's right hand and loosened the straps round her knees.

Ditzen, shocked and dismayed, scrambled to her feet and, to her horror, watched Forssmann push the catheter further inside his arm. She yelled at Forssmann, angry at having been deceived, but Forssmann remained calm and directed her to accompany him to the X-ray suite in the hospital's basement. There, Forssmann's friend and colleague Dr. Peter Romeis shouted at Forssmann to stop what seemed sure to be a fatal mistake. Romeis attempted to grab the end of the catheter to pull it out. "I had to give him a few kicks in the shin to calm him down," Forssmann recalled.

The X-ray machine was turned on. Forssmann saw the tip of the catheter had reached the head of the humerus. He then pushed the catheter further in, almost to the two-foot mark, and told the X-ray technician to take a picture. The X-ray showed the tip of the catheter in the right atrium of the heart.

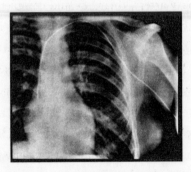

X-ray taken of Werner Forssmann's chest showing the catheter he threaded from his left arm into the right side of his heart

To the surprise and awe of all present, Forssmann appeared completely unruffled. Nothing had changed. There was no chest pain. No fatal arrhythmia. No fainting. For the first time, a doctor had shown that the inviolate inner sanctum of the heart could be plumbed without killing the patient. Now, using catheters, medicines might be delivered directly into the heart, imaging dyes could display blood flow through the heart, and cardiac abnormalities could be visualized in life. Forssmann catheterized himself eight more times over the course of the next two years, injecting radio-opaque dye into his heart without ill effect. In 1956, Werner Forssmann shared a Nobel Prize for his audacious work.*

"WE'VE KILLED HIM!"

Forssmann's breakthrough would eventually change medicine forever, but access to the heart alone did not diminish the scourge of heart attacks that were and remain the leading cause of death worldwide. Combating this silent killer would require somehow visualizing and accessing the heart-nourishing coronary arteries. But how? Conventional wisdom held that any catheter entering one of these tiny coronary vessels would block the flow of blood and therefore result in an immediate myocardial infarction. Moreover, if contrast dye were injected into a coronary artery, then it would completely fill the vessel, prohibit oxygen delivery by normal blood flow, and likely also cause death.

On October 30, 1958, about thirty years after Forssmann's self-experimentation, Dr. Mason Sones, Cleveland Clinic's larger-than-life chief of pediatric cardiology, was catheterizing a twenty-six-year-old man with a history of rheumatic heart disease. To visualize the aorta just above the heart, Sones intended to introduce a large amount of contrast dye (30 ml) quickly over just a couple seconds using an automatic, mechanical "power injector." When his catheter tip was positioned right where he wanted it, Sones brusquely gave the order to "fire!" and inject the dye. But the initial surge of dye made the

---

* In 1932, a few years after Forssmann's successful self-operation, he became a member of the Nazi Party. During WWII, he was a medical officer serving on the Eastern Front. Preferring to be captured by Americans instead of Russians near the war's end, he left his post and swam across the Elbe River under fire from an SS guard. Forssmann was put in an American POW camp and released in 1945. He subsequently worked as a lumberjack, since prior Nazi Party members were prohibited from practicing medicine for a number of years. Later, he was able to resume his work as a urologist in West Germany. Upon learning he would share the Nobel Prize, he told a newspaperman that he felt like a village pastor who had suddenly been informed he'd been made a cardinal.

tip of the catheter whip about wildly like the end of a fire hose and, to Sones's horror, the fluoroscope image showed the tip of the catheter lodge directly into the right coronary artery.

"Pull it out!" Sones shouted, but it was too late. The dye had gone in. Power injecting a large dose of toxic contrast dye into a tiny coronary artery was certain to be fatal—and sure enough, the patient's EKG line went flat (a condition termed *asystole*), signaling cardiac arrest. Sones reportedly cried, "We've killed him!" In his own words, from a 1982 recounting:

> When the injection began I was horrified to see the right coronary artery become heavily opacified and realized that the catheter tip was actually inside the orifice of the dominant right coronary artery . . . [I] ran around the table looking for a scalpel to open his chest in order to defibrillate him by direct application of the paddles . . . it was evident that he was in asystole . . .

In panic mode, Sones yelled, "Cough!" to the patient, hoping pressure from the lungs might squeeze the aorta and push the contrast dye through the vasculature more quickly. With a scalpel, he intended to open the man's chest to apply a defibrillator and shock the heart back to life—at the time, only possible by direct contact with the heart. But before he could cut, the EKG line blipped. Then it began to pulse. And then the jagged line started to race. "After three or four explosive coughs his heart began to beat again," Sones recalled. Finally, the heart settled back into a normal rhythm.

The patient lived. But Sones had made a terrible mistake. He had almost killed a young man, a fact that would naturally inspire soul-searching humility in most physicians. However, Sones's reaction was quite the opposite.

He realized that, for the first time in history, he had shown that it was possible for patients to survive injection of imaging dye into their coronary arteries—vessels whose character and anatomy had never been previously visualized in life. Of course, Sones did not wish to relive this harrowing experience, but he correctly surmised that injecting less-concentrated solutions, and lower volumes, of contrast dye would be significantly safer. Later that day, back at his office, Sones declared to his staff, "We just revolutionized cardiology!" Elaine Clayton, his secretary, cheerfully replied, "Again?"

But he had. Sones's subsequent pioneering work with coronary angiography helped map the coronary arteries and permitted direct visualization of the atherosclerotic plaques within them that lead to heart attacks, an essential milestone in the quest to conquer heart disease.

DOTTERING

At the time of Sones's achievement, treatment for heart attacks was almost nonexistent. In September 1955, while on vacation in Colorado, President Dwight D. Eisenhower developed chest discomfort while playing a round of golf, a symptom he attributed to indigestion. He went to bed early, but at 2:30 a.m., awoke with severe chest pain. His physician was summoned, and morphine was administered for pain relief. For the next nine hours, the doctor did little more than sit at the bedside and observe. An EKG, confirming the president had suffered a major heart attack, was not performed until 1 p.m., almost an entire day after his initial symptoms began. Eisenhower spent the next six weeks in the hospital on bed rest. He ultimately suffered seven heart attacks in his lifetime. The final one killed him in 1969, at just the time when major innovations in cardiac care were about to emerge.

We now know that the key causative event that triggers a heart attack is not simply stenosis of a coronary artery, nor a clot from elsewhere in the body that unluckily becomes lodged in one of the coronaries. It all starts with an atherosclerotic plaque. Plaques form when the endothelium, which lines the inside of an artery, is damaged, often due to years of hypertension, smoking, or high cholesterol. Contrary to what one might think, the biggest plaques aren't the most dangerous. Sudden occlusion of a coronary artery doesn't occur from a steadily growing plaque that slowly shrinks an artery's aperture. Instead, *plaque rupture* is the key event that triggers heart attacks—manifesting as hemorrhage, erosion, ulceration, or fissuring of a plaque. A ruptured plaque exposes its necrotic contents to the bloodstream, immediately attracting potent clot-forming factors and chemicals like platelets, thromboxane, and serotonin. It only takes a few minutes for these ingredients to aggregate and form a thrombus that completely occludes an artery. Once this occurs, myocardial cell death can begin within sixty seconds. The entire heart wall normally supplied by the occluded artery begins to die, with the extent of the damage usually dependent on the duration of ischemia. Impaired heart function also triggers lethal arrhythmias like ventricular fibrillation that can lead to sudden death.

The most fortunate heart attack victims benefit from the prompt arrival of emergency medical personnel who are able to defibrillate a patient out of an arrhythmia and speed the patient to the emergency room. There, the patient will receive morphine, oxygen, aspirin, nitroglycerin (a vasodilator), and a thrombolytic agent that helps dissolve blood clots (such as TNK-tPA, tissue plasminogen activator). Up to the early 1960s, treatments such as these were the best that medicine could offer. At the time, the heart attack mortality rate was 30 to 40 percent. Then, a revolution began that changed everything.

In 1963, a flamboyant radiologist at the University of Oregon named Charles Dotter was advancing a catheter inside the right iliac artery of a patient's leg when he inadvertently opened a partial obstruction in the artery by simply pushing the catheter through it. He quickly realized the implications of this serendipitous event. He and his fellow-in-training, Dr. Melvin Judkins, next developed a series of stiffer catheters, with incrementally increasing diameters. The plan was to first insert the smallest catheter through an arterial obstruction, and then a larger one which would dilate the vessel further, and so on with increasingly larger catheters until the obstruction was completely relieved and blood could flow freely once again. After testing their method on cadavers, they performed their technique for the first time on a patient named Laura Shaw on January 16, 1964. Shaw was an eighty-two-year-old diabetic whose poor circulation had resulted in non-healing, gangrenous toe ulcers that were extremely painful even at rest. Her doctors recommended toe amputations, but she refused. She was eventually sent to Dotter, who saw she had an isolated, segmental stenosis in the superficial femoral artery—the perfect case on which to test his catheters. The procedure went well, and almost immediately Shaw's foot grew warmer and less pale. Her pain soon improved, and within a matter of weeks the ulcer was healed.

With a showman's dramatic flair and a distaste for convention, Dotter reveled in exhibiting his catheterization skills. At one conference, while lecturing on catheterization of the heart, he suddenly announced, "I've been standing here and talking to you for about twenty minutes, and all this time I have had a catheter in my heart." Dotter then rolled back his sleeve and showed the audience where the catheter was inserted in his arm. "Now I'll show you what a normal heart reading looks like," he said, as he hooked himself up to an oscilloscope and proceeded to move the catheter into different chambers of his heart to show how and why the waveforms changed on the display.

Dotter's brash personality and successful procedure irked his surgical colleagues, who hated losing patients to the new technique, which became known as "Dottering." However, they still needed Dotter's radiographic help to plan their vascular bypass operations. One surgeon wrote Dotter the following order regarding his patient's femoral artery obstruction: "Visualize but do not try to fix!" The admonition did not stop Dotter from clearing the patient's obstruction anyway. Dotter never failed to delight in relating this story, and a year after the procedure, he climbed Mount Hood's 11,250-foot summit with that patient, bringing along a cameraman shooting 16 mm movie film to capture the moment.

Charles Dotter had developed an innovative new way to treat a serious disease. What he did not realize, however, was that his work was merely a preamble that would set the stage for arguably the most consequential medical development of the twentieth century.

A SIMPLE BALLOON

In the late 1960s, an East German physician named Andreas Grüntzig attended a lecture about Charles Dotter's techniques. Not every attendee was impressed.

One drawback of Dotter's method was that sometimes the forceful jabbing of catheters into obstructions dislodged the atheromatous clots and sent them further downstream to cause acute ischemia, creating a worse problem that sometimes required urgent surgery. But Dotter's approach sparked a new idea in Grüntzig's mind. Instead of using serial catheters of increasing diameters to bulldoze through atheromas, why not try opening up an artery by inflating a simple balloon inside it? This could improve the technique considerably.

Grüntzig became obsessed with the notion of an intravascular balloon. For years, he worked with his wife, Michaela, and assistant, Maria Schlumpf, at the kitchen table inside his small Zurich apartment, experimenting with ways to attach a balloon to the end of a catheter. Using glue and thread, latex and rubber, Grüntzig tried numerous techniques to fashion a usable prototype. He sought out manufacturers of shoelaces and ribbons for ideas, and chemists who worked with plastics. When testing his early prototypes on cadavers, the balloons often burst, or, if not, failed to compress a narrowing and instead simply expanded on either side of an obstruction in a dumbbell shape. He needed what he called, "a sausage-shaped distensible segment," one that would inflate evenly over its entire length and produce enough force to compress an atheroma.

The breakthrough came when an emeritus chemistry professor working at a local junior high school introduced Grüntzig to polyvinyl chloride (PVC), the perfect material for Grüntzig's needs. This type of plastic proved strong and rigid enough to prevent overinflation, and exerted pressure evenly along the balloon. Switching materials, Grüntzig developed a catheter with a balloon glued to the end and a hole through which imaging dye could be emitted. The catheter would travel to the aorta over a stiffer guide wire that could be pre-maneuvered to the desired location, important because the catheter itself was too flimsy and flexible to be threaded the requisite distance on its own.

Grüntzig first tried his technique in the coronary arteries of dogs, with success. In 1974, he performed what would thereafter be termed *balloon angioplasty* on a human for the first time, successfully treating a severely stenosed femoral artery in the leg of a sixty-seven-year-old man who suffered debilitating pain whenever he walked. On his monitor, Grüntzig watched the slowly expanding balloon dilate the patient's artery, with immediate and obvious widening of the channel and increased blood flow after balloon deflation and removal. Ultrasound tests confirmed that blood supply to the lower leg was significantly improved. Before long, the patient was walking the halls of the hospital pain-free.

In 1976, Grüntzig traveled to the annual meeting of the American Heart Association in Miami with a poster displaying the success of his experiments on dogs. He also intended to describe his early experience treating stenoses in human leg arteries. To his chagrin, very few of the many thousands of cardiologists attending the gigantic meeting stopped to look at his poster or listen to his presentation, and most who did didn't believe Grüntzig's idea had a chance of working in a human heart.

But one man did—Dr. Richard Myler, who invited Grüntzig to his hospital in San Francisco to try the procedure in a coronary artery of a human patient. In May 1977, Grüntzig and Myler performed the world's first successful coronary balloon angioplasty on a patient undergoing coronary bypass surgery. Later that same year, back in Zurich, Grüntzig offered "percutaneous" angioplasty (through the skin and not during an open heart surgery) to a thirty-eight-year-old businessman named Adolf Bachmann, who suffered from severe chest pain, or *angina*, with the slightest exertion. The recommended treatment for this was coronary bypass surgery, with its many risks and long recovery. When Grüntzig suggested that Bachmann could instead be treated with a much less invasive catheter procedure, Bachmann agreed.

On September 16, 1977, Grüntzig anesthetized Bachmann's groin and placed a needle into the femoral artery. Then he inserted a thin metal guide wire through the needle, up and into the artery. Removing the needle, he next slid a guiding catheter over the wire. Using intermittent puffs of contrast dye to visualize the anatomy on a fluoroscope monitor, he advanced the catheter up to the patient's aorta and into the left anterior descending coronary artery. Then he threaded his balloon catheter up through the center of the guide catheter and advanced the balloon into the coronary artery until it was positioned at the point of stenosis.

At this moment, Grüntzig and his many assembled onlookers still did not know for sure if blocking coronary blood flow by inflating a balloon might induce a heart attack or fatal arrhythmia. How long could Grüntzig safely occlude coronary blood flow? Was it better to expand the balloon slowly or a little more rapidly?

Grüntzig gently inflated the balloon. He later reported:

> To the surprise of all of us, no ST elevation [EKG finding typical of heart attack], ventricular fibrillation or even extrasystole [abnormal heart beats] occurred and the patient had no chest pain . . . After the first balloon deflation, the distal coronary pressure rose nicely. Encouraged by the positive response, I inflated the balloon a second time to relieve the residual gradient. Everyone was surprised about the ease of the procedure and I started to realize that my dreams had come true.

Grüntzig witnessed immediate increased blood flow coursing through the artery. Observers in the room gasped. It seemed miraculous. In a matter of minutes, Grüntzig had unblocked a coronary artery using a simple balloon and catheter, allowed Bachmann to avoid open heart surgery, and most likely saved his life. Bachmann, who had been awake and breathing under his own power the whole time, enjoyed immediate relief of chest pain.

When Grüntzig returned to the annual American Heart Association meeting one year after his largely ignored poster debut, he delivered a lecture and showed before-and-after angiogram images of the first four patients who had undergone coronary angioplasties. To Grüntzig's astonishment, in the middle of his presentation a spontaneous round of applause arose from the vast audience of cardiologists, which soon morphed into a standing ovation. Everyone present knew they had watched medical history being made.

By the early 1980s, many leading doctors around the world had adopted Grüntzig's technique as a treatment for coronary artery disease. Now a patient suffering a heart attack is often rushed from the emergency room to the cardiac catheterization lab, where an interventional cardiologist can perform balloon angioplasty, or more commonly, use the balloon to deploy a miniature wire stent that is even more successful at keeping the artery open long-term. After the procedure, the typical patient will be monitored for a few days and then be discharged to home. Grüntzig's invention of coronary angioplasty has saved millions of lives to date.

Yet despite this triumph, there are many patients with such severe heart disease that they cannot be effectively treated using angioplasty. These individuals may have multiple severely narrowed coronary vessels, or they might have suffered multiple heart attacks that have permanently damaged the heart muscle so badly that congestive heart failure results. For these patients with the worst heart disease, cardiac surgery still offers hope. And it is easy to see why many consider the development of modern cardiac surgery to be the most inspiring story of all.

OPEN HEART

To the non-surgeon, cardiac surgery seems nothing short of miraculous. It is violent, bloody, and beautiful. Cardiac surgeons work deftly, their movements precise and elegant, yet the violence is immediate: a motorized sternal saw is used to split the sternum in half in order to gain access to the heart. A stainless steel C-clamp is placed in the gap and a screw is turned, just like in woodshop, to draw the clamp's metal arms apart and create a small, square opening—a window to the heart.

Before the heart can be operated on, it must be stopped. To do this, bypass conduits are placed into the superior and inferior *venae cavae*—the large veins that return blood from the body to the heart. These conduits carry blood to a heart-lung machine that oxygenates the blood, and then returns it to the aorta. Once this bypass of blood flow around the heart and lungs is complete, a bucket of ice-cold saline solution is poured into the chest cavity and potassium-rich cardioplegia solution is instilled within the coronary arteries to slow the heart, stop its beat, and reduce its oxygen and metabolic demands.

With the heart still, the work can now begin. The most common cardiac surgery is the coronary bypass operation. The aim is to bypass obstructions in coronary arteries by sewing new conduits for blood flow to the diseased arteries past the point of blockage. This new supply of blood can come from another artery, often one of the internal mammary arteries which reside nearby behind the sternum, or via a vein harvested from the leg that can be used to convey blood from the aorta to the coronary artery just beyond the obstruction.

As you can imagine, all of this is difficult to do. Using impossibly delicate stitches, and with the aid of loupes attached to their glasses that magnify the view three to four times, surgeons adroitly sew these watertight connections. The risks are great, for if a connection that is not watertight goes unnoticed, internal bleeding will occur after the chest has been closed up—just one of dozens of complications that can occur and kill a patient.

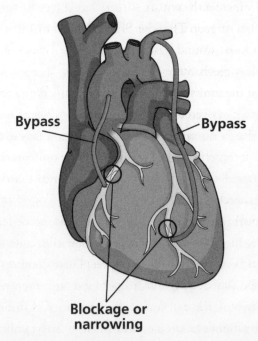

Coronary bypass grafts convey oxygenated blood from the aorta to coronary arteries distal to points of narrowing in order to replenish blood supply to heart muscle

HEART DISEASE: *The Mavericks*

After the work on the heart has concluded, it must be restarted. The heart is refilled with blood and small metal paddles are used to apply a jolt of electricity that stimulates the sinoatrial node. The heart begins to beat again. Pumping power from the heart-lung machine is slowly reduced until the surgeon is sure the heart alone will adequately meet the demands of the body. Finally, the patient is disconnected from the heart-lung machine and the chest is closed.

To the newly initiated, witnessing these miraculous events generates a powerful sense of awe. It confers the surreal sense that this may be something God did not intend for us to see or do. And yet, the industry and intrepidity of man has made it possible.

The story of mankind's journey into the heart is one of great drama and courage. For centuries, the idea of operating on a constantly moving organ, hidden behind the rib cage within the chest cavity, seemed preposterous. Noted giants of nineteenth-century surgery left choice dictums for posterity. The great Austrian surgeon Theodor Billroth stated in 1883, "A surgeon who tries to suture a heart wound deserves to lose the esteem of his colleagues." In 1896, English surgeon Stephen Paget decreed, "Surgery of the heart has probably reached the limits set by Nature to all surgery: no new method, and no new discovery, can overcome the natural difficulties that attend a wound of the heart." Clearly no sane surgeon would ever propose to operate on a beating heart. But sometimes, fate and serendipity unite to change history.

In September 1896, a twenty-two-year-old man in Frankfurt, Germany, was stabbed in the chest after fleeing a bar fight. Found unconscious in a park, he was brought to the nearest hospital, where his case report noted: "He was deathly pale, his pulse was barely palpable with frequent skipped beats, and his breathing was labored." A doctor explored the wound with a metal probe. The probe slid deeper and deeper, and to everyone's astonishment when it came to a stop the external end of the probe bounced up and down in time with the patient's heartbeat. It was clear that the knife had punctured the heart. Death was certain.

Except that a prominent surgeon named Ludwig Rehn happened to work at this particular hospital. When Rehn examined the languishing patient, a thought entered Rehn's mind—a mental leap that few, if any, of his peers would have conjured, much less acted upon. With his astute knowledge of anatomy, Rehn believed it might be possible there was only a small puncture in the heart and that blood from this wound was slowly pooling in the pericardial sac that tightly surrounds it. Today we call this condition *pericardial tamponade*.

Rather than concede the patient's certain demise, Rehn brought the patient to the operating room, where he opened the chest and found the pericardial sac tense and distended, compressing the heart just as he'd suspected. When he punctured the sac with his scalpel, a geyser of blood arose, but then the bleeding diminished and the heart beat more strongly, no longer constrained by the vise-like grip of high-pressure tamponade. Rehn next found the wound itself, a one-and-a-half-centimeter gash in the right ventricle; with each heartbeat, blood oozed out of it. Rehn applied pressure with his finger to stop the bleeding, and the heartbeat grew a little stronger still. Then, timing each movement with the diastole (relaxation between each beat) of the heart, he skillfully placed three sutures to close the wound. Rehn's subsequent recounting of the case conveys his sense of wonder:

> The sight of the heart beating in the opened pericardial sac was extraordinary . . . during systole, the heart muscle was hard as stone, the right ventricle disappearing under the sternum . . . I passed the needle quickly during the diastolic phase, as this was the only time when the right ventricle was exposed . . . In the succeeding diastole, the suture was tied.

For the first time in history, a surgeon had sutured a full-thickness laceration of a beating heart. The patient survived, and two weeks later Rehn even introduced him at a surgical meeting.

Still, Rehn's miracle could only be considered a one-off. It was the perfect confluence of serendipity and skill that was impossible to re-create or plan for—a patient with a tiny heart laceration (that did not result in immediate death) presented to a hospital where one of the most daring, expert surgeons in the world happened to work. Thus, it was not surprising that, for the next half century, little to no further headway in cardiac surgery was made. The heart remained inviolate in the minds of surgeons.

Until the Second World War.

"THANKS FOR THE CHANCE"

In 1944, Dwight Harken was a thirty-five-year-old, Harvard-trained U.S. Army surgeon treating wounded soldiers at London's 160th General Army Hospital. Confronted with all manner of horrific injuries, including young men whose chests had been blown apart by bullets or bomb blasts, Harken quickly became acclimated to the randomness of war and the indiscriminate nature of its victims' injuries. No two wounds were exactly the same. The war supplied Harken with a plethora of mind-boggling cases that no amount of stateside training could have ever prepared him for. More than once, a dying soldier was brought to him with a piece of shrapnel lodged in the heart wall. These hearts still beat with the metallic fragments plugging the lacerations they had made, but not well, and it was only a matter of time before such men died from exsanguination, infection, or strokes, caused by blood clots forming on irregular metal surfaces and then traveling to the brain. Chaplains were brought to these bedsides. Friends helped write letters home. These men were certain to die.

What could Dwight Harken, or any surgeon, possibly hope to do for these soldiers? If he grasped a metal fragment with a clamp and pulled it out, it would certainly result in an uncontrollable torrent of blood which would make it nearly impossible to see the heart wound, much less fix it. These wounds were not like the neat and tiny laceration Rehn had successfully sutured; Harken would merely be hastening death. Though Harken was a good surgeon, maybe even a great one, he had no reason to expect his hands were fast enough to suture a beating heart in time to prevent exsanguination. At most, he would have only a few minutes, operating deep in a hole, in a pool of blood.

Unless he could think of a better way to sew the heart.

On June 6, 1944, the date of D-Day itself, Harken examined a man with a severe chest wound. The patient's ribs and sternum were shattered. Not knowing if or how the heart might be affected, Harken brought him to the operating room. There he discovered a large metal fragment lodged in the right ventricle, bobbing up and down with each heartbeat. The situation appeared hopeless, but Harken had devised a new approach. He removed the fragment and immediately plugged the hole with his finger, to stem the bleeding. Then he passed sutures one by one to close the hole as he gradually removed his finger. In a letter describing the incredible event to his wife, Harken recalled, "The only moment of panic was when we discovered that one suture had gone through the glove on the finger that had stemmed the flood. I was sutured to the wall of the heart! We cut the glove and I got loose."

Harken had wisely increased his chance of success by using "swedges," little rolls of cloth placed between the silk sutures and heart wall to prevent the sutures from cutting through the tissue when tightened. He also timed his knots with diastole when the heart was at its fullest, so that subsequent contractions of the heart did not put additional strain on his sutures. Harken's surgery worked. For the first time since Ludwig Rehn, a more versatile method of repairing a wound in a beating heart had been devised. During the war, Harken successfully removed metal fragments from the hearts of over a dozen soldiers.

Compared to modern surgery, Harken's finger-plugging method was undeniably primitive. But it was a beginning. He later learned to pre-place sutures around the metal in a purse-string configuration so that he could more rapidly close the hole after pulling out the fragment. After the war, this method proved the foundation for the next leap forward, the first attempts to venture inside the heart itself to treat one of the greatest killers of the day—mitral stenosis.

The mitral valve separates the left atrium from the left ventricle. Composed of two overlapping leaflets, viewed from above it resembles a bishop's hat, or mitre, hence the name. During the relaxation of diastole, the mitral valve opens and blood rushes from the left atrium into the left ventricle. Then, during systolic contraction, a burst of pressure from the contracting left ventricle causes the mitral valve to slam shut and prevent back flow into the atrium while blood is rapidly ejected through the aortic valve into the aorta.

During the first half of the twentieth century, before the widespread use of penicillin, rheumatic fever was a common childhood illness. One of rheumatic fever's deadliest complications was mitral stenosis, a narrowing and stiffening of the mitral valve's leaflets. These scarred leaflets no longer opened or closed properly. A narrowed mitral valve opening resulted in a bottleneck that prevented the left atrium from fully emptying before the next heartbeat, which in turn caused blood to back up into the lungs. In time, as their lungs filled with more and more blood, patients became unable to breathe and incessantly coughed up blood. To its victims, mostly young men and women in the prime of their lives, mitral stenosis was invariably a death sentence. It was only a matter of time.

After the end of World War II, Dwight Harken returned to Boston. Determined to make his mark by inventing new surgical techniques, he set his sights on devising a way to treat mitral stenosis. His idea was to use his purse-string technique to pre-place a ring of sutures in the left atrium just above the mitral valve. Then he would make an intentional incision in the center of the ring and immediately stick his finger through the hole, while an assistant held the purse-string tight to prevent blood loss. With his index finger now inside the heart, he theorized he could literally poke it through the scarred and narrowed mitral valve orifice to open it wider, separate the leaflets that had grown fibrosed together, and hopefully regain normal valve function. The risks were enormous and largely unknown.

What if his finger inside the heart upset the heart's rhythm and caused ventricular fibrillation and death? What if he was too forceful and damaged the leaflets, or inadvertently tore them away? That might cause the opposite problem—mitral insufficiency, wherein blood would be shot from the ventricle backward into the atrium, instead of out through the aorta to the rest of the body. There were so many ways he might kill patients. And yet without trying, they were certain to die.

Harken was determined to succeed. But to his chagrin, the first surgeon to attempt this daring operation was actually his rival, another World War II veteran named Charles Bailey, from Philadelphia. In November 1945, Bailey operated on a thirty-seven-year-old man named Walter Stockton. Anyone could see Stockton was close to dying from mitral stenosis. He could barely walk or breathe, and frequently coughed up blood.

In the operating room, Bailey made an incision in Stockton's chest, spread the ribs, and cut a slit in the pericardium to expose the beating heart. He placed a ring of sutures in the left atrium above the mitral valve. Then he made a small incision in the center of the ring and a fountain of blood burst forth. Bailey inserted his finger and tried to tighten the purse-string suture around it, but the suture tore through the heart muscle and would not hold. In a report, Bailey later wrote, "The purse-string suture was pulled upon and tore out . . . Severe bleeding occurred and a large clamp was hastily applied . . . it was impossible to get mattress sutures to hold." Working furiously, yet in vain, Bailey watched his patient's blood drain away. The heart went still within a few minutes. "The patient died on the operating table of hemorrhage, no valvulotomy having been performed," the report concluded.

Undeterred, Bailey determined to try again. In another patient, seven months later, he tried using a new instrument he'd fashioned—a tiny, hooked metal probe he termed a "backward cutting punch," which he hoped might prove better than his blunt finger at dissecting and separating the mitral valve leaflets. Though the patient survived the operation, she died two days later. For these two deaths, and for insisting he be allowed to continue his experimental surgery, Bailey's operating privileges at Hahnemann Hospital in Philadelphia were revoked. Dr. George Goeckeler, the chief of cardiology, told him, "It is my Christian duty not to permit you to perform any more such homicidal operations." To this Bailey responded, "It is my Christian duty to perfect this operation. Nothing could be worse than what mitral stenosis does to people."

Bailey was tenacious. He took his procedure to Wilmington Memorial Hospital in Delaware, where he operated on another patient in 1948. Bailey succeeded in separating the fused mitral valve leaflets, but this time he was too forceful and pushed too hard—he damaged the leaflets so badly that they lost their normal capacity to prevent backflow from the ventricle to the atrium. The patient's disease had been converted from mitral stenosis to mitral insufficiency, and he also died, within a week after the surgery. In the aftermath, Wilmington Memorial Hospital revoked Bailey's operating privileges. Behind Bailey's back, some colleagues and students called him "the Butcher."

Bailey still had operating privileges at two more Philadelphia hospitals—Philadelphia General and Episcopal—but with word of his failures spreading, he was worried he might have only one more chance to attempt his procedure before these institutions banned him as well. To maximize his chance of success, he decided to schedule two operations on the same day, one in the morning at Philadelphia General and another in the afternoon at Episcopal Hospital. This way, even if his first patient died, he reasoned he should still be able to attempt the second case before the news traveled across town and someone stopped him. If both operations failed he would likely be finished—his reputation ruined with no hospital willing to accept him.

The first patient, at Philadelphia General Hospital, died on the table. In the wake of this disappointment, no doubt feeling the stress and gravity of the moment, Bailey went to Episcopal Hospital and started his operation on a twenty-four-year-old housewife named Constance Warner. Warner's childhood rheumatic fever had led to such severe heart failure that she was no longer able to care for her young child.

This time, Bailey used both a metal instrument and his finger to gently separate the leaflets. To his relief, the operation seemed to go well. His sutures held and no major hemorrhage occurred. Constance Warner survived the surgery. Her breathing improved. By the fourth post-operative day she was walking.

Bailey felt elated. His reputation instantly restored, he became determined to capitalize on the moment. He broadcast news of his success to his colleagues and even took Warner on a thousand-mile train ride to Chicago just one week after the surgery to parade her in front of the audience at a meeting of the American College of Chest Physicians. Warner went on to live a full life, had a total of four children, and died at the age of sixty-two.

Although Bailey would go down in history as the first to perform this feat, six days after Warner's operation, Dwight Harken also successfully performed mitral valve surgery on a patient in Boston. Prior to this, Harken had attempted the operation on six patients, all of whom had died. At this point, the worldwide success rate for this surgery was two out of twelve. After the death of one of these patients, Harken returned home despondent. That evening there was a knock at the door. The caller was a woman who handed over a note from the patient, her friend, which she'd promised to deliver if her friend did not survive the surgery. The note read: "Dear Dr. Harken: Thanks for the chance. A small portion of my estate has been left to see that this doesn't happen again."

Dwight Harken and Charles Bailey were pioneers of cardiac surgery, but they failed far more often than they succeeded. To press on in the face of so many deaths required supreme courage. Both harbored deep reserves of self-confidence and the ability to bounce back after heartbreaking disappointments. The cost of innovation was great, but through their efforts the door to the new universe of cardiac surgery had finally been shoved open.

TO BREATHE OUTSIDE THE BODY

Operations on the heart would remain a rare event—the parlor trick of medicine with its terrible risks to be confronted only in the face of certain death—until surgeons somehow found a way to still the beating heart without killing the patient. Surgeons like Harken or Bailey could clamp the flow of blood to the heart temporarily, but doing so only granted them about four minutes to complete an operation before the onset of permanent brain damage. Surgeons later learned to use hypothermia to extend this time; cooling the body to seventy degrees or lower reduced the body's metabolic rate and need for oxygen, a method that could extend the operating window to approximately twelve minutes. Any operation that took longer than this was impossible. Doctors could arrest and restart a heart, but until someone devised a way to provide oxygenated blood to the body without using the heart, these time constraints would consign the nascent field of cardiac surgery to only the most simplistic of maneuvers.

A Canadian surgeon named William Mustard took a novel, though not particularly imaginative, approach to extracorporeal oxygenation of blood. He tried using monkey lungs. In 1951, Mustard removed the lungs of four monkeys, placed them in jars pumped full of oxygen, diverted a child's venous circulation through rubber tubing to the lungs, and then returned the (hopefully) oxygenated blood to the patient. He tried this on twelve children, all with life-threatening cardiac abnormalities, over a period of three years. All twelve died. But some of them survived their operations—one for almost two weeks—thus proving that it was at least possible to oxygenate a patient's blood outside the body.

Another surgeon named Walt Lillehei from the University of Minnesota also tried to harness nature's own designs by exploring the idea of "cross-circulation" in children needing operations for congenital heart defects. This idea stemmed from his knowledge of fetal circulation. A fetus in the womb does not breathe. Its blood cross-circulates with its mother's through the placenta and is therefore oxygenated via the mother's lungs. In similar fashion, Lillehei concluded that it might be possible to connect a child's circulation to that of its mother or father. The child's blood would run through the parent's circulation, be oxygenated by the parent's lungs, and then be returned to the child, whose heart had been temporarily stopped to permit surgery.

Lillehei first tried this on dogs, and it appeared to work. He could take up to thirty minutes, an eon in the realm of cardiac surgery at the time, to close a dog's ventricular septal defect (a hole connecting the right and left ventricles). All the while, the dog was kept alive by connecting its circulation to another, healthy dog. In March 1954, Lillehei was ready to try the procedure on a person. Across the country, there were many children dying from congenital heart defects, each of them desperate for a cure. Yet Lillehei's inventive procedure also prompted unusual ethical concerns. In addition to the risks assumed by the patient undergoing surgery, Lillehei would also be subjecting the healthy parent to risk. Theoretically, this was a surgery that could result in 200 percent mortality if both individuals died.

After much discussion and debate, the University of Minnesota Hospital allowed Lillehei to proceed. His first patient was Gregory Glidden, a thirteen-month-old boy with a ventricular septal defect; through a hole between the ventricles, Gregory's heart shunted blood from the left ventricle to the right ventricle, which then pumped the blood back into the lungs in a doomed and futile cycle. Gregory survived the surgery—connected to his father's circulation—but died from pneumonia eleven days later. Undeterred, Lillehei soon tried operating on four-year-old Pamela Schmidt. Like Gregory, Pamela was dying from a ventricular septal defect. In the operating room, Pamela and her father lay supine, side by side, with Pamela's circulation connected to that of her father. Lillehei stopped her heart, opened it, and sutured shut the hole between the ventricles. Then he closed the heart, restored blood flow to it, and restarted it with an electrical pulse. Pamela's heart beat vigorously.

The surgery was a success.

Pamela and Lillehei became national celebrities. Pamela and her family were profiled in *Cosmopolitan* magazine. The state of Minnesota and the American Heart Association officially dubbed Pamela "The Queen of Hearts." Lillehei was inundated with interview requests and referrals of other children dying from heart defects.

And yet, the procedure's triumph was premature. Lillehei successfully operated on a handful of other children after Pamela, but later endured a series of seven cases in which six of the children died. In one of them, the child's mother suffered a devastating stroke from an air embolism that lodged in her brain. Lillehei's reputation fell. He had made a step forward in the quest for extracorporeal oxygenation, but his was not the crucial advance that would truly revolutionize cardiac surgery.

Meanwhile, another surgeon named John Gibbon had been pondering the extracorporeal oxygenation of blood since 1931, when, as a research fellow in Boston, he became haunted by the case of a young woman who had undergone routine gallbladder surgery but developed chest pain and trouble breathing fifteen days after the surgery. She had developed an embolism, a blockage from a clot, in her pulmonary artery. At the time, the only effective treatment was surgery to remove the clot directly; however, this surgery was so risky that it had never been successfully performed in the United States and had only been done nine times in the world to that point. Gibbon's boss, Dr. Edward Churchill, decided that they should operate only as a last resort—if the pulmonary artery became completely occluded and the patient was about to die. To be prepared for this, they brought the patient to the operating room. Gibbon was assigned to sit beside her in an all-night vigil, checking her vital signs every fifteen minutes.

At 8 a.m., the patient lost consciousness. She stopped breathing and there was no pulse. Within six and a half minutes, Churchill opened her chest, removed the large clot, and closed the pulmonary artery incision with a clamp. But even this was not quickly enough. The patient never woke up, and she died.

Gibbon was profoundly affected by this tragedy. He later recalled:

> During that long night, helplessly watching the patient struggle for life, the idea naturally occurred to me that if it were possible to remove continuously some of the blue blood from the patient's distended veins, put oxygen into that blood . . . and then inject continuously the now red blood back into the patient's arteries, we might have been able to save her life. We would . . . perform part of the work of the patient's heart and lungs outside of the body.

Gibbon became obsessed with this idea. He spent decades endeavoring to build a machine that could oxygenate blood. The challenges were numerous.

For one thing, red blood cells are very delicate and susceptible to trauma when moving through any manner of tubing or machinery. Another problem was blood's propensity to clot upon encountering any foreign material, such as the metal surfaces inside a machine. And, of course, the ideal way to convey oxygen to the blood, and remove carbon dioxide, would have to be devised.

Using a pump, tubing, plastic casing, an oxygen supply, and heparin to prevent clotting, Gibbon built a prototype machine to try on cats. This design spun the blood along the inside of a large, revolving, oxygen-filled cylinder, using centrifugal force to spread the blood out into a very thin film to increase the surface area available for gas exchange. With this machine, Gibbon managed to keep cats alive on complete heart-lung bypass for up to twenty-five minutes in 1939. However, it did not provide adequate oxygenation for use in larger animals or humans.

A breakthrough came when he discovered a patron: Thomas Watson, the CEO of IBM. Watson learned about Gibbon's vision and supported him with funding and engineers to improve Gibbon's prototype. The result was an innovative design in which venous blood from the patient was spread out in a film and sent cascading down a series of eight vertical, stainless steel mesh screens, a method that optimized both surface area and turbulence—Gibbon had learned that introducing gentle turbulence maximized gas exchange. The whole system was sealed in a clear plastic casing pumped full of oxygen.

At Jefferson Medical College in Philadelphia, on May 6, 1953, Gibbon performed his first successful case using what would become the modern heart-lung machine. His patient was Cecelia Bavolek, an eighteen-year-old college student with an atrial septal defect. She had been hospitalized three times in the previous six months due to heart failure and could no longer carry out normal activities without symptoms. By running Bavolek's circulation on bypass for twenty-six minutes using his machine, Gibbon was able to stop her heart and take his time repairing the hole between the atria. The surgery was a success.

The heart-lung machine set the stage for the meteoric growth of cardiac surgery. Within eight years, the first coronary artery bypass graft operation

(CABG) had been performed. Was there any medical procedure more glamorous, more audacious than this? The proposition that mankind could rearrange the circulatory plumbing designed by God in order to cheat death was heady news indeed.

Yet, it should be noted that the path leading to a successful surgical treatment for heart disease and angina had been littered with failed ideas. One doctor tried dissecting the nearby internal mammary artery from the back of the sternum and tunneling it directly into heart muscle. Another thought chafing the surface of the heart might induce increased blood flow to that area, so he opened patients' pericardial sacs and coated their hearts with talcum powder. Another tried to infuse oxygen through the veins in an attempt to "retroperfuse" the heart muscle. All these methods failed.

The surgeon typically credited with doing the most to improve CABG surgery techniques was Dr. René Favaloro, an Argentinian surgeon working at the Cleveland Clinic in the late 1960s. He advanced the idea of using a leg vein as a graft to convey blood from the aorta to the diseased coronary artery past the point of obstruction. His patients enjoyed prompt relief of angina, and their post-operative angiograms showed that cardiac perfusion had been restored to a remarkable degree. Today, bypass surgery is the mainstay of the cardiac surgeon, used in patients with very severe or multivessel coronary disease who cannot be successfully treated by angioplasty.

THE GREATEST PRIZE

On December 3, 1967, surgeon Christiaan Barnard performed the first heart transplant in history at Groote Schuur Hospital in Cape Town, South Africa. No one could deny that this marked the zenith of surgical achievement to date. But the story of this remarkable accomplishment begins with the pioneering work of Dr. Norman Shumway, a surgeon from Stanford University who labored for over a decade to develop the ideas and techniques that made Barnard's surgery possible.

Beginning in the 1950s, Shumway and his colleague Dr. Richard Lower confronted several problems related to transplantation. One question they sought to answer was whether a heart could still function after being completely removed from the body—that is, after all the nerves connected to it had been transected. To answer this, they experimented on dogs by removing their hearts and reimplanting those hearts back into the same animals. The dogs survived.

Another difficult challenge was to determine what was the best and most effective way to implant a donor heart into a recipient. The heart is connected to the circulation via two veins (the venae cavae) that deliver blood to the right atrium, the pulmonary artery that conveys blood from the right ventricle to the lungs, four veins that return blood from the lungs back to the left atrium, and the aorta, which connects to the left ventricle. There are therefore eight vascular connections in total—quite a lot from the viewpoint of a surgeon when one considers that each new connection must be painstakingly and precisely sewn together to form a perfect, watertight seal.

Shumway developed a better way. Instead of cutting out the heart by severing all the tubular vessels attached to it, what if he left most of the recipient atria behind—the sections of each atrium that received the two veins in the right atrium, and the four veins in the left atrium? These sections of recipient atria could simply be sewn into a matching void sectioned out of the donor heart. This eliminated the need to reconnect six of the heart's eight vascular connections, reducing the length of surgery and risk of failure.

Just as important as Shumway's surgical innovations was his study of how to overcome rejection of the donor organ by the recipient's immune system, termed *transplant rejection*. The discovery of powerful immunosuppressive medicines like cyclosporine aided his quest to surmount this obstacle. But there was another problem Shumway couldn't fix—the definition of death.

Today, we recognize that death may occur in the form of brain death, which is characterized by irreversible loss of brain function and can happen even when a patient is still being kept alive using a ventilator. But in the United States in the mid-1960s, brain death was not a recognized form of death; death was defined only as the complete cessation of both heartbeat and respiration. This meant it would be practically impossible for Shumway, or any other American cardiac surgeon, to find a donor heart to transplant, which would require having a recipient patient prepped and ready to go in the operating room the moment a donor heart happened to become available, such as in the aftermath of a prospective donor's sudden death in a motor vehicle accident. It was impossible to predict when such an event might occur.

This conundrum prompted contentious ethical debates in many hospitals across the United States . . . until the day that news of Christiaan Barnard's successful heart transplant surgery in South Africa sped across the globe. After Barnard's breakthrough, debate in America related to brain death evaporated almost overnight.

Unlike the eminent and highly regarded Norman Shumway, Barnard was practically unknown—a non-academic surgeon working at a suburban Cape Town hospital. He had successfully used Shumway's techniques to remove the heart of Denise Darvall, a young woman declared brain dead after being struck and killed by a car while walking in a crosswalk. The world's first heart transplant recipient was Louis Washkansky, a fifty-seven-year-old diabetic with severe heart failure after suffering multiple heart attacks.

In South Africa, a patient was deemed dead when a physician declared it so. Guidelines concerning the definition of death were more permissive than in the U.S., and this gave Barnard the freedom to use a brain-dead donor. Darvall and Washkansky were brought to adjacent operating rooms. Barnard disconnected Darvall's ventilator and waited for her heart to stop beating. Then, in a matter of minutes, he extracted Darvall's heart and brought it to Washkansky.

In his autobiography, Barnard wrote, "Inserting my hand, I removed Washkansky's heart from his body and placed it in a basin . . . Below this was the hole and it seemed immense. I had never seen a chest without a heart."

Washkansky's heart had become greatly enlarged over many years as it had adapted and struggled to supply adequate blood flow to the body in its diseased state. In contrast, Darvall's heart was far smaller, and it looked tiny in Washkansky's chest cavity. Barnard hoped the heart would be strong enough to meet the circulatory demands of such a large man.

Once he had finished implanting the heart, Barnard restarted it with a small electric shock and it began to contract normally, though not strongly enough to wean the patient off the heart-lung machine. It appeared that the new heart could not provide sufficient blood pressure to perfuse Washkansky's body. Barnard tried to reduce the power from the heart-lung machine a second time, but again found the donor heart would not pump strongly enough.

He decided to wait a little longer. Tense minutes passed.

Barnard checked all the suture lines; there were no leaks. Was the heart beating just a little more strongly now?

Yes, it looked like it was.

It definitely was.

The third time Barnard reduced the pumping power from the heart-lung machine the blood pressure stayed high enough. In Afrikaans, he exclaimed, "Dit lyk of dit gaan werk!" *It looks like it's going to work!*

Barnard gazed across the table at his team. He later recalled, "Eyes over masks blinked back—moist with joy and wonder. In the theater, the tension of silence broke with mixed sighs, mumbled words, and even a little laugh. All of us, like the heart itself, were once again sure of ourselves." All told, the operation had taken about five hours.

News of the world's first heart transplant electrified citizens around the world, and Christiaan Barnard became an international celebrity. Unfortunately, Louis Washkansky died from pneumonia eighteen days after the surgery, but an autopsy confirmed the transplanted heart displayed no signs of failure or rejection.

Barnard's second transplant patient lived for nineteen months after surgery. His third patient lived twenty months. His fifth and sixth patients lived thirteen and twenty-three years, respectively. Barnard later became even more famous for his glamorous lifestyle and for dating movie stars; he was married three times. Meanwhile, Norman Shumway continued to labor to perfect the post-operative care that would allow transplant recipients to live longer and better lives. In 1981, he was a key member of the surgical team that performed the world's first combined heart-lung transplant.

At present, about 3,800 cardiac transplants are performed annually in the United States. The system of organ donation is tightly regulated by the United Network for Organ Sharing (UNOS), which pairs donors and recipients according to numerous factors including priority granted to the sickest patients and the likelihood of organ rejection. Every day, thousands of patients remain on the waitlist for a heart that they desperately hope might become available to prolong their lives.

TOMORROW

Today, the most effective tactic in the battle against heart disease is prevention. Smoking cessation, exercise, and weight loss campaigns have significantly reduced mortality from coronary artery disease. Few medicines in history have matched the worldwide benefits of statin drugs to lower cholesterol. Nowadays, congenital heart disease can be surgically corrected early in life, even in utero. Survival rates for patients undergoing every type of cardiac procedure are at all-time highs.

And yet, despite our advances, coronary artery disease remains the world's leading killer. Each year, about 805,000 Americans have a heart attack—approximately one every forty seconds. What other weapons can we deploy in the fight against this relentless adversary?

Stem cells may be a tantalizing part of the answer. Stem cells are unique because they have the potential to develop into many different cell types in the body. Doctors have developed a way to extract stem cells from a patient's heart, replicate them in culture, and then inject them back into the heart to support damaged tissue or augment the function of healthy tissue. This foray into the world of regenerative medicine has already shown positive results by improving cardiac function and reducing the size of scars in patients who have previously suffered heart attacks. Stem cells have also been used to seed

3D-printed, biodegradable scaffolds to create functional human heart valves and blood vessels. For more complex structures, cadaver organs can gain new life as scaffolds themselves. In 2008, scientists decellularized a rat heart using a variety of solutions and detergents that stripped all the original cells away, but left intact the extracellular matrix that formed the organ's structure. They then seeded this organic framework with cardiac cells from newborn rats and, after eight days, the cells had formed a beating heart that could conduct electrical impulses and pump blood.

In 2016, this concept was replicated to produce functional cardiac tissue in a full-sized human heart—a decellularized cadaveric heart was populated with stem cells, infused with nutrients, and grown in a bioreactor that simulated conditions inside the body like the stress of high ventricular pressure. When stimulated with electricity, the new cardiac muscle exhibited contractions. Thus, we are moving ever closer toward attaining the futuristic goal of fabricating replacement hearts, so that one day, transplant patients like Dick Cheney will no longer need to spend years desperately waiting for a donor heart—or worse, dying before one becomes available.

Not every high-tech advance is years or decades away. Using an ingenious intraoperative stabilizing system, it is now possible to operate on a beating heart, eliminating the added risk and expense of the heart-lung machine. Interventional cardiologists, who continue to improve and perfect what can be accomplished using catheters, can now deploy artificial heart valves and even repair atrial and ventricular septal defects, allowing more patients to avoid invasive cardiac surgery. A downside of Cheney's implanted defibrillator was its reliance on wires and batteries with limited lifespans (the type of ICD that Cheney received employs a battery that must be surgically removed and replaced every three to six years). Today, doctors are implanting wireless pacemakers directly into the heart via catheter. These tiny capsules contain batteries that last between eight and thirteen years. A future alternative to batteries could be the use of a small radiofrequency generator implanted in the chest wall, which can transmit energy to the pacemaker in the same way a wireless station charges a smartphone.

The human heart remains a unique and powerful muscle with an incredibly long and reliable battery life all its own. After a remarkable century of discoveries and innovation, the heart is no longer the mystery it once was. No doubt humankind will succeed in decrypting more of its secrets in decades to come; but, no matter how many are uncovered, the heart will always inhabit a position of majesty and mystique in the human psyche. It will always be regarded with awe.

2

DIABETES

The Pissing Evil

In 1918, on an autumn day in New York City, an eleven-year-old girl named Elizabeth Hughes relished the chance to eat cake and ice cream at a friend's birthday party. Her parents, Charles and Antoinette, found it strange that, after the party, Elizabeth seemed extremely thirsty. She drank water constantly and appeared unable to quench her thirst. In the coming days, they also noticed she had a very strong appetite and was eating much more than usual, but at the same time, she appeared tired, sometimes even listless. This was very unusual for Elizabeth, who was normally highly energetic, playful, and adventurous. Her parents were perplexed. What had brought about this change?

That winter, Elizabeth contracted the flu. This frightened Charles and Antoinette. The Spanish flu was killing indiscriminately—young and old, rich and poor. No family was safe. They prayed for Elizabeth's recovery, and mercifully, her condition improved. For a time, it seemed logical that Elizabeth's symptoms had been caused by the flu. After all, she had always been a very healthy child.

But by the spring of 1919, no one could deny that Elizabeth's persistent and disturbing symptoms were the mark of something more sinister. In addition to perpetual thirst and frequent urination, her weight loss became alarming. Standing four feet, eleven inches tall, Elizabeth went from seventy-five to

DIABETES: *The Pissing Evil*

sixty-five pounds, though she ate ravenously. The Hugheses took Elizabeth to see a doctor, who suspected she had diabetes mellitus. Tests of the blood and urine confirmed the diagnosis.

It was a death sentence.

Charles and Antoinette soon learned that doctors did not know why children like Elizabeth seemed unable to metabolize food properly. In fact, they learned that food, normally essential to life, would literally kill their daughter. The more she ate, the sicker she would be. There was no cure, no treatment. The average child lived for eleven months after diagnosis.

Their only option to prolong Elizabeth's life would be to practically starve her. A child Elizabeth's age typically ate 2,000 calories a day. If they wanted Elizabeth to live as long as possible, they would have to restrict her to 750 calories on most days, 300 calories on some days, and one day per week she should fast completely. What little food she ate would have to be carefully measured and meted out. Her diet would be exclusively proteins and fat. No carbohydrates; she could never eat bread or sweets again. Elizabeth's "starvation treatment" diet looked like this: one egg for breakfast; half an orange mid-morning; five olives and a few brussels sprouts for lunch; and an egg, two spoonfuls of spinach, and a half pat of butter for dinner.

On such a diet, Elizabeth might at best add a year or two to her life. Or, like many children, she might starve to death far sooner.

Why bother? some parents of diabetic children wondered. The cruelty of a starvation diet—watching their beloved children waste away, suffering the pain of constant, agonizing hunger—could be too much. Some considered it more humane to treat their children to a smorgasbord—a final feast of their favorite foods—and let them literally eat themselves to death. Wasn't it more merciful to end things quickly, when death was inevitable anyway?

But others continued to hope and pray for a cure that might arrive in time to save their loved ones. Elizabeth's parents made a difficult decision and agreed to give her up to the care of Dr. Frederick Allen, a leading diabetologist who sequestered his patients at a sanitarium called the "Physiatric Institute," which was first established in Manhattan and later moved to a larger facility in Morristown, New Jersey. At the Institute, patients were closely monitored. Trained cooks and nurses carefully prepared and recorded each child's dietary intake. Urine output and amount of exercise were meticulously tracked. For every child, the goal of treatment was to determine the minimum amount of food required to stay alive. Dr. Allen and the nurses were forced to harden their hearts and grow numb to the children's pleas for food. They constantly reminded themselves that, without starvation, all of the children would already be dead.

Elizabeth lived at the Institute intermittently from 1919 to 1922. By 1921, her weight had dropped to fifty-two pounds. Severely weakened, she became emaciated and her hair started to fall out. Still, she remained amazingly compliant and cheerful. Her self-control and ability to stick to her diet greatly impressed her doctors and nurses, who grew very fond of her. She tried to distract herself from constant hunger by keeping busy; her activities included going on walks, writing short stories and letters, and reading.

Elizabeth was also devoted to her parents, who lived busy lives and traveled quite often due to her father's work. In addition to letters, she was able to keep up with her parents' activities because they were commonly reported in the newspapers. This was because her father was the United States Secretary of State, Charles Evans Hughes.*

Hughes's signature achievement was to organize the Washington Naval Conference of 1921–22, which culminated in a disarmament treaty between the United States, Britain, and Japan. Hughes was credited with a diplomatic masterstroke that convinced these nations to agree to limit their production of battleships and maintain, respectively, a 5:5:3 ratio of total warship tonnage, preventing a new arms race in the aftermath of the Great War.

And yet, for all his accomplishments, and despite being one of the most powerful and famous men of his time, Hughes was powerless to help his youngest

---

* In 1916, Hughes had taken the unprecedented step of resigning from the Supreme Court to run for the presidency. He narrowly lost to Woodrow Wilson. He would return to the Supreme Court as Chief Justice in 1930.

daughter, Elizabeth, whose skeletal appearance wrenched his heart. She was perilously close to death, and there was nothing he could do but pray for a miracle.

"THE PISSING EVIL"

Humans have three sources of energy: fat, protein, and carbohydrates. The body breaks down carbohydrates into glucose, which is readily used by our cells to provide the fuel needed to move, breathe, think, and do everything else we require to live our lives. Insulin is a hormone that regulates glucose metabolism. A hormone is a chemical that is produced by a gland in one part of the body and then travels to another part of the body to accomplish its purpose. Insulin is made and secreted by clusters of cells in the pancreas called the *islets of Langerhans*. In every cell of the body, insulin acts as a gatekeeper that permits glucose to enter the cell so it can be used. Insulin also promotes muscle growth and enables storage of glucose into glycogen, which is akin to stockpiled glucose that can be called upon when regular glucose levels are low.

Because of insulin, our bodies normally manage to keep the fasting glucose level in our blood in an ideal range between 70 and 100 mg/dL. If the body's insulin is insufficient, then glucose cannot be used by cells and remains in the bloodstream, causing hyperglycemia. If one's glucose level increases beyond 180 mg/dL, it overcomes the kidney's ability to extract it from the bloodstream and the excess glucose begins to appear in the urine.

Diabetics either cannot make enough insulin (type 1 diabetes) or have tissues that cannot utilize insulin to its full effect (type 2 diabetes). The classic symptoms of this metabolic disease include:

1. Frequent urination and thirst: Glucose escaping in the urine osmotically draws more fluid out of the blood, which in turn increases thirst.
2. Hunger and fatigue: These symptoms occur because the body can't get energy from glucose.
3. Weight loss: Without insulin to help build muscle, one's muscles at-

rophy. And without energy from glucose, the body breaks down fat for energy (though it must be noted that obesity is a major risk factor for type 2 diabetes).
4. Vaginal infections: High blood glucose elevates the sugar in one's sweat and other secretions, creating perfect conditions for growth of yeast and bacteria.
5. Blurry vision: Fluctuating glucose levels can cause swelling of the lens in the eye.

Diabetes' effects have been recorded since antiquity. Around the fifth century B.C., Indian physicians recorded that the urine of affected patients tasted sweet and attracted ants. Because overwhelming thirst and frequent urination made it seem like fluids passed right through the body, second-century physician Aretaeus of Cappadocia named the condition *diabetes*, derived from a Greek word meaning "sieve" or "to pass through." In the seventeenth century, British physician Thomas Willis added the term *mellitus*, or "from honey," to denote the sweet-tasting urine and to differentiate it from *diabetes insipidus*, another condition with frequent urination. Willis called diabetes the "pissing evil" and reported the case of a noble earl who "became much inclined to excessive pissing . . . in the space of twenty-four hours, he voided almost a gallon and a half of limpid, clear, and wonderfully sweet water, that tasted as if it has been mixed with honey."

THE PANCREAS

Picture the pancreas. It's not a glamorous organ. A medical student's first opportunity to hold a pancreas in his or her hands in pathology class, or in a live patient during an operation, might spur a one-word description: slimy. It isn't entirely facetious to liken the organ to Jabba the Hutt. Same color. Same texture. And like that giant slug-like creature from *Star Wars*, the pancreas even has a head and a tail.

DIABETES: *The Pissing Evil*

The pancreas is located in the upper left abdomen, behind the stomach and in front of the left kidney. Its head is tightly nestled into a bend of the duodenum (the first portion of the small intestine) like a fist pressed hard into the palm of your other hand. The pancreas is a "hidden" organ, buried so deep in the abdominal cavity that it's almost impossible to palpate clinically, and diseases associated with it often only become evident when far advanced—for example, when a tumor becomes so large that it begins to impress upon neighboring organs.

The pancreas has two general functions. First, it emits an *exocrine* secretion of digestive enzymes produced by acinar cells. These digestive enzymes (with names like trypsin, chymotrypsin, lipase, and amylase) ooze down the main pancreatic duct and are deposited into the duodenum. Approximately 85 percent of the pancreas is comprised of acinar cells that generate 2 to 2.5 liters of enzymes per day. A much smaller portion of the pancreas is devoted to its second function: *endocrine* secretion of hormones like insulin, glucagon, and somatostatin produced within the islets of Langerhans. These unique cell clusters are named after Paul Langerhans, a German doctoral student who discovered them in 1869. He saw that they were distinct from the abundant acinar cells, but he did not ascertain their function.

Still, nineteenth-century physicians began to suspect the pancreas was integral to the pathophysiology of diabetes, partly because deceased patients sometimes displayed damage to the organ at autopsy, and individuals with pancreatic injury developed the disease. This link was confirmed in 1889, when German physicians Oskar Minkowski and Josef von Mering removed a dog's pancreas and reported the dog developed diabetic symptoms along with high blood and urine glucose. Reimplanting a small piece of pancreas back into a depancreatized dog reversed this.

By the turn of the twentieth century, scientists knew the pancreas influenced glucose metabolism but did not understand how. Many theorized that the pancreas, and specifically the islets of Langerhans, must secrete something, perhaps a hormone that stayed in the bloodstream and regulated blood sugar (termed an "internal" or endocrine secretion)—but how to identify and isolate it?

Why not grind up some pancreas, inject it into animals or patients, and see what happened? The mysterious hormone was sure to be somewhere in that mix, scientists thought.

Some did try this, but most attempts to treat diabetics with extracts of pancreatic tissue failed or proved inconclusive. In particular, adverse side effects, such as infection at the site of injection or severe allergic reactions, discouraged investigators. Many doctors blamed their inability to isolate the hormone produced by the islets of Langerhans on the coexistence of the pancreas's powerful digestive enzymes (termed an "external" secretion), which they thought must be destroying the more delicate internal secretion, whatever it was.

If only there was a way to isolate the internal secretion all by itself.

THE "INTERNAL" SECRETION

The story of insulin's discovery is a tale of perseverance and triumph, but also envy, paranoia, and bitter resentment. In October 1920, Dr. Frederick Banting was a twenty-eight-year-old decorated Canadian veteran of the Great War[*] . . . and a failure. He'd established a surgery practice in London, Ontario; but, even after several months, he had not attracted more than a handful of patients. He was in love with a girl who didn't want to marry him. Money got so tight that he decided to sign on as a doctor for an oil drilling expedition journeying to the Northwest Territory—until the expedition said they didn't want him. He tried to make ends meet by giving general medical lectures for two dollars an hour at the University of Western Ontario, a humbling assignment for a trained surgeon.

One day, Banting was assigned to give a lecture on carbohydrate metabolism to a class of physiology students. He had no special interest in carbohy-

---

[*] Banting served as a medical officer in World War I. At the Battle of Cambrai he suffered a severe shrapnel wound to his right arm. Despite being ordered to the rear, he continued to treat wounded soldiers for twelve hours. For this action he was awarded the Military Cross by the British government.

drates, physiology, or endocrinology in general, so he read up on it. On the night of October 30, 1920, he perused a surgical journal and came across an article titled "The Relation of the Islets of Langerhans to Diabetes with Special Reference to Cases of Pancreatic Lithiasis." The author was an American pathologist who had performed an autopsy on a patient with a pancreatic stone (pancreatic lithiasis) that had blocked the pancreatic duct and prevented the flow of digestive enzymes. The digestive enzyme-producing acinar cells of the patient's pancreas atrophied away. But the islet cells survived.

For Banting, this article sparked an epiphany, one that kept him up most of the night. Conventional wisdom held that scientists' inability to isolate the mysterious internal secretion was due to the damaging effect of digestive enzymes produced by pancreatic acinar cells. But, Banting thought, what if there was a way to replicate what had happened to the patient in the article? What if he could eliminate the secretion of digestive enzymes? Then the pancreas would secrete the internal secretion alone, and Banting could isolate it.

But how could he replicate the effect of a pancreatic duct stone?

He was a surgeon. He could block the pancreatic duct by tying it off.

That night, Banting wrote a message to himself in a notebook:

Diabetus

Ligate pancreatic ducts of dog. Keep dogs alive till acini degenerate leaving Islets.

Try to isolate the internal secretion of these to relieve glycosurea.

If Banting were to ligate, or tie off and close, the pancreatic duct that delivered digestive enzymes to the duodenum, then this should cause the acinar cells to atrophy over the course of the following seven to ten weeks. Then, if he succeeded in isolating the internal secretion of the islet cells, he could inject it into a depancreatized dog with induced diabetes and see if the internal secretion relieved diabetes.

Banting took his idea to John Macleod, a professor at the University of Toronto and a leading expert on carbohydrate metabolism. Macleod was Scottish, forty-four years old, and had come to Toronto after serving as chair of the physiology department at Western Reserve Medical School* in Cleveland, Ohio. At their meeting, Banting did not impress Macleod. The young surgeon was not an articulate speaker and had only a superficial understanding of metabolic physiology. Banting's enthusiasm was obvious, but so was his inexperience and somewhat amateurish approach to a scientific problem that had befuddled scientists around the world for decades. Still, Macleod thought enough of the idea to grant Banting lab space for his experiments.

In May 1921, Banting gave up his medical practice and moved to Toronto to begin the research. After helping Banting get organized, Macleod left to spend the summer vacationing in Scotland. A twenty-one-year-old medical student, Charles Best, became Banting's assistant. Banting and Best spent the summer operating on dogs. They removed the pancreases of one group of dogs, rendering them diabetic. In another group, they ligated the pancreatic ducts, waited about seven weeks for the digestive enzyme–producing acinar cells to degenerate, and then harvested the pancreases, hoping they contained only the internal secretion free from the external secretion.

An immediate puzzle was how to best prepare the pancreatic extract. They decided to slice up the pancreas, partially freeze it in chilled Ringer's solution (a mixture of salts in water), grind up this slurry with a mortar and pestle, and filter the remains. The product was a pinkish liquid.

On July 30, 1921, at 10 a.m., Banting and Best injected 4 ml of their extract into the vein of a white terrier whose blood sugar percentage was 0.20 (normal range 0.08 to 0.13 percent in dogs). At 11 a.m., the dog's blood sugar had fallen significantly, to 0.12 percent—a strong effect. The dog's clinical condition also improved—previously listless, the dog now stood up, moved about, wagged its tail, and even resumed showing affection to the two investigators. Two days later, they gave their extract to a depancreatized dog that was

---

* Now Case Western Reserve University School of Medicine.

in a coma. Amazingly, within an hour the dog awoke and was able to "stand & walk." Positive results were also obtained with a third dog. In addition to being stronger and more energetic, Banting and Best noted that the animals' wounds began to heal more rapidly as well.

There remained numerous challenges. Surgery to ligate the pancreatic duct was difficult and sometimes led to the dogs' deaths due to blood loss or excessive anesthesia. Other dogs suffered severe infection or anaphylactic reactions to the extract. Giving too much extract rendered dogs hypoglycemic, sometimes resulting in coma and death. The young researchers went through so many dogs they were forced to venture into the streets of Toronto to buy dogs for one to three dollars each. Moreover, waiting for the ligated pancreases to degenerate took seven weeks or more, and it seemed like there was never enough extract to test with.

Frederick Banting (right) and Charles Best, photographed in the summer of 1921 on the roof of their laboratory building at the University of Toronto

Yet the rookie scientists knew they were on to something. In a letter sent to Macleod on August 9, 1921, Banting wrote, "I scarcely know where to begin." Banting reported that their extract, which he and Best chose to call "isletin," lowered blood sugar and improved the dogs' clinical symptoms, sometimes dramatically. Macleod wrote back and cautioned that much more work would be needed to ensure "no possibility of mistake." Macleod wrote, "It's very easy often in science to satisfy one[']s own self about some point but it's very hard to build up a stronghold of proof which others cannot pull down. Now, for example, supposing I wanted to be one of these critics I would say that your results on dog 408 were not absolutely convincing because . . ."

He went on to suggest new experiments Banting and Best should pursue, some of which the duo had already completed by the time Macleod's letter reached them in early September. They had performed a control experiment by depancreatizing two dogs, one of which received extract while the other did not. Within two days, the untreated dog was lethargic and could barely walk. Two days later, it died. In contrast, the dog treated with extract had markedly reduced blood sugar and remained in "excellent condition," running around the room, "frisky." This dog, number 92, and others who were being kept alive using the extract, became quite dear to Banting. But there was always a limited supply of extract, and after running out, dog 92 died three weeks after it had been depancreatized. Banting later wrote, "I have seen patients die and I have never shed a tear. But when that dog died I wanted to be alone for the tears would fall despite anything I could do. I was ashamed then [sic] I hid my face from Best."

After returning from Scotland in late September, Macleod met with Banting. This encounter portended a difficult relationship between the pair. Banting, who had worked all summer without any pay and had been living a meager, frugal lifestyle subsisting off his savings, asked Macleod for a salary, a boy to care for the dogs, a dedicated room for Banting and Best to perform

DIABETES: *The Pissing Evil*

their work in, and repairs to the floor of the operating room, which was quite dilapidated. To Banting's dismay, Macleod resisted. He implied that Banting and Best's research was no more important than other projects under his supervision and that giving them more resources would detract from other research. This response was highly offensive to Banting, who felt he had made tremendous sacrifices in order to pursue the project. Banting later wrote this account of the meeting:

> I told him that if the University of Toronto did not think that the results obtained were of sufficient importance to warrant the provision of the aforementioned requirements I would have to go some place where they would. His reply was, "As far as you are concerned, I am the University of Toronto." He told me that this research was "no more important than any other research in the department." I told him that I had given up everything I had in the world to do the research, and that I was going to do it, and if he did not provide what I asked I would go some place where they would. He said that I "had better go."

Banting turned to leave, at which time Macleod appeared to soften, and then relented. He ultimately conceded to all of Banting's demands.

After this encounter Banting vented his frustration to Best, who recalled that Banting "began to froth at the mouth." In time, Banting's resentment toward Macleod, whom Banting felt had been insufficiently supportive from the start, would grow to outright hostility. For many months, Macleod remained oblivious to the severity of Banting's anger. In his view, Banting and Best were amateur investigators with no real experience in physiology or research. If anything, it was he who was making the sacrifice to support such an unproven pair. A lack of mutual understanding would continue to plague the discovery team, with severe repercussions in the months to come.

John Macleod, photographed in 1928

The experiments resumed, and Macleod asked Banting and Best to present their work at a meeting of the university's Physiological Journal Club on November 14, 1921. Meetings such as these are ubiquitous at medical schools everywhere. They are considered informal and are generally attended by doctors and students from within the institution, not the general public. It was the ideal place for Banting and Best to present their work for the first time, since neither had any significant prior experience presenting scientific research.

Banting disliked public speaking. He tended to get extremely nervous and had a penchant for mumbling. Still, he carefully prepared his presentation for the journal club. He asked Macleod to introduce him and planned for Best to display posters showing their data while Banting did the speaking.

But to Banting's chagrin, Macleod gave a lengthy introductory speech that summarized practically everything Banting was planning to say. As a result, Banting—already anxious and ill at ease—became extremely flustered and did a poor job presenting the research.

Burning with humiliation, Banting directed his anger at Macleod. It bothered him that Macleod had given the appearance of leading the research. He had stolen Banting's thunder by explaining the whole project himself during the introduction. Moreover, Macleod had used the word "we" when he described the work, which rankled Banting and fueled his resentment. After the presentation, "students were talking about the remarkable work of Professor Macleod," Banting later wrote ruefully. Macleod was supposed to be their mentor—a world-renowned expert who should be helping them—but in Banting's view, any support from Macleod had only been given grudgingly and the professor had been completely absent during the main period of discovery. Now it seemed clear to Banting that, with their promising results, Macleod was keen to jump on the bandwagon and assume credit for their findings.

Banting did not share his feelings with Macleod, who remained oblivious. Much later, Macleod would write, "Had I been told of this attitude of Banting at the time, it would have served to warn me of his peculiar temperament and of his entirely unwarranted suspicions that I was trying to receive the credit for the results he had obtained."

Meanwhile, the most problematic and rate-limiting aspect of their research had become the requirement to wait seven weeks after pancreatic duct ligation for acinar cells to atrophy, before harvesting the pancreases to obtain usable extract. In mid-November, Banting thought of a possible way to speed their production of extract. He had read that newborn or fetal animals had a greater percentage of islet cells compared to digestive enzyme–producing acinar cells. This made sense, particularly in fetuses, because these animals had not yet ingested any external food and would therefore never have needed to produce digestive enzymes. Banting reasoned such pancreases might contain the internal secretion free of the external secretion. He knew that many farmers impregnated their cattle just before slaughter because pregnancy made them fatter and better feeders, so there were plenty of calf fetuses available at slaughterhouses. He and Best went to a local abattoir and excised the pancreases of nine calf fetuses. From these, they prepared their pancreatic extract, and to their delight, found that this extract worked, successfully reducing the blood sugar of depancreatized dogs. No longer would they need to perform complicated duct ligation surgeries, nor wait seven weeks for dog pancreases to atrophy before harvesting. All they had to do was go to the slaughterhouse, where they could obtain all the fetal pancreases they wanted.

In late November, Banting and Best wrote up their first paper detailing their results. This was titled "The Internal Secretion of the Pancreas," and would be published in the February issue of the *Journal of Laboratory and Clinical Medicine*, a well-regarded American journal. Banting asked Macleod if he would like to be a co-author on the paper, but Macleod declined. Macleod later wrote, "Banting asked me if I wished my name to appear along with his and Best's, and my reply was that I thanked them but could not do so since it was their work and 'I did not wish to fly under borrowed colours.'"

In early December, Banting and Best made another important discovery when they decided to use alcohol, instead of saline, as the medium with which to concentrate, and hopefully purify, their extract. This new method, which had been initially suggested by Macleod, yielded a more efficacious extract that reduced canine blood sugar more consistently. Then they wondered, if the mysterious active ingredient was soluble in alcohol, maybe they could use this technique to obtain the internal secretion from whole pancreas, not degenerated pancreas or fetal pancreas. The next time they removed a pancreas from a dog to render it diabetic, instead of throwing the organ away, they used it to prepare an extract by their new alcohol method and injected the extract back into the same dog. The dog's blood sugar dropped. It worked. This revealed that Banting's original hypothesis—that the internal secretion could only be isolated in the absence of digestive enzymes produced by acinar cells—was incorrect. With the right purification techniques, it was possible to isolate the internal secretion from whole pancreas—a vital discovery that ultimately made production of the active ingredient far quicker and easier than before.

A TEAM OF RIVALS

By now, the team knew that their discoveries could lead to beneficial treatments in human patients far sooner than anyone had previously dared hope. But to accomplish this, they would have to improve the purity of their extract's active ingredient even further. To date, large quantities of the extract—usually injections of 5 to 10 ml or more—were required to achieve significant reductions in blood sugar. And the potency of each batch remained variable. Sometimes a small amount of extract dropped the blood sugar dramatically. Other times it took serial, large-dose injections to obtain the desired result.

In mid-December, the team invited a biochemist named James Bertram Collip to join them. Collip was a young professor from the University of Alberta who had been awarded a Rockefeller Foundation Traveling Fellowship to fund a period of research work at the University of Toronto. He had prior experience working on the isolation of biological compounds and blood chemistry, so he was well suited to help the team purify the active ingredient.

Collip worked around the clock, experimenting with various methods of refining the internal secretion. He could change the concentration of alcohol, alter the temperature at which various steps were performed, vary the rate of evaporation using fans, use filters to aggressively remove lipids and other impurities, rapidly heat or cool the product, or tinker with a number of other factors, all in an effort to produce the purest, most potent extract possible. In late December, he performed an important experiment in which he sacrificed a depancreatized dog that had been treated with extract and removed the dog's liver. He tested the liver and, to his delight, found that it contained a high degree of glycogen. Since only the internal secretion could induce the liver to store glucose as glycogen, this provided further evidence that the extract was a successful substitute for the body's natural internal secretion. Collip settled on a strategy of dissolving the extract in higher and higher concentrations of alcohol. When the active ingredient remained soluble in alcohol, lipids and other impurities would precipitate out as solids after being centrifuged. By this method, he was able to remove more contaminants than ever before and concentrate the final product much more effectively. As the calendar turned to 1922, he was getting tantalizingly close to success.

Unfortunately, however, just as their scientific progress was accelerating rapidly, relations between the team members began to deteriorate. On December 30, 1921, Banting, Best, and Macleod traveled to New Haven, Connecticut, to present their research at the American Physiological Society Conference, a major scientific meeting. Many important figures in diabetes research were present, including Frederick Allen, from the Physiatric Institute in New Jersey, and Elliott Joslin, a physician from Boston who had written an important textbook on diabetes. This was the big leagues, and Banting was extremely nervous. Regrettably, his performance at this venue was even worse than it had been at the journal club in Toronto the month prior. Before a large and august assembly, Banting mumbled and could hardly be heard by many in the audience. He failed to make a convincing case for the isolation of the pancreas's internal secretion. In his own words, he later recalled: "When I was called upon to present our work I became almost paral[y]zed. I could not remember nor could I think. I had never spoken to an audience of this kind before. I was overawed. I did not present it well."

When Banting finished, experts in the audience began to ask questions and raise concerns about the investigators' assertions. Macleod found himself in the very difficult position of watching his mentee floundering. Banting's research came from Macleod's lab and therefore reflected on Macleod's own reputation. As the project's senior scientist, he had no choice but to enter the discussion and try to salvage Banting's presentation, which he largely succeeded in doing.

However, instead of being grateful, Banting grew angry and envious that Macleod was so articulate—whereas he had failed so miserably. He blamed Macleod for hijacking the discussion and for using "we" and "our work" freely in his discourse. Banting's insecurity and festering anger at Macleod's initial lukewarm enthusiasm for the work now evolved into full-blown paranoia that Macleod was trying to claim credit for research he had barely participated in at all. In an unpublished memoir, handwritten in 1940, Banting described his state of mind while riding the overnight train back to Toronto: "I did not sleep a wink on the train that night—I did not even go to my berth but sat up in the smoker condemning Macleod as an imposter & myself as a nincompoop . . . I knew Macleod for what he␣was[,] a talker & a writer. Apart from his pen & his tongue he would not even be a lab. man for he had no original ideas, he had no skill with his hands in an experiment."

Banting's paranoia also began to spread beyond Macleod alone. He became jealous of Collip's success in isolating a purer form of the internal secretion and realized that history would best remember the person whose pancreatic extract was first used to successfully treat a human patient. It was Collip's job to purify the active ingredient, so it made sense that his formulation would be used first; but Banting now wanted his extract to be the first. Banting asked Macleod to help him convince the hospital to let him use the extract he and Best had produced on the first human subject. Perhaps against his better judgment—Collip's product was almost certainly purer than Banting's—and perhaps to placate Banting's increasingly demanding manner, Macleod agreed to do this, and permission from the hospital was obtained.

The first patient to officially receive Banting's pancreatic extract was a fourteen-year-old boy named Leonard Thompson. By following Frederick Allen's starvation diet therapy, Leonard was emaciated and just sixty-five pounds. Clinically he was lethargic and pale; his hair was falling out and his breath smelled of acetone. He lay in bed almost all day and resembled a famine or concentration camp victim—mere skin and bones, with a distended belly. Banting and Best prepared their extract from whole beef pancreas using the same technique they had employed in December to produce extract that was successful in dogs. Its appearance was a viscous brown fluid. On January 11, 1922, 7.5 ml was injected into each of Leonard Thompson's buttocks by a resident physician named Ed Jeffery. They had chosen an amount that was half the dose expected to successfully lower blood sugar in a dog of equal weight. When the blood sugar results came back, Banting and Best learned the treatment had had only a modest effect. The blood sugar dropped from 440 mg/dL to 324 mg/dL. Sugar was still found in the patient's urine. No clinical improvement was seen, and unfortunately, an abscess formed at the site of one of the injections.

No doubt Collip must have found Banting's behavior extremely irritating. Banting appeared to have rushed to treat a patient with an inferior product out of self-interest. Meanwhile, Collip continued to work day and night. Just a few days later in mid-January, in the middle of the night, Collip succeeded in producing what he considered nearly pure extract for the first time. Of this moment, he later wrote: "I experienced then and there all alone in the top story of the old Pathology Building perhaps the greatest thrill which has ever been given me to realize." At this point, James Collip became the first person in history to actually see what the team ultimately decided to call "insulin" in its physical form.

This was a triumph, but one that was soon marred by an unforgettable altercation between Collip and Banting. In Banting's own words:

> The worst blow fell on[e] evening toward the end of January. Collip had become less & less communicative and finally after about a week's absence he came into our little room about 5:30 one evening. He step[p]ed inside the door and said "Well fellows I've got it." I turned and said, "Fine[,] congratulations. How did you do it[?]" Collip replied, "I have decided not to tell you." His face was white as a sheet. He made as if to go. I grabbed him in one hand by the overcoat where it met in front and almost lifting him I sat him down hard on the chair. I do not remember all that was said but I remember telling him that it was a good job he was so much smaller other-

wise I would "knock hell out of him." He told us that he had talked it over with Macleod and that Macleod agreed with him that he should not tell by what means he had purified the extract.

Collip never wrote of this incident, but Best, who was present, later wrote that Collip, "announced to me that he was leaving our group and that he intended to take out a patent in his own name on the improvement of our pancreatic extract . . . Banting was thoroughly angry and Collip was fortunate not to be seriously hurt . . . I can remember restraining Banting with all the force at my command."

Relations among the discovery team had reached a nadir. On January 23, patient Leonard Thompson received Collip's extract. It was far more effective than Banting's first treatment twelve days earlier. Thompson's blood sugar dropped from 520 mg/dL to 120 mg/dL. There was almost no glucose in his urine. He felt much better, with greater strength and energy.

As more patients were treated, it became clear the team had created a medical miracle. A scientist named George Clowes, who led research and development for Eli Lilly & Company in Indianapolis, Indiana, had attended the group's December presentation in New Haven and realized what insulin might mean for the world's diabetics. Through great persistence, Clowes eventually convinced the discovery team and the University of Toronto to grant Eli Lilly & Company the rights to manufacture and sell insulin. This proved to be a crucial step because the Torontonians soon experienced difficulty producing consistently potent insulin in large quantities, something Eli Lilly & Company was better equipped to do using the industrial-sized equipment at their large facility in Indianapolis. Eli Lilly & Company was soon ordering massive quantities of pancreases from abattoirs across the Midwest to be delivered to their plant in Indiana.

Banting, who set up a diabetes clinic in Toronto, became inundated with letters and telegrams from doctors and desperate patients. Because insulin was at first in very short supply, he was forced to turn away the vast majority of those seeking his help. Yet nothing could stop news of the miracle medicine from spreading quickly around the world. Insulin could bring emaciated, listless boys and girls back to life, enabling them to gain back fifty or a hundred pounds, eat formerly forbidden foods, and laugh and play like healthy children.

"SIMPLY TOO WONDERFUL FOR WORDS"

On July 3, 1922, Antoinette Hughes, spouse of U.S. Secretary of State Charles Evans Hughes, wrote Banting a letter about her daughter Elizabeth. By this point, Elizabeth Hughes was five feet tall but weighed less than fifty pounds fully clothed. Her diet had been reduced to less than 300 calories per day.

Antoinette asked if Banting would be willing to see Elizabeth if she brought Elizabeth to Toronto. "My daughter," she wrote, "has had Diabetes for a little more than three years. Her case was, and still is, severe and, though it has been handled skillfully and with the utmost care, her tolerance is still very low and she is pitifully depleted and reduced." She alluded to Elizabeth's uncommon willpower and resolve: "She is a model patient having never once, in these three years, gone over her diet. The steps backward—when they have come—have invariably been due to no fault of hers and we feel that this strength of character—which her nurse considers quite unusual—is—of course—much in her favor."

Although Banting's initial reply to Antoinette was his standard, polite regrets that he was unable to see new patients due to the shortage of insulin, he later agreed to see Elizabeth in mid-August. Banting's clinical notes from August 16, 1922, describe her like this:

> wt 45 lbs. height 5 ft. patient extremely emaciated, slight aedema of ankles, skin dry & scaly, hair brittle & thin, abdomen promment [sic], shoulders drooped, muscles extremely wasted, subcutaneous tissues almost completely absorbed. She was scarcely able to walk on account of weakness.

Banting immediately started giving Elizabeth insulin, which allowed her to increase her diet. After just one week on insulin, her daily caloric intake was 1,220. A week later, she was eating a normal diet of 2,200 to 2,400 calories. In a diet journal, Elizabeth noted eating her first slice of white bread in three and a half years. Four days later came her first taste of corn in three years; nine days later, macaroni and cheese; six days after that, grapes, and then bananas and plums.

In a September 24 letter to her mother, who had departed on a state visit to Brazil, Elizabeth wrote, "I can't express my gratitude for the chance I'm having, in being up here to take advantage of this wonderful discovery for it is truly miraculous... Dr. Banting... brings all these emminent [sic] Doctors in from all over the world who come to Toronto to see for themselves the workings of this wonderful discovery, and I wish you could see the expression of their faces as they read my charts, they are so astounded in my unheard of progress and change in my looks." At the beginning of October, she wrote in another letter, "Oh, it is simply too wonderful for words this stuff and you just wait and see how it[']s changed me . . . Wouldn't Dr. Allen have ten fits if he knew what I was on now."

When Frederick Allen traveled to Toronto to see Banting and attend a conference of prominent diabetologists, Banting brought him and several other clinicians to see Elizabeth. Of the moment Dr. Allen saw her, Elizabeth wrote, "Dr. Allen said with his mouth open—Oh!—and that's all he did. He just kept saying over and over again that he had never seen such a great change in anyone and he actually cracked a joke as he was leaving saying he was glad to have been introduced to me or he wouldn't have known who it was."

Soon physicians across North America were celebrating miraculous results in patients just like Elizabeth—children on the brink of death, now suddenly resurrected. Boston's Dr. Elliott Joslin related his experience in biblical terms, saying, "By Christmas of 1922 I had witnessed so many near resurrections that I realized I was seeing enacted before my eyes Ezekiel's vision of the valley of dry bones."

(Left) A five-year-old boy named Teddy Ryder in July 1922, before receiving insulin.
(Right) After insulin treatment, one year later

In one unforgettable account, Banting shared the following story at the conclusion of his unpublished 1940 memoir:

A man carried his wife into the office under his arm ... he deposited in the easy chair 76 lbs of the worst looking specimen of a wife that I have [ever] seen. She snarled and growled and ordered him about. I felt sorry for him. I placed her in hospital more in pity for him than in regard for her.

She was one of the most uncooperative patients with which I have ever delt [sic] ... She would steal candy or any kind of food she could lay hands on. She demanded that her poor husband come to the hospital early in the morning and every night he must not leave until she was asleep, yet she scolded him[,] cursed him and treated him like nothing all day long ... She was a terrible looking specimen of humanity with eyes almost closed with aedema, a pale and pasty skin, red hair that was so thin that it showed her scalp . . . Above all she had the foulest disposition that I have ever known . . . She was in hospital some weeks and improved considerably and then he took her home. I was frankly glad to see the last of her. For his sake I had been kind . . .

A year later I was at my desk early on[e] morning when the phone rang. A cheerful chuckling voice asked if I would be there for ten minutes. I said I would. The receiver was hung up. I went on with my correspondence.

In a few minutes I heard the outer door open and a moment later my office door was thrown wide open as in rushed one of the most beautiful women I have ever seen. She was a stranger. I had never seen her before yet she threw her arms around my neck and kissed me before I could move from where I stood. Over her beautiful head I saw the laughing face of the patient husband. I stood back. The three of us stood hand in hand. I looked at them. The husband said, "Doctor I wanted you to see her now. This is the girl I married—before she had diabetes . . ."

Months later I received a tiny envelope with the name and a pink ribbon. A daughter. And I wondered if the little one had red hair, and I prayed she would never have diabetes.

Sadly, the triumph of the Canadian team remained tarnished by internal conflict among its members. There is no photograph of all four discoverers together. In 1923, Banting and Macleod were awarded the Nobel Prize in Physiology or Medicine.* Incensed that Macleod had been recognized as an equal with him, and furious that Best had been excluded, Banting considered refusing the honor. He later agreed to accept it, but announced that he would split his share of the prize money with Best. Shortly thereafter, Macleod announced that he would split his share with Collip.

Banting's grudge against Macleod never abated. He always resented Macleod for showing little interest in the project at the outset, felt Macleod had

---

* Banting was thirty-two years old and he remains the youngest recipient in history to receive this honor.

contributed little, and believed the professor had taken undue credit once they succeeded. Macleod considered Banting incredibly oversensitive, hopelessly paranoid, and impossible to placate. Debate and argument over who had done what and when—indeed who deserved credit for discovering insulin itself—persisted for decades. In truth, all four investigators made significant contributions to the discovery. Banting had the original idea, drive, and passion to start the project. He and Best did the crucial early experiments that established the proof of concept. Collip's expertise allowed the team to isolate and purify insulin quickly, which undoubtedly saved lives by making insulin available more promptly. Macleod did contribute ideas; he provided direction, sometimes prevented Banting and Best from pursuing unproductive lines of research, and helped demonstrate the merit of the research to the scientific community.

Macleod left the University of Toronto in 1928, possibly, in part, because Banting's public animosity toward him had made life in the community very unpleasant. Macleod returned to Scotland, where he became a professor at the University of Aberdeen, which was thrilled to welcome a Nobel laureate to its roster. Macleod died in 1935.

Charles Best replaced Macleod at the University of Toronto as professor of physiology; he lived the longest of all the discoverers, until 1978.

James Collip became a professor at McGill University and later served as Dean of Medicine at the University of Western Ontario. During his career he performed important research on parathyroid hormone. He died in 1965.

Frederick Banting became a national hero and enjoyed many accolades. He was awarded lifetime annual stipends from the Ontario and federal governments of $10,000 and $7,500, respectively. The University of Toronto established the Banting and Best Department of Medical Research, and Banting spent the rest of the 1920s and 1930s working on various research ideas that he hoped would lead to a cure for cancer. During the discovery period, he had turned down lucrative offers to bring his research to American universities, as well as enormous sums to patent the production of insulin for commercial gain. Though close to destitute, Banting considered the notion of profiting from his discovery ethically repugnant; his greatest desire was for insulin to be produced inexpensively and made available to all patients, rich and poor.

Ironically, Banting's relationship with Best took a downturn in later years. Just as surprising—Banting came to like and appreciate Collip, whom he befriended and served with on the Canadian National Research Council in the late 1930s. In 1934, Banting was knighted by the Canadian government. He married, divorced, then married again. In the Second World War, he served as chairman of Canada's medical research unit. On February 20, 1941, Banting was a passenger in a Hudson bomber that departed Gander, Newfoundland, en route to England. The plane developed engine failure and, in heavy snowfall, crashed in the wilderness next to a frozen lake. Banting survived the crash but died from his injuries twenty hours later. He was forty-nine years old.

Elizabeth Hughes went on to enjoy her life as a wife, mother, and grandmother but was always very secretive about her diabetes. She remained so private that her fiancé did not learn she had diabetes until she told him a week after their engagement. She died of a heart attack in 1981 at the age of seventy-four, after having had 42,000 lifesaving insulin injections in her lifetime.

COMBATING A PUBLIC HEALTH CRISIS

Today, type 1 diabetics like Elizabeth Hughes who cannot make enough insulin comprise only about 5 to 10 percent of diabetic patients worldwide.* This reflects the fact that the prevalence of type 2 diabetes, in which an indi-

---

* Why can't patients with type 1 diabetes make insulin? There are three possible reasons: a genetic defect that inhibits proper cell function, an autoimmune disorder in which the body's own inflammatory cells attack and disable the islets of Langerhans, or a viral infection that causes damage to these cells.

vidual's tissues cannot use insulin to its full effect, has skyrocketed in recent decades.*

Why does insulin work less effectively in type 2 diabetics?

Genetics plays a role. For example, a study from Finland showed a 34 percent concordance rate for identical twins to develop type 2 diabetes. But lifestyle remains the most important factor, particularly with regard to obesity. Approximately 80 to 90 percent of type 2 diabetics are overweight or obese. It is not entirely understood why fat cells reduce the effect of insulin on the body—a predicament termed *insulin resistance*—but it's believed that increased fat metabolism leads to a reduction of insulin receptors on cell surfaces. Without the ability to use glucose, the body breaks down fat for energy, but this leads to more problems because acidic ketone bodies (a product of fat breakdown) accumulate in the bloodstream. Unchecked, this derangement of the body's metabolic system can lead to ketoacidosis, coma, and death.

For the first time in history, instead of dying from eating too little, humans are now dying from eating too much, and an epidemic of obesity has become a slow-moving public health crisis. Diabetes now affects 422 million people worldwide, including an astounding 37.3 million Americans—approximately 10 percent of the population. It is the eighth leading cause of death in the United States. Over one-third of American adults, about ninety-six million, suffer from "prediabetes," with blood glucose levels that are higher than normal but not yet high enough to be labeled type 2 diabetes. It is estimated that up to 70 percent of prediabetics will eventually develop diabetes.

Thankfully, physicians now have an expanding number of sophisticated pharmacological options in their fight to overcome the disease. In the 1950s, the development of sulfonylureas, oral medications that lower blood glucose by stimulating insulin release from the pancreas, was a major advance. Another mainstay of modern treatment is metformin, an oral drug that reduces the release of glucose by the liver, decreases intestinal absorption of glucose, and increases sensitivity to insulin's actions in peripheral tissues of the body.

---

* Note: many type 2 diabetics still benefit from, and require, insulin treatment.

DIABETES: *The Pissing Evil*

In this century, medicines termed GLP-1 *agonists* and DPP-4 *inhibitors* have been developed to mimic or boost hormones called *incretins* that bind to the pancreas's GLP-1 receptors and stimulate increased production of insulin. Another class of beneficial medications is termed SGLT-2 *inhibitors*. These oral drugs enhance the kidney's ability to excrete glucose, which not only lowers blood glucose but also promotes weight loss and has been shown to improve the health of patients with heart and kidney disease.

The production and use of insulin have become far more sophisticated than in the past. For almost fifty years, the process of making insulin from cow or pig organs remained time-consuming and laborious. It took about eight thousand pounds of animal pancreas to yield one pound of insulin. It wasn't until 1967 that a scientist named Dorothy Crowfoot Hodgkin used X-ray diffraction to determine insulin's precise chemical structure. It became clear that insulin is virtually identical across animal species, from humans to fish to cows and pigs, which explained why animal insulin worked in humans just as well as endogenous insulin. In 1978, Genentech, the pioneering biotechnology company, was the first to produce synthetic insulin using recombinant DNA technology. Scientists successfully introduced a copy of the human insulin gene into bacteria, each of which then began to act as a tiny factory that produced a great deal of insulin autonomously. Synthetic insulin was approved by the U.S. Food and Drug Administration (FDA) in 1982 and was known by the brand name Humulin, to differentiate it from pig or cow insulin. Today, patients use short-acting and long-acting formulations of insulin to meet their needs, as well as wearable insulin pumps, invented in the 1970s, which are small, computerized devices that deliver insulin through a tiny cannula placed under the skin.

The more recent advent of continuous glucose monitoring (CGM) promises to further revolutionize diabetic care and is a significant step toward realizing the promise of an "artificial pancreas." A CGM monitor is a sensor that is commonly applied to the belly or arm with an adhesive patch. It contains a tiny filament that measures the glucose level in the body's interstitial fluid—fluid that fills the spaces between cells in our tissues. The sensor takes readings every few minutes and transmits the data to a pump, smartphone, or other device. This allows patients to monitor blood glucose levels day and night, allowing for more precise and effective insulin dosing.

Scientists and medical device companies are working hard to combine CGM sensors with insulin pumps in a closed-loop design that would detect blood sugar levels and respond by automatically delivering insulin if the blood sugar is high, and glucagon—which raises blood sugar—if low. Companies like Medtronic and Tandem already manufacture insulin pumps that integrate sensor readings with automatically dosed insulin to some extent, but a true "artificial pancreas" that requires no patient input remains elusive.

Which brings us to the present, and a potentially game-changing development: the implantation of stem cells, induced to become functioning islet cells that produce insulin. Transplantation as a strategy to cure diabetes is not new. Pancreas transplants have been performed since the mid-1960s. Today, transplantation of islet cells alone can be performed as well—but these treatments remain uncommon due to limited supply of available donor organs. In November 2021, it was revealed that the first patient who had received stem cell–derived islet cells in a new clinical trial had been essentially cured of type 1 diabetes. His new islet cells functioned properly and completely regulated blood sugar levels. This breakthrough was the result of a two-decade quest by laboratory scientists to discover how to induce stem cells to become islet cells. It will be years before such an advance might manifest as treatment for greater numbers of patients, but the news, reported on the front page of the *New York Times*, has raised hopes that a true cure for type 1 diabetes may someday be found.

For now, the battle against diabetes continues to be waged by many: physicians and nurses, dietitians and school administrators, trainers and coaches, parents and friends. For most type 2 diabetics, efforts to lose weight and curb obesity remain essential goals. The miracle of insulin prolonged lives, but increased patient longevity also served to uncover diabetes' myriad sequelae, including retinopathy, nephropathy, and peripheral neuropathies. A century after insulin's discovery, diabetes is no longer a death sentence, but it remains one of our most serious health challenges.

3

BACTERIAL INFECTION

The Magic Bullet

November 30, 1940
Southampton, England

Southampton was reeling. A German bombing raid on November 23 had destroyed hundreds of buildings and caused seventy-seven deaths, as well as over 300 additional casualties. The port's dockyards and factories were prime Luftwaffe targets, particularly the Supermarine Spitfire plant located in a suburb of the city. Now, a week later, the enemy was sending an even larger group of warplanes to pulverize the city.

The recent carnage had prompted the reassignment of policemen from other cities and towns to Southampton. One of them was forty-three-year-old Albert Alexander, a constable from Oxford. On November 30, Alexander watched as 120 German aircraft dropped 800 bombs on the city over a six-hour period. It was Southampton's worst bombing raid of the war. One hundred and thirty-seven people died. Over a thousand homes were destroyed.

During the onslaught, as Alexander manned his post at the police station, a bomb scored a direct hit on the building.

Alexander survived.

He was one of the lucky ones. His only injury was a small cut near the corner of his mouth, from a piece of flying shrapnel. Unlike so many other poor souls, he would live to return to his family; he and his wife Edith had two young children.

After the bombing, Alexander wanted to resume his normal duties. The cut on his face seemed trifling. It was unsightly, but he hoped it would simply heal on its own.

Except that it didn't. The annoying wound began to fester. The pain and redness grew to more than he could bear. It worsened to involve more than just the skin of his face; the infection spread upward, across his eyes and up to his head and scalp. After several days, even he had to admit he had no choice—to the professionals he must go. Alexander checked himself into the Radcliffe Infirmary, the primary hospital in his hometown of Oxford.

The doctors cultured samples from the sore, which yielded two types of bacteria: *Staphylococcus aureus* and *Streptococcus pyogenes*. They treated him with sulfapyridine, one of the famed "sulfa drugs," or sulfonamides, being mass-produced for the boys in the field. Unfortunately, an eight-day course of the drug did not help Alexander. In fact, it made things worse by causing an allergic skin rash.

By January, Alexander had developed multiple abscesses upon his head and face. The doctors incised them to drain the pus. When he developed right shoulder pain, an X-ray revealed osteomyelitis of the head of the humerus—the infection was spreading. Even worse, Alexander's left eye became very painful and badly infected—so badly that he went blind in that eye. Then, that eye's cornea grew so damaged and weak that it spontaneously perforated.

The doctors had no choice but to remove his eye.

Now desperation set in. By early February, Alexander's entire face and head were covered with abscesses. He was emaciated and in constant pain. His lungs became infected, threatening his ability to oxygenate his blood. A blood transfusion did not help.

It became obvious that he was going to die.

How had this happened? All from a tiny cut? Even during wartime, with bombs falling all around and tragedy so familiar, this months-long ordeal seemed an exceptionally cruel way to die. What would become of his family? How would his wife be able to raise both children on her own?

In mid-February, Alexander's doctor, Charles Fletcher, offered a small ray of hope. An unproven, untested Hail Mary. Fletcher had just received an experimental new medicine from a team of researchers at Oxford University. There was a chance this drug had antibacterial properties.

Would Alexander be willing to try it?

What choice did he have?

"ANIMALCULES"

Infectious diseases caused by bacteria killed more people than heart attacks, cancer, or any other affliction from the beginning of recorded history until the mid-twentieth century. Death from a simple cut, like the one suffered by Albert Alexander, was commonplace.

The first person to identify the bacterial microorganisms responsible for such tragic, premature deaths was a Dutchman named Antonie van Leeuwenhoek who took up microscopy as a hobby in the 1640s. A draper by trade, he regularly employed magnifying glasses to count the thread density of cloth. With an affinity for lenses and their refractive properties, he learned to manufacture his own lenses and used them to build a microscope. Van Leeuwenhoek soon saw that even a single drop of water teemed with microscopic life. He carefully scrutinized the tiny organisms that colonized a sample of rainwater left stagnant for days in a barrel and called them "animalcules." Because these animalcules appeared to move, he presumed they were alive and described them as "a thousand times smaller than the eye of a full-grown louse." After examining a saliva sample, he marveled, "All the people living in our United Netherlands are not as many as the living animals that I carry in my own mouth."

Bacteria* are single-celled microorganisms that are typically classified by their shape—for example, rod-, sphere-, spiral-, or comma-shaped. They are protected by a cell envelope (comprised of a cell membrane with or without a cell wall), lack a nucleus, and contain free-floating DNA, often arranged in circular units called plasmids. A supplementary way of classifying bacteria is by their appearance when stained with various dyes—the most important being the "Gram stain," devised by Hans Christian Gram in 1884. Gram staining colors bacteria that have a thick cell wall purple (Gram positive), while bacteria that either possess a thin cell wall or lack a cell wall counterstain pink (Gram negative).

The number of bacteria on Earth is estimated to be five million trillion trillion (5×10^{30}). A gram of soil may contain more than forty million bacteria, and there are more bacteria in our bodies than there are human cells. Most bacteria cause no harm, and many are helpful, particularly ones that reside in our gastrointestinal tract and promote digestion. But some bacteria are pathogenic and responsible for a number of humanity's worst afflictions, including bubonic plague, leprosy, syphilis, anthrax, cholera, tuberculosis, diphtheria, and many more.

Prior to the broader understanding of pathogenic microorganisms that would grace the twentieth century, there existed two competing beliefs that claimed to explain the spread of disease. These were known as the contagion and anti-contagion theories. Contagionists believed that disease was passed via personal contact by an unknown source of pestilence, sometimes called an "invisible bullet" or "animalcule," and later labeled "germs." Contagionists felt the best way to combat disease was to separate the infected from the uninfected—hence their reliance on strict quarantines of affected populations, and often the cessation of trade with foreign areas that were blamed as the source of the pestilence.

In contrast, anti-contagionists believed the environment was more blame-

---

* "Bacteria" is derived from the Greek word *bakterion* meaning "small staff." It is the diminutive form of *baktron*, meaning stick, rod, or staff.

worthy than human contact. They attributed disease to environmental factors such as miasmas, or "bad air," emanating from decaying organic material like rotting food or animal corpses. According to anti-contagionists, the best way to combat an epidemic was to clean up squalid, overcrowded slums and foul-smelling, sewage-laden streets in order to properly cleanse the air. It is largely thanks to anti-contagionists that cities constructed proper sewage systems. Not surprisingly, anti-contagionists believed the best way to protect oneself during an epidemic was to leave for the countryside, or perhaps retreat to a sanitarium, where the air was certain to be fresh and clean. Good personal hygiene was also of paramount importance.

There appeared to be ample evidence to support both theories, but neither was definitive in all situations. Contagionists pointed to diseases like smallpox, where personal contact was obviously the primary conduit for disease spread. But other conditions, such as yellow fever and malaria—both mosquito-borne illnesses—did not fit this mold. Meanwhile, the anti-contagionist viewpoint seemed adequate to explain why some disease outbreaks affected poverty-stricken areas the most, but it could not account for rural epidemics. And then there were diseases like cholera that failed to fall into either camp. Cholera was not spread by human contact. Nor was it due solely to squalid conditions because epidemics sprang up indiscriminately among rich and poor, urban and rural, and in winter, when "bad air" from stifling, humid conditions in cities was considered reduced.

The excellent detective work of British surgeon John Snow helped to demonstrate that cholera was, in fact, most commonly a water-borne illness. During an 1854 epidemic in London, Snow plotted the residences of cholera-stricken patients on a map. He noticed that most patients drew water from a pump at the corner of Broad and Cambridge Streets in Soho. He took pains to seek out and interview as many families as possible, and the deeper he investigated, the more conclusive his initial assumption became. Even cases that did not at first seem related to this particular pump were eventually shown to be—for example, one sick fifty-nine-year-old woman who lived far away from the pump had no apparent connection to it, until her son, who did live near it, revealed to Snow that his mother visited him often and drank from it frequently because she preferred the taste of the water from that specific pump. Snow theorized that cholera was spread through contaminated water, not by personal contact or miasma. In this case, the city's crumbling sewer system had allowed waste from an old cesspit to seep into the water supply. Snow convinced officials to remove the pump handle, after which the epidemic rapidly receded.

Snow's work was later used to support contagionists' "germ theory" of disease. But even decades later, not everyone was convinced. Florence Nightingale was one die-hard anti-contagionist who considered germs hogwash. She devoted her career to improving living conditions, cleanliness, and sanitation. Another well-known anti-contagionist named Max von Pettenkofer, a Bavarian chemist, dramatically demonstrated his conviction that cholera was not contagious by drinking a broth laden with cholera in front of witnesses in 1892. Amazingly, von Pettenkofer only suffered mild symptoms and did not contract cholera, seeming to validate his theory. Partly due to the actions of authorities like von Pettenkofer, the germ theory of disease was not widely adopted until the turn of the twentieth century. And even mid-nineteenth-century contagionists like John Snow had to admit that, though they believed something in contaminated water was responsible for causing diseases like cholera, they could not fathom, nor identify, what that something was.

A SAVIOR OF FRENCH WINE AND SILK

In 1856, Louis Pasteur was a thirty-three-year-old professor of chemistry in Lille, an industrial city in northern France. The father of one of his students, a brewer, came to him with a problem. The brewer used fermented beet sugar to make alcohol. Although most of his wine production came out perfectly, clear and tasting great, there were always some barrels in which the wine went sour and couldn't be sold. This had become an increasingly common problem

BACTERIAL INFECTION: *The Magic Bullet*

among winemakers in France, with adverse economic consequences. Why was this happening? The brewer hoped Pasteur could help find answers.

Up to that point, Pasteur had focused primarily on a topic befitting the interest of a chemist: the study of crystallography. He had never delved into the life sciences, but he agreed to inspect the brewer's wine casks. Sure enough, some of the wine had spoiled. When Pasteur looked at samples of the good and bad wine under a microscope, he saw a significant difference between them. In the good, clear wine he saw small, round yeast cells; but in the sour wine samples the yeast cells were inundated with tiny, black, rod-shaped specks. He did not know what these were, but because of their absence from the good wine he concluded that they were the reason why the bad wine had spoiled. These microorganisms were overwhelming the yeast cells and as a result, the wine was being fermented into sour-tasting lactic acid instead of alcohol. Pasteur told the brewer that every barrel of wine should be inspected and any contaminated samples should be thrown out, and certainly never mixed with the contents of other casks, which would contaminate the whole lot. Pasteur's method was adopted by winegrowers throughout France and, some believe, effectively saved the wine industry from ruin.

Pasteur became more interested in these mysterious microorganisms. He learned they also soured milk, but that he could kill them by heating them. This heating process became known as "pasteurization" and was adopted as a way to preserve milk and other foods and beverages.

In the 1860s, Pasteur made additional groundbreaking discoveries. To debunk the commonly believed theory of "spontaneous generation," which held that microbes originated spontaneously out of the air rather than as the offspring of parent microbes, Pasteur designed an experiment using a swan-necked flask and showed that microbes did not colonize boiled water exposed to sterile air. He also aided France's silkworm trade by identifying microbes that sickened silkworms and threatened to ruin the industry.

For these efforts, Pasteur gained the gratitude and admiration of many Frenchmen. He was well on his way to establishing himself as the leading figure in the new field of microbiology. But then, in 1876, incredible news reached France about the discoveries of a much younger German doctor that no one had ever heard of.

THE UNKNOWN GERMAN

Robert Koch was the last kind of physician one might expect to become a leading bacteriologist. Twenty-one years younger than Pasteur, he wasn't a professor at a prestigious university or the author of dozens of published papers. He was a mere country doctor in a village called Wollstein, which is now in western Poland. But he would soon prove to be one of medicine's greatest pioneers—and Pasteur's nemesis.

Koch spent most of his time taking care of farmers' aches and pains in an exam room located inside his own house. After his wife gave him a microscope as a birthday gift, he converted part of the exam room into a rudimentary laboratory and began to discover things.

In 1873, an anthrax outbreak struck the livestock around Wollstein. This highly contagious disease felled animals without warning and regularly decimated herds of sheep and cattle. A cow that appeared healthy one day would begin moving slowly the next, behaving listlessly. It would refuse to eat. Shortly thereafter, it might sink to the ground, a grotesque, sickened beast with blood streaming out of its mouth and nostrils. Soon it would be found dead on the ground, supine with its legs sticking upright in the air. Farmers had no idea what caused this plague and cast blame on evil spirits, excessive feeding, or poisoning. Even worse, humans could contract the disease and die as well.

Koch set out to investigate this pestilence. Using his microscope, he examined the blood of diseased animals and saw dark rods. He watched them grow and divide. There were also curious, small spheres that could grow into rods and link themselves together. Koch found that he could transfer the disease to rabbits and mice by injecting them with blood from a diseased cow. Within a day, these animals developed anthrax-type symptoms and died.

When Koch examined their blood, he saw the same dark rods again. When he injected this blood into another animal, that animal also died. Over the course of the next three years, his careful investigations allowed him to recognize the tiny spheres were spores of dormant bacilli that could later transform into active, pathogenic rods. He could kill the microscopic rods by subjecting a sample to heat, but the spores did not die. They retained their ability to produce new rods and more spores. This finding explained how the bacteria could survive through the winter, and why some farmers found certain fields "plagued"—animals that grazed upon them died year after year. With little to no training in microbiology, Koch had isolated *Bacillus anthracis*, the causative microorganism of anthrax, and single-handedly elucidated its full life cycle.

Koch yearned to share his discovery with others and possibly publish it in an academic journal, but he was filled with self-doubt. How could he be sure his conclusions were correct? He did not have the resources of an established university, nor were there any local doctors or scientists he could discuss his findings with. He decided to make a bold move and write to a world-famous botanist at the University of Breslau,* Ferdinand Cohn, who had also studied microbes. Koch humbly asked Cohn if he would be willing to review Koch's experiments and results before he tried to publish them. One can imagine Cohn rolling his eyes at Koch's request, as he later admitted, "I had been receiving such communications from dilettantes, regarding alleged discoveries, in this yet undeveloped field. I anticipated little of value from this request from a completely unknown physician from a rural Polish town." But, perhaps taken by Koch's earnest letter, Cohn invited Koch to come to Breslau that very Sunday, April 30, 1876, to show Cohn his work.

It is easy to picture Koch's eagerness and nervous excitement upon receiving this invitation. Tellingly, he chose to leave Wollstein by train in the middle of the night, at 1 a.m., to guarantee his punctual arrival at the university in Breslau. Koch brought along a cornucopia of scientific paraphernalia, including microscopes, slides, vitreous fluid samples, chemical reagents,

---

* Breslau is now Wrocław, Poland.

rabbits, frogs, and mice—some of which were infected with anthrax. Over the course of three days, he humbly and respectfully performed multiple demonstration experiments for the assembled professors and students, methodically showing how he had isolated the bacilli and transmitted it from one animal to another.

The university professors were stunned.

Cohn recalled, "Within the very first hour I recognized that he was the unsurpassed master of scientific research." Another well-known professor named Julius Cohnheim declared, "I consider this the greatest discovery in the field of bacteriology." It seemed impossible that an unknown country doctor had solved the mysteries of anthrax as an amateur microbiologist working in isolation without any sophisticated equipment or even a medical library at his disposal.

In his diary, Koch wrote, "My experiments were well received." In one afternoon, he had done more than anyone to prove that the germ theory was true. Up to that point, although Pasteur's experiments had done much to advance the germ theory, there remained many skeptics. Many believed that microorganisms were a byproduct of infection, not the cause of it. Others were willing to accept microorganisms were responsible for some infectious diseases, but not all. Now, for the first time, a microorganism had been isolated and unequivocally proven to be the cause of an important disease.

Koch's discovery would launch a veritable gold rush of prospectors on the hunt for the pathogens behind many of the world's deadliest diseases. The late 1800s became an exhilarating time when new discoveries by daring microbe hunters could hit the headlines at any moment. These scientific innovators were motivated to help humanity, certainly, but also by desire for personal glory. It was an age of technological wizardry—the last quarter of the nineteenth century saw the advent of marvelous inventions such as the telephone, phonograph, electric light bulb, radio, and internal combustion engine. Inventors like Thomas Edison and Alexander Graham Bell became rich, as well as famous. Technology and science were the watchwords of the day.

In Europe, the impetus to discover was also shadowed by another powerful impulse: nationalism. In an era where colonial success in Africa and Asia depended, in large part, on advances in tropical medicine, scientists were treated as national figures, and each nation cheered its champions.

Meanwhile, after the conclusion of the brief, but devastating, Franco-Prussian War (1870-1871), the perpetual rivalry between France and Germany intensified, touching practically every aspect of public life, including the nascent field of microbiology. Soon Europe's two greatest medical investigators, Pasteur and Koch, would face off as adversaries on the battlefield of science.

Both men were fiercely patriotic. Koch had proudly served as a field surgeon during the war, which Germany won decisively. After France's surrender, Pasteur returned an honorary degree he had been awarded from the University of Bonn and wrote to one of his students, "Every one of my works to my dying day will bear the epitaph: Hatred to Prussia. Vengeance. Vengeance."

THE RIVALS

Louis Pasteur did not rest on his laurels—it was important to establish French dominance in all things, and especially in the race for scientific discoveries. In 1879, he took on a new investigation—an epidemic of chicken cholera had killed 10 percent of the fowl in France. Pasteur isolated the causative microorganism* and learned to grow it in chicken broth. He could inject the microorganism into a chicken, and it would die within a couple days. Then, a serendipitous mistake produced a breakthrough. During a period while Pasteur was away on a trip, a sample of culture broth was inadvertently forgotten and left stagnant. A commonly told version of this event blames Pasteur's assistant, Charles Chamberland, who had either been instructed to promptly inoculate chickens with this pathogenic fluid or to regularly add fresh broth to the mixture to keep the microorganisms growing well—tasks he evidently

---

* The bacteria that caused fowl cholera was named *Pasteurella multocida* in Pasteur's honor.

failed at. Whatever the reason, Pasteur and Chamberland found that when this old, stale culture fluid was finally injected into a chicken, the chicken had only a mild case of the disease and did not die.

This was a surprise.

Rather than discard both the chicken and the ineffective culture, Pasteur had an idea. He next injected the same chicken with freshly cultured microbes—a solution that had always killed chickens within a couple days. To his amazement, the chicken still did not die. What was so special about this chicken? Was it something they had done, or was there some other, unknown factor at play?

Pasteur theorized that when the pathogenic microorganisms became old and exposed to oxygen in the environment over time, they lost their ability to cause disease. But how had the old sample conferred protection to the chicken against fresh, virulent microbes? Pasteur wasn't sure. Without fully understanding why, he had discovered a new method of vaccinating animals—by employing a weakened, or "attenuated," microbe that would not cause significant disease but could stimulate the immune system to produce protective antibodies.

Pasteur went straight from this novel accomplishment to another, even more important one. He turned his attention to the disease that Koch had so famously elucidated in 1876: anthrax. Hoping to outdo Koch by creating an anthrax vaccine via the same method of attenuation that had worked with chicken cholera, Pasteur isolated the anthrax microbe and weakened it using potassium dichromate and heat. On May 5, 1881, he initiated a public test at the village of Pouilly-le-Fort, thirty miles southeast of Paris. This highly publicized event was attended by hundreds of spectators and several newspaper reporters. Pasteur inoculated twenty-four sheep, six cows, and one goat with his anthrax vaccine. A control group consisting of approximately the same number of animals was observed. Two weeks later Pasteur returned to give the treated animals an additional, slightly stronger dose of the vaccine; and two weeks after that, he inoculated all fifty-odd animals with virulent anthrax.

The animals were closely watched. Within two days, every unvaccinated sheep and goat in the control group had died. Every vaccinated animal re-

BACTERIAL INFECTION: *The Magic Bullet*

mained healthy. Pasteur's experiment was a triumph. Journalists broadcast news of this accomplishment around the world. In 1882, 85,000 animals were vaccinated in France, decreasing mortality due to anthrax from 9.01 percent to 0.65 percent.

Pasteur's reputation soared to new heights. With this feat, he appeared to have regained his position as the world's foremost scientist.

But then, less than a year later, Robert Koch answered with an astonishing new discovery of his own.

Since appearing seemingly out of nowhere with his remarkable inaugural presentation at the University of Breslau in 1876, Koch had made numerous methodological improvements in the field of bacteriology, including using glass slides and coverslips, examining live microbes by employing a "hanging-drop" technique, culturing bacteria on solid media rather than only in live animals, and advancing microphotography. In 1880, he accepted a research position in Berlin at the Imperial Health Office. Now, armed with a staff of assistant researchers, he decided to turn his microbe-hunting sights on tuberculosis, the number one killer of the nineteenth century.

Tuberculosis was a mysterious disease that often caused mild initial symptoms, such as a cough or low-grade fever, but then could hide latent in the body for months or even years before reappearing as a serious malady. When such relapses occurred, there was no mistaking their severity. A patient's cough could become violent and produce bloodstained sputum as infected lung tissue became inflamed, necrotic, and scarred. Overwhelming fatigue and listlessness, accompanied by fever and night sweats, were followed by late-stage bone pain and pus-filled tubercles or abscesses. A patient's pallor and increasing emaciation gave the impression that the body was being consumed by some evil force from within—features that inspired two of the affliction's common nicknames: "consumption" and "the white death." Because of its variable and long-term clinical course, many physicians believed tuberculosis was not contagious and might instead be an inherited or spontaneous disease. It is estimated that tuberculosis caused at least one-quarter of deaths worldwide in the second half of the nineteenth century.

Koch's search for the bacteria that caused tuberculosis did not begin well. He obtained samples of sputum from tuberculosis patients, and tested tissue from tubercles, the bumpy lesions patients developed under their skin, but he could not see any bacteria using his microscope. This had proven to be a problem in many diseases—though suspected to be present, microbes were not visible under normal conditions using natural light. In order to see them, scientists realized they must devise a way to color them so they would stand out. Koch had befriended a younger doctor named Paul Ehrlich who was fixated on the idea that some organic chemicals, including dyes, have specific affinity for various tissues and microorganisms. After testing hundreds of dyes that failed to elucidate the cause of tuberculosis, Koch tried using a methylene blue dye that Ehrlich had left him. He incubated a tuberculosis sample with the dye and a dash of caustic potash (potassium hydroxide) overnight. The next morning, Koch examined the sample under a microscope and was amazed to see little blue rods. This was the bacteria that would be known as *Mycobacterium tuberculosis*.

Through his efforts to isolate *Mycobacterium tuberculosis* and prove that it was the one and only cause of the disease, Koch and his assistant Friedrich Loeffler devised and later promoted a repeatable method of investigation that became known as "Koch's postulates." To unequivocally show that a microorganism was responsible for an infectious disease, one must:

1. Show that the microorganism is present in every case of the disease.
2. Isolate the microorganism from the host and grow it in culture outside the body.
3. Inject the cultured microorganism into a healthy animal and show that it causes disease.
4. Re-isolate the microorganism from the inoculated animal to confirm it is identical to the original, causative microorganism.

Koch fulfilled each of these postulates using *Mycobacterium tuberculosis*: he isolated it, grew it in culture, injected it into rabbits and guinea pigs to cause the disease, and isolated it again from these animals.

BACTERIAL INFECTION: *The Magic Bullet*

In a lecture that has been called "the most important in medical history," Koch presented his findings at a meeting of the Berlin Physiological Society on March 24, 1882. For the crowd's perusal, he placed a microscope and more than two hundred items—microscopic specimens, test tubes, flasks, and culture plates—across a large table. The title of the talk was simply "On Tuberculosis." Anticipation was high; tuberculosis was the worst killer of the age and everyone was curious to see if Koch might reprise the sort of bombshell discovery he had made regarding anthrax.

Koch stated, "If the number of victims is a measure of the significance of a disease, then all diseases, even the most dreaded infectious diseases, such as plague or cholera, must rank far behind tuberculosis. Statistics show that one-seventh of all human beings die of tuberculosis, and that if one considers only the productive middle-age groups, tuberculosis carries away one-third and often more."

Koch went on to note that no one had yet determined the cause of tuberculosis. Many believed it to be a microorganism, but many others didn't. He described his efforts to concoct a stain that would show the culprit, and then described what he had seen under the microscope: "The color contrast between the brown tissues and the blue tubercle bacilli is so striking that even isolated bacilli can be clearly seen and identified." Koch described his postulates and explained how he had fulfilled each of them using *Mycobacterium tuberculosis*. He demonstrated that the microorganism was present in every case of the disease and proved it was responsible for the mysterious affliction that had befuddled physicians for centuries.

The audience was stunned. After Koch's last words, there was complete, awe-filled silence. The attendees had witnessed an earth-shattering speech, full of revelations. The cause of the world's most deadly disease had been identified. A method of investigating further diseases and the microorganisms that caused them had been established. Paul Ehrlich later wrote, "All who were present were deeply moved and that evening has remained my greatest experience in science."

OPEN WAR

Pasteur and Koch met for the first time at the Seventh International Medical Congress in London in August 1881. They were cordial to one another. Koch had developed an improved method of growing bacteria, on a film of gelatin on a glass plate rather than in broth, and Pasteur complimented him on this.* But this civility did not last. A few months later, perhaps piqued that Pasteur's vaccine appeared to supplant his own anthrax research, Koch published a caustic attack on Pasteur's work in a German medical journal. In this unprovoked opening salvo, Koch declared, "The assumptions from which Pasteur proceeded were not correct, and the whole experiment was, therefore, worthless." He discounted Pasteur's claims, suspected him of using tainted cultures, and criticized his work as unreproducible, all in a dismissive and insulting tone that shocked the European scientific community. "Only a few of Pasteur's beliefs about anthrax are new, and they are false," Koch asserted.

When a translation of Koch's article was reprinted in France, the French public became enraged. It was hard to believe that the author was the same modest country doctor who had written so humbly to Ferdinand Cohn only five years earlier. His concurrent, and undeniably groundbreaking, work on *Mycobacterium tuberculosis* may have imbued him with a newfound sense of imperiousness.

Somewhat reluctantly, Pasteur concluded he had no choice but to respond to Koch's attack. French honor, in addition to his own, was at stake. In February 1882, he sent a trusted and capable assistant, Louis Thuillier, to Berlin to demonstrate the anthrax vaccine's effectiveness to German sci-

---

* Two of Koch's assistants soon made further improvements. One, Walther Hesse, explained to his wife and assistant, Fanny Angelina, that Koch's gelatin medium unfortunately liquefied in warm weather. She suggested trying agar, a different type of gelatin which she used to make jelly that stayed solid in the summer. Agar has been the preferred bacterial growth medium ever since. Another Koch assistant, Julius Petri, improved upon the flat glass plate by making a round one with a slightly raised wall and a lid to cover it. This became the petri dish.

entists. The animal demonstration was a success, but Koch still refused to publicly acknowledge Pasteur's achievement.

The next clash occurred in September 1882, at the International Congress of Hygiene and Demography in Geneva, Switzerland, a conference both Koch and Pasteur attended. Pasteur delivered a speech, with Koch sitting in the front row, in which he defended his work concerning chicken cholera and anthrax. He pointedly stated, "Yet however blazingly clear the demonstrated truth, it has not always had the privilege of being easily accepted. I have encountered, both in France and abroad, obstinate objectors . . . Dr. Koch, who finds nothing remarkable in this experiment, wishes to know whether the chilled chickens that contracted the disease were not chickens capable of getting it naturally . . . The author does not believe that I operated as I said I did."

To the amazement of the crowd, a fuming Koch jumped up during Pasteur's speech and tried to disrupt his nemesis's oration. When Koch got his chance to take the podium, he said:

> When I saw in the program of the Congress that M. Pasteur was to speak today . . . I attended the meeting eagerly, hoping to learn something new about this very interesting subject. I must confess that I have been disappointed, as there is nothing new in the speech which M. Pasteur has just made. I do not believe it would be useful to respond here to the attacks which M. Pasteur has made on me . . . I will reserve my response for the pages of the medical journals.

Part of Koch's pique reportedly stemmed from a simple error in translation. In his oration, Pasteur repeatedly spoke of Koch's publications as *"recueil allemand,"* or a "collection of German works." However, both Koch and his interpreter misheard this as *"orgueil allemand,"* or "German pride." This was the reason why Koch jumped out of his chair and tried to interrupt Pasteur's speech. He was terribly insulted that Pasteur had called his entire body of work the product of German arrogance (even though Pasteur hadn't). Pasteur, oblivious to the reason for his adversary's petulant behavior, angrily motioned for Koch to keep quiet and sit down. After Koch's diatribe in Geneva, Pasteur responded that he would gladly wait for Koch's response. In a private letter to his assistant Émile Roux, Pasteur wrote, "Koch acted ridiculous and made a fool of himself." To his son, he wrote, "It was a triumph for France; that is all I wanted."

Not surprisingly, Koch's highly anticipated response heaped more fuel on the fire in a dispute that had become increasingly embarrassing for both men, though neither seemed to recognize this. Three months after the conference, Koch published a pamphlet titled "On Anthrax Vaccination: Response to a Speech Given at Geneva by M. Pasteur." He believed that microbe attenuation was not possible and reiterated that Pasteur had not presented any new information. Worse than Koch's message was his insulting, scathing tone, which seemed purposely intended to humiliate Pasteur and, by proxy, France. For example, he wrote:

> I was anxious to hear valuable scientific results from Pasteur's work in attenuating anthrax bacilli . . . But the Congress heard none of this . . . only the worthless fact that as yet so-and-so many thousand animals had been inoculated . . . Pasteur's polemic did not involve a factual refutation, but consisted of generalities. It was mostly directed against me personally, and was delivered in an emotional tone . . . Pasteur's method must be rejected as defective . . . After all, Pasteur himself is not a physician, and one cannot expect him to make sound judgments about pathological processes and the symptoms of diseases . . .

The pamphlet's message ran to ten thousand words of gratuitous derision, smear, and pomposity.

Pasteur's vigorous response came as a letter titled "Response to M. Koch, private counselor to the government of Berlin," in the January 1883 issue of the *Revue Scientifique*. He attacked Koch for being unwilling to acknowledge how indebted he was to French achievements in science. "You do not acknowledge that you are mistaken about the very principle of the attenuation of the bacterium," Pasteur wrote, adding, "You ascribe to me errors that I

BACTERIAL INFECTION: *The Magic Bullet*

have not committed . . . you denounce them and make a lot of noise with your triumph . . . You are wrong, sir, you are setting yourself up for another foiled expectation in which you will be forced to change your opinion."

This war of words between the world's two greatest scientists, and their nations by extension, continued with no signs of abating. In August 1883, a cholera outbreak in Alexandria, Egypt, created a new front in the conflict. Both Germany and France sent scientific missions to aid the response and, hopefully, isolate the cause of the disease. The German team was led by Koch himself, while the French team, named "Le Mission Pasteur," was led by Louis Thuillier, Pasteur's trusted adjutant. In September, the twenty-seven-year-old Thuillier tragically contracted cholera and died. It was a grim reminder of the risks all the investigators took. The German team respectfully attended Thuillier's funeral. Koch served as a pallbearer.

Ultimately, Koch and the Germans would win this round. After Thuillier's death, the French team abandoned the work and returned home. Koch was unable to definitively determine the causative microbe in Egypt, but, undeterred, he next traveled to Calcutta, India—the presumed source of the pandemic. There, from a twenty-two-year-old man who had died just ten hours after first displaying symptoms, Koch successfully isolated a comma-shaped bacillus that would become known as *Vibrio cholerae*. He had found the cause of cholera.

In May 1884, Koch returned home to a hero's welcome. The medical detective who had uncovered the secrets of anthrax, tuberculosis, and cholera was deemed a national treasure. But once again, Koch's time atop the throne was brief. In 1885, news of an almost unbelievable triumph from France rocketed around the world.

It would become Louis Pasteur's greatest achievement.

TRIUMPH AND DISASTER

Pasteur had turned his attention to rabies, a terrible disease contracted from the bite of an infected animal. Two to eight weeks after a person was bitten, symptoms would present, including incredible thirst, paralysis, and severe throat constriction that made swallowing impossible and caused most victims to suffocate to death.

Pasteur first tried to isolate the causative organism by looking for it in dog saliva, but he couldn't find anything. He did not know that rabies was caused by a virus, which he could not visualize using his microscope. Seeing that rabies was a disease that caused convulsions, paralysis, and madness, he deduced that the offending microbe might reside in the spinal cord and brain. He minced up the spinal cords of infected rabbits and injected this into a healthy dog. The dog soon contracted rabies and died.

Now that he could be certain this tissue contained the virulent microorganism, he sought to weaken it by letting it dry in a sterile flask, a method conceived by Émile Roux. Pasteur experimented to determine how long it would take for the weakened rabies sample to no longer cause disease in a healthy animal. He found that it took two weeks. He then devised a careful experiment in which he injected successively more virulent samples into a healthy dog: first, harmless fourteen-day-old dried nervous tissue from a rabid dog; then the next day, thirteen-day-old tissue; the next day, twelve-day-old tissue, and so on—all the way down to the point where he finally injected the dog with fresh, virulent tissue that had not been weakened at all.

The dog survived.

He repeated the experiment several times in other dogs with success. Pasteur had created a vaccine to protect animals against rabies.

On July 6, 1885, a nine-year-old boy named Joseph Meister was brought to Pasteur in Paris from a nearby village. Joseph's arms and legs had been ravaged by a rabid dog that had bitten him fourteen times. Death from rabies seemed certain, but the boy's doctor had heard Pasteur had devised a way to protect animals from rabies. Would it work on a human?

Hoping rabies' long incubation period might give his vaccine time to confer immunity before the disease took root and killed the boy, Pasteur agreed to try. Since he was not a physician himself, he supervised a doctor who

BACTERIAL INFECTION: *The Magic Bullet*

injected Meister with increasingly virulent spinal cord tissue from a rabid rabbit, daily for ten days.

Pasteur anxiously watched the boy for any sign of rabies.

None ever came. The boy gradually recovered over the course of four months. Pasteur's vaccination method was an effective treatment for rabies in humans. Three months later, Pasteur saved another boy who had been bitten by a rabid dog. In 1886, nineteen Russians traveled to Paris for treatment after being attacked by rabid wolves. Despite the prolonged period of time between injury and treatment, Pasteur was able to save all but three of them. By October 1886, 2,500 people had been treated with Pasteur's rabies vaccine. Once again, Pasteur had proven himself worthy of society's greatest accolades, which he received in abundance through scores of awards, honorary degrees, medals, and other tokens of gratitude from around the world.

Louis Pasteur

Perhaps feeling more pressure than ever to match his rival blow for blow, Robert Koch claimed in 1890 to have discovered an incredible new breakthrough—a cure for tuberculosis. He vaguely referred to his treatment as "lymph," and stated that it appeared to cure the disease in guinea pigs. Rather than kill microbes directly, he believed lymph attacked infected tissue itself, rendering it necrotic. This presumably destroyed the bacterial habitat, explaining the treatment. News of Koch's cure instantly spread across the globe and was met with public euphoria. Consumptive patients flocked to Berlin in hopes of receiving treatment. Thousands of doctors came to learn more about the medicine. Koch officially named it "tuberculin," but remained strangely secretive about what the panacea was or how he had made it.

Unfortunately, Koch did not subject tuberculin to the same strenuous and methodical testing that he had become known for. He eventually revealed that tuberculin was merely a filtered sample of *Mycobacterium tuberculosis* grown in glycerine broth. Over the following year, numerous physicians in many countries reported that Koch's cure did not work. Clinical trials in Berlin revealed that rather than helping patients, the medicine often caused severe allergic reactions and sometimes even made the disease worse.*

As evidence of tuberculin's ineffectiveness accumulated, Koch retreated in disgrace. Many suspected he had been pressured into making a premature announcement about the potential treatment by government ministers anxious to tout another German scientific victory. Others believed his lack of transparency stemmed from a plot to profit off the cure. Koch, the once humble country doctor, had become a medical Icarus. Though his reputation never fully recovered, he was awarded a Nobel Prize in 1905 for his research on tuberculosis. He outlived Pasteur, who died in 1895, and succumbed to a heart attack in 1910, at the age of sixty-six.

---

* Tuberculin would later find use as an effective method of testing for tuberculosis.

BACTERIAL INFECTION: *The Magic Bullet*

Robert Koch

MAGIC BULLETS

Koch and Pasteur had done more than anyone to identify disease-causing microorganisms and prove the germ theory correct. But Koch's failure with tuberculin served to highlight one unchangeable fact—that it was one thing to identify the causes of infectious diseases, and quite another to treat and cure them. If Koch, the genius of Germany, could not combat infection, what chance did other scientists have?

Koch's friend and protégé Paul Ehrlich would devote his career to succeeding where Koch had not. He would carry his mentor's work into the twentieth century at the vanguard of what would emerge as a therapeutic revolution. In 1889, after partially recovering from a case of tuberculosis he probably contracted from working in his lab, Ehrlich joined Koch's team in Berlin. Koch assigned him to work with Emil von Behring, a scientist with an interest in diphtheria. Von Behring and a Japanese scientist named Shibasaburo Kitasato had discovered that the blood of diphtheria patients contained a toxin produced by bacteria. The toxin, not the bacteria, was what actually caused the disease. Animals infected with diphtheria produced an "antitoxin" (later recognized as antibodies) that could be isolated from the serum component of the blood. When von Behring injected the antitoxin into an animal infected with diphtheria, it helped the animal fight the infection, sometimes to the point of cure. Working together, Ehrlich, von Behring, and Kitasato next developed a diphtheria antitoxin for use in human patients.* In order to produce antitoxin on a large scale, they kept stables full of horses that had been inoculated with the toxin of the causative bacteria, *Corynebacterium diphtheriae*. The horses underwent regular blood draws, from which the antitoxin was harvested. The first children saved using the antitoxin were treated at the University of Leipzig in 1892.†

In 1902, Ehrlich's expertise with chemical dyes inspired him to embark on a new quest. He had sometimes seen that bacteria that took up dye were not just stained; they also were killed. He realized this could be a remarkable tool. If Ehrlich could find a dye that killed a pathogenic microorganism, it could prove an effective treatment. But how to find the right one? There were thousands of dyes and dozens of known infectious pathogens. It would be like finding a needle in a haystack. Nevertheless, Ehrlich began to look. He called this holy grail "the magic bullet," and didn't stop searching for seven years. He and his assistants methodically tested hundreds of dyes, constantly

---

* Von Behring and Ehrlich later became adversaries because von Behring developed a diphtheria antitoxin commercially on his own, reneging on a prior agreement with Ehrlich to do so together.

† A famous diphtheria outbreak involving twenty children occurred in January 1925 in Nome, Alaska. The only way to supply the town with urgently needed diphtheria antitoxin was by a dog sled relay involving twenty mushers and 150 sled dogs that took 5.5 days to cover 674 miles. This "Serum Run" relay captured the imagination of people around the world. Balto, the final leg's lead sled dog, became world-famous, though Togo, the dog leading the penultimate leg, covered a much greater distance, 260 miles, compared to Balto's 55 miles.

tinkering with their composition by adding or removing various chemicals. For years, no headway was made.

Then, in 1909, a Japanese scientist named Sahachiro Hata came to work with Ehrlich. Hata began to retest numerous dyes from Ehrlich's lab that had previously failed to show any usefulness. Hata hoped to find a weapon against *Treponema pallidum*, the spiral-shaped bacteria that caused syphilis. One solution, titled "Number 606," had already been evaluated by Ehrlich's assistant a year before and deemed ineffectual. However, upon retesting, Hata found that Number 606 killed *Treponema pallidum*! When given to animals with syphilis, their sores healed and their bloodstreams became clear of bacteria. It was a discovery borne of diligence, dogged persistence, and luck.

Ehrlich then tested Number 606 on a human patient. To his delight, it worked. He renamed the chemical "salvarsan" and gave 65,000 doses of it to hospitalized patients in Germany throughout the year 1910. Salvarsan became the first chemotherapeutic drug—the first chemically synthesized medication—in history.*

FROM A MOLD

As the first man-made offensive weapon against infectious disease, salvarsan represented an important medical milestone. But its significance would soon be eclipsed by its heirs—antibiotic medications that would revolutionize healthcare and save hundreds of millions of lives in the twentieth century. It is not hyperbole

---

* In 1908, Ehrlich was awarded a Nobel Prize for his work on immunity in general, but whereas most Nobel laureates receive the prize as a capstone earned in retirement, Ehrlich would be the rare scientist whose best work would come after the honor. When Ehrlich died in 1915 at the age of sixty-one, he was admired as one of the world's greatest scientists. However, when the Nazis gained control of Germany in 1933, they seized and burned books about him in an attempt to erase his legacy because he was Jewish. Ehrlich's wife and family escaped to the United States in 1938.

to label the invention of antibiotics as the greatest medical accomplishment in human history, because from time immemorial, until the 1940s, more humans died from infectious diseases than any other cause. This changed due to a discovery so unlikely, so dependent on a seemingly impossible sequence of serendipitous moments and mistakes, that a novelist telling the tale might be rightly criticized for excessive fantasy. Many medical discoveries recounted in this book would probably still have been made, eventually, by someone else if the real-life discoverer had never existed. Not so with penicillin. Though other antimicrobials would certainly have been discovered, this all-important one might have been lost to the ages were it not for a fleeting insight—an unusual observation by a shy and diffident British microbiologist who noticed something peculiar about a tiny mold that had annoyingly contaminated his latest experiment.

And yet, the story is not so simple. Because a discovery is nothing without the recognition of what has been discovered.

In 1921, Alexander Fleming was a forty-year-old microbiologist in the inoculation department at St. Mary's Hospital in London. He was by nature introverted and reserved. A friend once said, "He was not a conversationalist and awkward silences were sometimes broken by awkward remarks . . . talking to him was like playing tennis with a man who, whenever you knocked the ball over to his side, put it in his pocket." But one thing Fleming excelled at was growing and studying bacteria.

One day in November 1921, Fleming came to work with a cold. He decided to put a drop of his own nasal mucus on a nutrient plate, curious to see what, if anything, might grow from it. Two weeks later, he saw that the plate had been colonized by bacteria, not originating from the mucus itself, but seeded from the air. Interestingly, there was a clear halo around the drop of mucus—a circular zone where bacterial growth had been inhibited. Further investigation allowed him to discover that mucus, as well as tears, sputum, plasma, and even breast milk, contained an enzyme that helped protect the body against bacteria. He named it "lysozyme." Although Fleming never found lysozyme to impede any bacteria that caused illness in man, he nevertheless considered it fascinating and devoted significant time to studying it.

BACTERIAL INFECTION: *The Magic Bullet*

Seven years later, in 1928, Fleming embarked on a study of staphylococcus. He grew these bacteria on numerous petri dishes in his cramped laboratory space, which was quite messy and disorganized. To record growth, he regularly lifted the cover off the petri dishes to inspect the expanding bacterial colonies. In July, Fleming went on a five-week vacation to Scotland. Before leaving, he stacked his petri dishes, typically numbering in the range of forty to fifty, at the end of a table to make room for a lab assistant who would be using the space in his absence.

Fleming returned from vacation on September 3, 1928. Many legends would arise about what happened next. What we know for sure is that Fleming reinspected his plates and on one of them noticed the presence of a mold growing near the edge of the dish. It was dark green and felty, about twenty millimeters in diameter. In the area immediately surrounding the mold was a zone of dead, lysed bacteria. A little farther away, bacterial colonies appeared irregular and stunted in growth. On the opposite side of the plate, the staphylococcus appeared healthy and normal looking.

Fleming reportedly said, "That's funny."

He showed the plate to his colleague Merlin Pryce, who remarked, "That's how you discovered lysozyme."

Alexander Fleming's discovery plate showing large, fluffy penicillium mold at the top of the plate, lysed staphylococcal bacteria in the immediate vicinity, and healthy bacterial colonies at the opposite side of the dish

A commonly told version of this story claims that the mold blew in off the street through an open window and landed on Fleming's open petri dish sometime that summer. But most historians believe this is very unlikely because the furniture in Fleming's lab made it very difficult to reach the window, and microbiologists would not generally open windows that could, indeed, become a prolific source of contaminants. The more likely source of the mold was the mycology lab of C. J. La Touche residing on the floor below. Such a mold could have wafted up the staircase and into Fleming's lab. The repeated uncovering of the petri dish allowed multiple occasions for a contaminant to land on the nutrient agar, and La Touche confirmed that the mold was one of the species he had been working with, *Penicillium rubrum*. Later, it was found that La Touche initially misidentified this mold, which was actually a sample of *Penicillium notatum*.

Another account of the story maintains that Fleming inspected all his old petri dishes, deemed them unremarkable, and placed them in a sink to be washed. The discovery plate was mere inches above a Lysol bath, about to be disinfected and wiped clean, when Merlin Pryce happened to stop by. Fleming chose a few of the plates at random to show Pryce and, upon reopening the history-making one, looked closer and said, "That's funny."

Regardless of the exact circumstances of the moment, one serendipitous aspect of the environment is considered almost certain to have played a role in the discovery: the London weather. Under normal conditions, it would be very unlikely, if not impossible, for a wayward sample of mold to land on a field of agar teeming with staphylococcus and take root, much less thrive. The staphylococci would easily overwhelm and destroy a nascent mold. The only way the mold could have established itself was if it had landed on nutrient medium free of competing microorganisms and been given time to grow unopposed. How could this have happened? Fleming had already seeded this plate with staphylococcus.

The answer was the weather. The temperature in London from July 28 to August 6 was unusually cool, with highs generally between 61°F and 68°F, and a low temperature of 47°F on one day. This was followed by a more seasonably warm period in mid-August with temperatures reaching up to 79°F. Bacteria need warmth to flourish and do not grow well in the cold. The initial

BACTERIAL INFECTION: *The Magic Bullet*

cold period inhibited the staphylococci while giving the mold time to grow and become established. Then, the subsequent warm period encouraged the bacteria to grow. If the summer weather had been more typical, the mold would probably never have formed.*

Fleming was intrigued by his unexpected discovery, so uncannily similar to the way he had detected lysozyme years before. He learned to grow the mold atop the surface of nutrient broth, and realized it was not the mold itself that killed bacteria, but a yellow-colored fluid the mold produced. These highly potent droplets of fluid could be present on top of the mold and pipetted off, or found dripping into the underlying nutrient broth, which also became bactericidal and was nicknamed "mold juice." In a petri dish, Fleming streaked many different types of bacteria up to a furrow containing the mold juice and saw that it inhibited several kinds of bacteria, not just staphylococcus. He tried using other molds to see if they might also subdue bacteria, but none could. He gave a presentation about the penicillium mold at a Medical Research Club meeting on February 13, 1929, but no one in the audience expressed any interest; no questions were asked.

Fleming's interest in the mold subsided. He thought it might be useful as a way to inhibit unwanted bacteria from contaminating cultures of other microorganisms he wished to study, a kind of bactericidal "weed-killer," but he did not pursue further uses for it.

Despite his failure to pursue the penicillium mold's therapeutic potential, Fleming did do two invaluable things. First, he generously shared samples of *Penicillium notatum* with other labs that also found it useful for preventing unwanted bacteria from contaminating their everyday cultures. This perpetuated the original mold sample, which was disseminated to locations in Europe and North America. Second, he wrote a paper that was published in the *British Journal of Experimental Pathology*, titled "On the Antibacterial Action of Cultures of a Penicillium, with Special Reference to Their Use in

---

* The role of the weather in penicillin's discovery was elucidated by Fleming's colleague Professor Ronald Hare, who recreated Fleming's experiments and published his findings in 1970.

the Isolation of B. Influenzae." In it, he coined the name "penicillin," and, although he had already abandoned attempts to use it therapeutically, he stated in the discussion section, "It is suggested that it may be an efficient antiseptic for application to, or injection into, areas infected with penicillin-sensitive microbes." At the time, no one in medicine seemed to take note of it.

Nine years passed before the torch was picked up again.

"IT LOOKS QUITE PROMISING . . ."

In 1938, Howard Florey was the chairman of the Sir William Dunn School of Pathology at Oxford University. Australian by birth, he traveled to England for the first time as a Rhodes scholar after completing medical school. At Oxford, he impressed his teachers and gained a reputation as a brilliant investigator. He was known for a dry wit, for deadpan humor and gentle needling at others' expense, and for being sparing with praise—all attributes most subordinates eventually found endearing once they got to know their boss and learned how self-deprecating he could be. Introspective and reserved, Florey was not particularly sociable. His bluntness could rub sensitive colleagues the wrong way, but he offered no apologies. He was forthright, honest, and quite plainly, ambitious.

Intrigued by the effects of sulfonamides—moderately effective antimicrobial compounds that had already been developed into an oral medication*— Florey began to search for more powerful bactericidal substances. In 1935, he had hired a promising twenty-nine-year-old German Jewish chemist named Ernst Chain. Chain spoke English with a heavy German accent and looked uncannily like Albert Einstein. Unlike the more reserved Florey, Chain had an impetuous personality. He was voluble and friendly, but passion sometimes got the better of him, and it was not unusual for him to raise his voice and

---

* The first sulfa drug, prontosil, was discovered in 1932 by German scientist Gerhard Domagk. Like Paul Ehrlich, Domagk had tested hundreds of chemical compounds in search of a beneficial chemotherapeutic agent.

shout when his temper flared. He jokingly referred to himself as "a temperamental Continental." In short, Chain was brilliant, enthusiastic, occasionally egotistical, and, as time would reveal, highly sensitive to perceived slights.

Florey and Chain developed a close working relationship. They had common interests and often walked home together, discussing experiments on the way. In the summer of 1938, in the course of his library research, Chain came across Alexander Fleming's 1929 paper on the penicillium mold. He was immediately intrigued and couldn't believe his good luck when he was able to easily acquire a sample of *Penicillium notatum* from a colleague, Margaret Campbell-Renton, who worked in a lab down the hall. This mold sample had been obtained from Alexander Fleming in 1929 by Dr. Georges Dreyer, the man who preceded Florey as department chairman at the Dunn School. Dreyer and Campbell-Renton had kept the mold alive for use as a tool to kill undesirable bacteria in cultures of *Bacillus influenzae*.

Now Chain set out to procure his own bactericidal "mold juice," but he soon learned that inducing the mold to produce significant amounts of the fluid was difficult. Chain tried altering the temperature and pH of the fermentation broth. He repeatedly evaporated the mold juice in an effort to concentrate the active ingredient. His efforts were aided by a talented young biochemist named Norman Heatley, who functioned as Chain's assistant. Heatley had a knack for "microchemical" measurements, measuring minute quantities of elements such as carbon or nitrogen, in biological samples.

Unfortunately, Heatley bristled under Chain's supervision. Though Chain was humorous and fun-loving by nature, he had been trained at German universities where the atmosphere was traditionally rigid and strictly hierarchical—subordinates were expected to be exceedingly obedient and subservient to their superiors. Heatley resented Chain's brusque orders and rude manner, and soon found working with Chain to be intolerable. The pair often engaged in arguments, and sometimes shouting matches, over trivial matters such as the color of penicillin. Before long, Heatley decided to move on to a new position in Denmark. He intended to bid both Chain and Oxford adieu in September 1939.

But, at this critical moment, on September 1, 1939, German panzer divisions crossed into Poland at the vanguard of an overwhelming attack that destroyed the Polish military in little more than a month. The Second World War in Europe began, upending the lives of countless civilians, including Heatley. Travel to the continent was ill-advised, forcing Heatley to abandon his plan. He had no job, in research or otherwise.

As a result, Florey invited Heatley to stay on and continue working on penicillin, as Florey's assistant, not Chain's. To this arrangement, Heatley readily agreed. It was a fortuitous outcome that would prove crucial to the challenging road ahead—a years-long odyssey of such daunting complexity that the investigative trio's ignorance at this early stage could rightly be considered a mercy.

Heatley took on the job of growing the penicillin mold by fermentation. He made immediate progress. He determined that the mold grew best atop a pool of very shallow fluid, no more than 1.5 centimeters in depth. The greenish blue-colored mold grew in fronds upon the surface, and dropped spores into the underlying fluid that eventually become the source of more mold. Heatley soon collected every tray, pot, bowl, and baking dish he could find—any shallow container he could use to grow mold. The best receptacles proved to be bedpans. He experimented by adding chemicals, sugars, salts, yeast, and even meat to the underlying fluid to see if these might enhance the mold's growth.

While Heatley focused on producing more mold juice, Chain worked on purifying it. Reminiscent of James Collip in his quest to purify insulin, Chain endeavored to extract the purest form of penicillin possible from the fluid bath upon which the mold floated. He filtered the broth, altered its pH, and mixed it with various solvents such as ether; this enabled him to remove contaminants that were ether-insoluble. The work continued through 1939 and into 1940. Chain hoped to convert penicillin into a stable, solid form that could be easily preserved and transported. To do this, he freeze-dried samples of water-soluble penicillin and then extracted the water through sublimation at low pressure and temperature. "I obtained . . ." Chain later wrote regarding the final product, "a very nice brown powder."

BACTERIAL INFECTION: *The Magic Bullet*

By March 1940, Chain finally had enough penicillin to conduct preliminary tests in animals. The team injected it into mice to see if it was toxic. It was not. They learned it did not enter the bloodstream if given orally. When injected intravenously, it was rapidly excreted in urine but still retained much of its potency—drops of urine placed in a petri dish were bactericidal. This indicated that penicillin's strength was relatively undiminished by passage through the body and that it could therefore be an active systemic killer of bacteria—a crucial characteristic of any potential medication. Even when penicillin was diluted to one part per million, it remained bactericidal. The investigators did not know how it killed bacteria, only that it did. Under the microscope, bacteria mixed with penicillin seemed to grow larger, and then longer, before bursting. They stopped reproducing. They died.

In a simple yet groundbreaking experiment, Florey inoculated eight mice with lethal doses of streptococci on May 25, 1940. He gave four of the mice varying doses of penicillin. By the next day, all the control mice were dead, and all the treated mice were alive and healthy.

"It looks quite promising," the typically understated Florey said the next day, while beside him an ebullient Chain practically danced a jig and declared it "a miracle." They repeated similar experiments several more times, on up to seventy-five mice, using different dosages. These experiments were reported in the *Lancet* in August 1940, in an article titled "Penicillin as a Chemotherapeutic Agent."

This was a crowning achievement—a landmark discovery that would rank among the greatest accomplishments in the history of medicine.

No one seemed to notice.

THE POLICEMAN

To Florey's chagrin, he received no comments, questions, or requests for reprints from other scientists after the publication of his team's important article. What he thought had the potential to be a groundbreaking report appeared to make zero impact. It seemed like none of his countrymen paid any attention to it; though, to be fair, Florey would have understood why scholarly articles in medical journals were not at the top of British minds that summer.

They were all, himself included, under attack.

The Battle of Britain had raged over southern England all summer. During the Blitz, in September, the Luftwaffe bombed British cities nightly. Scientists, including members of the penicillin team, spent part of their days digging an air raid shelter on the grounds behind the Dunn School of Pathology. Newspapers printed daily casualty lists. No one knew when their number might be up, and normal life for all had been completely upended.

The war also served to stymie Florey's next goal—to find a pharmaceutical company willing to develop methods of mass-producing penicillin. He reached out to several companies, but none were interested. All were too consumed with producing urgently needed supplies for the war effort, including vaccines, antitoxin, and sulfa drugs. Those that took the time to listen to his presentation concluded that growing enough penicillin to treat a human patient would be too difficult. After all, it had taken Florey years to make enough to treat several dozen mice.

Florey, Chain, and Heatley reluctantly faced two deeply upsetting and seemingly insurmountable problems. First, they would never be able to grow enough penicillin to conduct a clinical trial on their own. It would take approximately forty gallons of mold fluid to produce enough penicillin to treat a single human patient for one day. They needed a commercial partner with the expertise and manufacturing equipment to handle vast amounts of temperamental mold fluid, but no pharmaceutical company in Britain was willing or able to help them. They were at an impasse.

On top of this was a second problem even more distressing than the first. It appeared very likely that Britain would soon be invaded by the German military. Florey, Chain, and Heatley concurred that it would be unacceptable for the enemy to benefit from their work, so they agreed that if and when the Germans came, they would destroy all their lab notes, data, and equipment. Yet the prospect of losing their precious penicillin mold, perhaps forever, was too terrible to contemplate, so they each rubbed penicillium spores into the pockets and inner linings of their coats, where they would be preserved as long as they remained dry. This way, if the unthinkable occurred, each man could take his chances and flee with the raw material to continue the research abroad. With any luck, one of them would make it.

The trio's sobering predicament sounds harsh by any measure, but at the time, nothing about their world was normal. Florey and his wife, Ethel, had already taken the extraordinary step of sending their two children, Paquita and Charles, away to North America along with 123 other children to be lodged with Canadian and New England families that had volunteered to take them in. The risk of a trans-Atlantic crossing through U-boat infested waters, and years of family separation, were deemed preferable to the chance of the children dying in a desperate battle to repel German invaders, or the nightmare of living under Nazi rule. Each day, the Oxford team dutifully continued their work, anxiously watching the skies and listening to the radio with a sense of foreboding.

A pivotal development occurred on February 12, 1941. A doctor named Charles Fletcher sought out Florey to tell him about a patient, forty-three-year-old Albert Alexander, a policeman who had suffered a cut on his face near the corner of his mouth during a bombing raid in Southampton. This seemingly trivial injury had morphed into a terrible infection over the course of several months, spreading to his scalp, eyes, and arm. Now Alexander was septic—bacteria had seeded his bloodstream; his lungs were infected and his body was covered with abscesses. Alexander was in constant pain and close to death. Sulfonamide drugs had had no effect. Dr. Fletcher understood Florey had an experimental medicine that might help.

Without hesitation, Florey gave Fletcher a sample of penicillin, which Fletcher injected into Alexander, first as a dose of 200 mg, followed by three hourly doses of 100 mg. Within a day, Alexander began to exhibit "striking improvement" according to Florey's subsequent case report. There was almost immediate "cessation of scalp-discharge, diminution of right eye suppuration and conjunctivitis." Alexander's fever cleared. He felt dramatically better and even his appetite returned. Florey, Chain, and Heatley were ecstatic. This was proof that penicillin could save countless lives, both military and civilian. Florey said to Fletcher, "This is the sort of thing that only happens to you once in a life."

Constable Alexander continued to improve over the course of five days. Infections of his face, scalp, and right orbit completely resolved. This was good because Florey had run out of penicillin. Already, Alexander had begun receiving less frequent dosing due to the diminishing supply. To gain more, the team collected the patient's urine and bicycled it back to the lab each day to extract the penicillin for reuse—a duty they nicknamed the "P-patrol."

Now the team hesitated. Since Alexander had improved so much, was it still necessary to give him more penicillin? The case of another patient had been brought to their attention—a fifteen-year-old boy suffering from sepsis after hip surgery. They decided to give their reclaimed penicillin to the boy, who made a complete recovery. But now there was no penicillin left. And to the horror of all, Alexander's condition began to worsen again. His infection recurred, and his doctors were powerless to act.

Three and a half weeks after his last dose of penicillin, Albert Alexander died from overwhelming sepsis.

This tragic experience left a lasting impression on all the investigators. Despite Alexander's death, they were now convinced that penicillin worked and that they had made a major medical breakthrough. But without far more penicillin, it would be impossible to help anyone. It was infuriating—in order to convince British pharmaceutical companies to make penicillin, they needed more of it to conduct a large study; but to perform such a study they needed quantities of penicillin that could only come from the help of pharmaceutical companies. It was an impossible conundrum, with no perceivable solution.

Except one.

They could leave Britain and try their luck in the United States.

This was no easy task. Travel to North America was risky and difficult. Even if they got to America, there was no guarantee that they would succeed in finding a manufacturing partner. The whole enterprise was based on faith . . .

BACTERIAL INFECTION: *The Magic Bullet*

and a sense of desperation. Luckily, Florey had one potential lead. In April 1941, he met with an American physician named Warren Weaver who was visiting London. Weaver was the head of the Rockefeller Foundation's Natural Sciences Division in New York City. When Florey described penicillin's enormous potential, Weaver was impressed. He promised to help Florey and reported to the Rockefeller Foundation: "This project... if it were indeed successful, would be more revolutionary than the discovery of sulfonamides... and must be recognized as a project of the very highest potential importance. We certainly ought to do all that we can to accelerate its progress."

At the same time, a new threat caused Florey much anxiety and lent even greater urgency to his mission. During the war, the transfer of scientific publications between Britain and Germany had ceased, but Florey received word from a friend in Switzerland that the Germans had acquired copies of the *Lancet* issue in which his team had described penicillin and its potential. Now the Germans were interested in penicillin as well. German scientists, physicians, and industrial capacity were known to be stellar. Florey had no doubt it was only a matter of time before Germany would have the advantage of penicillin, which could aid their military just as much as Florey hoped it would help the Allies.* Florey felt intense pressure to act quickly. He instructed his lab, and counseled other researchers, not to share samples of *Penicillium notatum* with anyone who could have an association with foreign scientists, including Swiss investigators who might have connections to German counterparts. Ultimately, the vital importance of penicillin's development made Florey's decision to travel to America an easy one.

Far less straightforward was the feat of reaching the U.S. This required numerous approvals from both sides of the Atlantic. Florey was instructed to keep his plans secret, and did not mention it to his laboratory staff or to Ernst Chain, who was surprised to see Florey's suitcases at the lab on the day before the planned departure. Chain asked where he was going. Florey explained

---

* Florey was correct to be concerned. The Germans were indeed interested in developing penicillin and even sent medical journals describing it to Japan via submarine in 1944.

that he and Heatley were going to America, but that Chain had not been chosen to accompany them.

Chain did not take this news well. He bitterly resented being left behind, especially when the more junior Heatley had been chosen to go. Chain regarded himself as Florey's equal partner and judged his own contribution to have outshined Heatley's. But Florey explained that they were going to America to increase production, and this was Heatley's area of expertise. In 1979, Chain recalled this moment in an interview that demonstrated his resentment had remained undimmed; he said, "I left the room silently but shattered by the experience of this underhand trick and act of bad faith, the worst so far in my experience of Florey. It spoiled my initial good relations with this man for ever."

Prior to this, the bond between Florey and Chain had already begun to fray due to an assortment of slights perceived by Chain—some trivial, others less so—culminating in the collapse of a relationship that had begun as a warm friendship. Florey had a propensity for making verbal jabs, which he considered innocuous, but that the sensitive Chain found hurtful.

Howard Florey (left) and Ernst Chain (right)

Furthermore, Chain's fury at being excluded from the trip to America came on the heels of another severe disagreement between the pair over patenting the process of making penicillin. Chain strongly urged Florey to file for a patent. He did not want royalties to enrich themselves; rather, Chain explained, they would be used to benefit the Dunn School or Oxford Uni-

BACTERIAL INFECTION: *The Magic Bullet*

versity, and to fund their future research. In Germany, where Chain had been trained, closer ties existed between academia and industry, and seeking patents for scientific discoveries was considered normal. But Florey remained unmoved. In Britain, the scientific community considered it anathema to be seen as conducting research for any measure of personal gain. Patents were simply not done and were considered unethical. Any discoveries belonged to the people, who, after all, had funded the charitable and government foundations that supported the work.

Chain fumed. He remained convinced that substantial funding would be crucial to support their future penicillin investigations, and that patents and royalties were the best way to secure that funding. He later wrote, "I saw a whole tremendous virgin field and we were the leaders and would remain so if we got enough money. I argued our position again and again with Florey and we had bitter fights." Chain also argued that if they didn't patent it, someone else would, and there might even come a time when they would have to pay royalties to make or use the very medicine they had discovered.

Reflecting their long-standing differences, Florey later wrote to Chain, "It is quite clear that I cannot carry on any further acrimonious conversations with you as they lead to no progress and waste time and energy which I cannot afford."

On June 27, 1941, Florey and Heatley flew from London to neutral Portugal. Three days later they boarded a Pan Am Clipper flying boat that landed in the Azores, and then Bermuda, before proceeding on to New York City.

AMERICAN ODYSSEY

In a briefcase that never left his side, Florey carried samples of the priceless penicillium mold, perpetually worried that changes in temperature or humidity over the course of their long journey might harm or kill the specimens. Fortunately, the samples survived, and Florey and Heatley arrived in New York on July 3, 1941. That very day they met with Alan Gregg, the director of the Rockefeller Foundation's Medical Sciences Division. All of Florey's hopes—the years of labor and sacrifice, the chance to save countless lives—had come down to this moment, a meeting with one of the foundation's highest-ranking scientists.

Speaking without notes, Florey explained all the work they had done with penicillin, its miraculous powers, the tragedy of running out of it and watching Albert Alexander die, the challenges to growing it, and the need for large-scale, industrial help. He captivated Gregg. Of the meeting, Heatley wrote in his diary, "The Professor spun the penicillin tale in a really expert way . . . I remember him best of all for that performance—and he was so tired after that long journey . . . it revealed the wide grasp of his scientific mind. Even though I knew the subject well, he showed me new facts, and I realized suddenly how great a man he was . . . I count that hour in Gregg's office as one of the great experiences of my life."

Recognizing penicillin's potential, the Rockefeller Foundation, Florey's acquaintances in the scientific community, and U.S. government officials moved quickly to support its production. The Bureau of Agricultural Chemistry and Engineering, a section of the Department of Agriculture, maintained a large laboratory facility in Peoria, Illinois, called the Northern Regional Research Laboratory. This was the ideal place to work on optimizing the production of penicillin in its purest possible form. Heatley went to Peoria to share their precious penicillin samples and all he knew about growing the mold. He would not return to Britain for a year.

One of the first innovations the Americans introduced was the use of corn steep liquor as a better fluid upon which to grow the mold. This single improvement multiplied the yield many times. An effort to seek out other, possibly more productive, strains of the penicillium mold resulted in a serendipitous discovery by lab member Mary Hunt, who visited numerous grocery stores and markets to bring back every moldy piece of fruit or vegetable she could find. Nicknamed "Moldy Mary," Hunt found a cantaloupe colonized with *Penicillium chrysogenum*, a mold discovered to produce six times more penicillin than *penicillium notatum*. This cantaloupe mold would become the parent source of all penicillin medication worldwide.*

Meanwhile, Florey sought the help of U.S. pharmaceutical companies to mass-produce the drug. It was a difficult task. Heatley recorded in his diary that Florey admitted to feeling like "a carpet bag salesman trying to promote a crazy idea for some ulterior motive." Ultimately, Florey succeeded, aided by the support of Dr. Alfred Newton Richards, chairman of the U.S. Committee on Medical Research and Development. Richards vouched for the promise of penicillin and told leaders at the pharmaceutical companies Merck, Squibb, Lederle, and Pfizer that the government was behind the effort because developing it was in the national interest. Further spurred by the December 7, 1941, Japanese attack on Pearl Harbor and subsequent U.S. entry into the war, all four companies signed on to manufacture the drug.

His mission accomplished, Florey returned to Oxford and resumed leading the team's work. Two British companies—Imperial Chemical Industries and Kemball, Bishop, and Co.—agreed to start manufacturing penicillin in substantial amounts. In March 1943, Florey published a report summarizing his team's experience treating over 180 patients with penicillin; most of these individuals were cured. In April 1943, U.S. soldiers who had fought in the Pacific Theater were the first servicemen to receive penicillin to treat aggressive or chronic infections. As with other aspects of U.S. wartime production, America's ability to mass-produce penicillin rose at an exponential rate. By 1944, twenty-two large-scale penicillin-producing plants were being built in North America. Penicillin was grown in giant vats containing tens of thousands of gallons of fluid. The U.S. military used penicillin to dramatic effect. The drug seemed to be a cure-all

---

* An American named Andrew Moyer, who worked with Heatley to mass-produce penicillin in Peoria, was eventually awarded patents in Britain for his system of production. Chain's premonition that he and his colleagues would end up paying royalties for the medicine they had discovered came true—such fees accompanied their purchases of the processed penicillin they needed for their experiments.

for myriad diseases, including pneumonia, syphilis, tetanus, meningitis, diphtheria, rheumatic fever, and more. Thousands of soldiers laid low from sexually transmitted diseases like gonorrhea became fit for duty after penicillin treatment; tens of thousands more owed their lives to the drug . . . and the team from Oxford, England.

But despite this triumph, there was still one, all-important aspect of the penicillin saga that had yet to be settled. It was a part of the story that none of the principals could have predicted would become so problematic.

THE CREDIT

Alexander Fleming had observed the progress of Florey, Chain, and Heatley with interest. After the team's initial publication in the *Lancet* in 1940, Fleming paid a visit to Florey's lab at Oxford to learn what they had done "with my old penicillin" and to learn more about how penicillin had been produced. Fleming's appearance particularly surprised Chain, who thought Fleming was dead. Florey graciously hosted Fleming and the whole team withheld no information about how they had succeeded. They even gave Fleming a sample of their purest penicillin, which Fleming later tested and determined to be far superior to sulfa drugs.

The twist in the story began on August 5, 1942, when Fleming called Florey to ask if Florey could supply penicillin to treat Fleming's fifty-two-year-old friend, a man suffering from streptococcal meningitis—a fatal diagnosis. Florey felt obligated to help the man who had discovered penicillin. He personally brought Fleming a sample of the drug. Fleming administered penicillin to the patient intramuscularly for five days, with little clinical effect; but then he injected it directly into the patient's spinal canal, which prompted a remarkable improvement, and soon a full recovery.

The news of this miraculous, Lazarus-like resurrection was picked up by the *London Times*, which ran a story on August 27, 1942, titled "Penicillium." No scientists' names were mentioned. Then, on August 31, a letter from Sir

Almroth Wright, the director of the Inoculation Department at St. Mary's Hospital (effectively Fleming's boss), was published in the same newspaper. It stated:

> Sir,
>
> In the leading article on penicillin in your issue yesterday you refrained from putting the laurel wreath for this discovery round anyone's brow. I would, with your permission, supplement your article by pointing out that, on the principle *palmam qui meruit ferat** it should be decreed to Professor Alexander Fleming of this laboratory. For he is the discoverer of penicillin and was the author also of the original suggestion that this substance might prove to have important applications in medicine.

The glaring spotlight of public attention was suddenly directed squarely upon the unassuming, taciturn, sixty-one-year-old Alexander Fleming, and reporters descended on St. Mary's Hospital to interview him. To the surprise of many, Fleming appeared to enjoy the limelight. He granted multiple interviews and allowed photographers into his lab. Thorough reporters took the time to learn of the contributions by the team at Oxford, but when they traveled there to interview Howard Florey, they encountered a man with a very different attitude toward journalists. Florey despised the press and had no interest in being interviewed or photographed. He considered it uncouth for scientists to be seen engaging in any behavior that could be regarded as advertising or self-aggrandizement. As a result, newspapermen gravitated toward Fleming, especially since his modest demeanor fit the image of the humble public hero perfectly. Before long, reporters stopped bothering to mention contributions made by Florey's Oxford team at all. It was not uncommon for articles to include incorrect information, from claiming that Fleming had developed penicillin entirely on his own, to asserting that he had directed its production at Oxford for use at St. Mary's. In 1945, the title of a *New*

---

* Translated from Latin as "Let whoever earns the palm bear it."

York Times article about penicillin lauded the work of Fleming and "two coworkers," referring to Florey and Chain.

It is hard to overstate the degree to which Fleming, or perhaps Fleming's myth, became adored by citizens across the globe. As the presumed inventor of the world's greatest wonder drug, he attained celebrity status and became one of the most famous figures of the twentieth century. Much of the last ten years of his life was devoted to traveling and receiving awards around the world. He was knighted and elected to the Royal Society. In 1944, *Time* magazine put a portrait of Fleming on its cover and started a fund for Fleming, which it encouraged readers to contribute to because Fleming had not received any financial gain from his discovery.

Alexander Fleming

In contrast, the contributions of Florey, Chain, and Heatley were, at best, little known and, at worst, completely forgotten. Though Fleming generally did not claim credit for anything he had not done, he could have done much more to set the record straight with reporters eager to portray him as the sole, heroic face of the penicillin miracle. When he did sometimes give credit to Florey's group, the press often simply ignored that aspect of the story.

Not surprisingly, Florey became highly irritated by Fleming's behavior, but said nothing publicly. To do so would break his rigid code of honor against self-promotion. Chain blamed Florey for not sticking up for the work of the Oxford group. He did not know that Florey had been advised by trusted luminaries in Britain's scientific community to remain silent. A private letter written on December 11, 1942, to Sir Henry Dale, scientific adviser to the British cabinet, revealed Florey's pique: "I have now quite good evidence . . . that Fleming is doing his best to see that the whole subject is presented as having been foreseen and worked out by Fleming and that we in this Department just did a few final flourishes . . . This steady propaganda seems to be having its effect even on scientific people in that several have now said to us 'But I thought you had done something on penicillin too.'"

While Florey hoped and assumed the idolization of Fleming would abate as time passed, it did not. Instead, it only grew. Florey and his Oxford team continued to grumble. A year and a half later, in a June 19, 1944, letter to Sir Edward Mellanby, the Secretary of Britain's Medical Research Council, Florey wrote:

> It has long been a source of irritation to us all here to witness the unscrupulous campaign carried on from St. Mary's calmly to credit Fleming with all the work done here. I have sufficient evidence of one sort and another that this is a deliberate and clever campaign. My policy here has been never to interview [with] the Press or allow them to get any information from us even by telephone . . . In contrast, Fleming has been interviewed apparently without cease, photographed, etc. . . . with the upshot that he is being put over as the "discoverer of penicillin" (which is true) with the

implication that he did all the work leading to the discovery of its chemotherapeutic properties (which is not true) . . . my colleagues here feel things are going much too far and, while for the most part do not want publicity or special credit for themselves, are getting quite naturally restive at seeing so much of their work going to glorify and even financially enrich someone else.

In 1945, the Nobel Prize in Physiology or Medicine was awarded to Fleming, Florey, and Chain. The rules prevented more than three individuals from being honored, and Heatley was left out. Today, Alexander Fleming's name remains closely associated with the invention of penicillin, while few recall the names of Florey, Chain, and Heatley. Author Eric Lax has highlighted an observation by the eminent neuroscientist William Maxwell Cowan, who made the point that, for all the lives that Florey's work on penicillin saved, the first life rescued was surely that of Alexander Fleming, whose career would otherwise have been forgotten as a footnote in the history of microbiology.

Fleming died in 1955, of a heart attack at the age of seventy-four. His passing was mourned around the world and he was laid to rest in St. Paul's Cathedral near Horatio Nelson and Christopher Wren. Florey died in 1968 and is buried in Westminster Abbey. Later in their careers, he and Chain had exchanged more positive, and somewhat conciliatory, letters. Chain was elected to the Royal Society in 1949, and was knighted in 1969. He died in 1979. In part to remedy his being overlooked by the Nobel Committee, Norman Heatley was awarded Oxford University's first honorary doctorate of medicine in 1990. He died in 2004.

RESISTANCE

Since the discovery of penicillin, more than 150 different antibiotic medications have been developed. Whereas penicillin works by inhibiting the proper maintenance of bacterial cell walls,* other antibiotics employ strategies such as attacking enzymes bacteria need to make proteins or disturbing bacterial DNA synthesis. These successor drugs were born out of necessity because, almost immediately, it became clear that bacteria had the capacity to evolve, mutate, and develop tactics to evade, dupe, and defeat penicillin and practically every one of its heirs.

As early as 1940, Ernst Chain and Edward Abraham, another member of the Oxford team, reported their discovery of a bacterial strain that could deactivate penicillin by producing an enzyme called penicillinase (also known as beta-lactamase) which disrupts the drug's structural beta-lactam ring. In his 1945 Nobel Prize speech, Alexander Fleming sounded a prescient warning that undertreatment using antibiotics would fail to fully kill bacteria, and worse, "educate them to resist penicillin." Within twenty-five years, more than 80 percent of *Staphylococcus aureus* samples were no longer sensitive to penicillin. In 2019, the U.S. Centers for Disease Control and Prevention (CDC) estimated that approximately 2.8 million Americans are infected with drug-resistant bacteria annually, and of these, more than 35,000 die.

How did this happen?

In any group of microorganisms, just like in any group of people, there are bound to be some individual bacteria, fungi, viruses, or protozoa that are stronger, heartier, or more resilient than the average microbe. These characteristics make some bacteria harder for antibiotics to kill, and after all the

---

* Penicillin's chemical composition, a beta-lactam ring, was elucidated by Dorothy Crowfoot Hodgkin of Oxford University in 1945. In 1957, penicillin was synthesized, obviating the laborious task of growing the mold itself. Penicillin worked by inhibiting an enzyme necessary for maintenance of bacteria cell walls. But it was only effective against gram-positive bacteria. The first antibiotic to kill gram-negative bacteria was streptomycin, discovered by Albert Schatz in 1943. Working in the lab of Selman Waksman at Rutgers University, Schatz tested hundreds of soil samples and found two actinomycete (a class of bacteria) strains that secreted a substance that killed gram-negative bacteria. Schatz and Waksman later engaged in significant battles over credit and patents. Waksman was awarded a Nobel Prize in 1952, but Schatz was not recognized.

other, average bacteria are wiped out, natural selection dictates that these stronger survivors will multiply and gradually command an ever-increasing share of the bacterial universe. Sometimes their advantageous characteristics arise from spontaneous mutations; bacteria's rapid rate of reproduction, in some cases occurring as quickly as every twenty minutes, yields inordinate opportunities for such beneficial mutations to develop and increase. Overuse of antibiotics dramatically hastens this process by wiping out the weak and the vulnerable, allowing survivors to gain increased dominance. Furthermore, each time we use an antibiotic, more bacteria are exposed to its killing method, and slowly but surely a few will learn to survive by developing novel ways to evade or defeat that method. Those few soon become millions, and then billions, of robust, resilient microorganisms that grow harder and harder to kill. Shockingly, in addition to passing resistant genetic information vertically from generation to generation, bacteria have also shown the ability to transfer genes horizontally, between bacteria, so that even bacteria that have never been exposed to antibiotics can become resistant to them.*

Antibiotic overuse was a rampant problem from the beginning. Once penicillin became plentiful, physicians were liberal in giving it to patients for all manner of maladies, including viral cold or flu conditions, for which antibiotics have no beneficial effect. This continues today. One study of antibiotic prescriptions written in the U.S. in 2010–2011 estimated that approximately 30 percent were unwarranted. In many European countries, antibiotics can be bought over the counter without a prescription.

Excessive antibiotic use is not confined to the field of medicine. An even larger problem is the widespread use of antibiotics in agriculture and for livestock—it is estimated that approximately 70 percent of antibiotics produced today are given to animals. The accidental discovery in the 1940s that antibiotics accel-

---

* Horizontal genetic transfer between bacteria occurs via (1) transduction, when viruses carry genes from one to another, (2) conjugation, when bacteria transfer genes through contact, and (3) transformation, when genes spilled into the surrounding milieu by dead, decomposing bacteria are absorbed by nearby bacteria.

erate animal growth prompted farmers to give their livestock small, intermittent doses—amounts not high enough to kill the most resilient bacteria but sufficient to eradicate weak microbes, leaving the strong ones to survive and thrive.

The methods by which bacteria evolve to resist antibiotics are numerous. They can attack the antibiotic itself by creating enzymes that impede the drug. They can alter the shape of binding sites to prevent an antibiotic molecule from latching onto, or penetrating, the bacterial cell wall. They can develop cell wall pumps that pump the antibiotic out of bacteria before they can be killed. When an antibiotic targets a specific step in a bacterium's metabolic pathway, the bacterium can alter that metabolic pathway to evade harm.

Though we endeavor to keep up with constantly evolving bacteria by inventing new antibiotics, there have always been some bacteria that have managed to thwart every drug ever produced. Sometimes this occurs with shocking rapidity, as with penicillin—within one year of its release, penicillin-resistant *Staphyloccus aureus* was identified. Methicillin was a drug introduced in 1960 to overcome penicillin-resistant organisms, but methicillin-resistant *Staphylococcus aureus* was reported that same year. Vancomycin was developed in anticipation of methicillin's limitations, but in 1988, vancomycin-resistant *Enterococcus faecium* was identified. And the list goes on. More recently, a new combination drug, ceftazidime-avibactam, was introduced in 2015, but a resistant sample of *Klebsiella pneumoniae* was discovered later that year. Although most bacteria remain susceptible to one or several antibiotics in our armamentarium, this isn't always the case, and increasing antibiotic resistance contributes to longer hospital stays, prolonged courses of treatment, higher expense, increased deaths, and often, the use of alternative medicines with more significant adverse side effects.

What can be done? How can we even begin to approach this overwhelming problem and stay one step ahead of our cunning enemies?

One thing that won't work is banking on the perpetual invention of new drugs. The rate of antibiotic development has slowed considerably in recent decades—primarily due to the cost; it takes approximately $1 billion to bring a new drug to market. Pharmaceutical companies view antibiotics, which patients need rarely, as far less lucrative than medicines that patients take every day, like drugs for arthritis or cholesterol, for example. In 2012, the U.S. Congress passed the Generating Antibiotic Incentives Now (GAIN) Act in an attempt to encourage greater antibiotic development. The legislation promised faster approval of new medicines and an additional five years of patent exclusivity. Unfortunately, in the decade after its passage, the GAIN Act did not significantly stimulate the creation of novel antibiotics.

On the other hand, there is no shortage of ideas for other ways to combat the problem. Awareness is important—both to inform physicians to prescribe antibiotics more judiciously (and certainly not for viral infections) and to instruct patients not to demand antibiotics. Every patient should finish their full course of medication even after beginning to feel better, because stopping too soon leaves more hearty bacteria alive. Reducing the need for antibiotics by promoting improved sanitation and access to clean water in places that lack it, and encouraging the use of vaccines, will do much to diminish our antibiotic footprint. More than anything else, government initiatives and international treaties that limit the use of antibiotics in farming and for livestock will help tremendously.

There are also novel, alternative treatment options that may contribute to our salvation. For example, scientists are studying genetically engineered bacteria that can produce bactericidal peptides that break down microbial defenses. It has also been noted that some metals have antibacterial effects, and the use of metal nanoparticles is being studied as a potential new method of combating bacteria. Finally, "phage therapy," a strategy that employs viruses termed *bacteriophages* that selectively attack bacteria, is another promising approach.*

---

* To describe phage therapy as novel is not entirely true. Bacteriophages were discovered in the early twentieth century. Whereas their therapeutic potential was largely ignored or even dismissed in the West, particularly after the advent of penicillin, bacteriophage therapy remained popular in the Soviet Union, which became scientifically isolated during the Cold War and lacked access to many antibiotics. Now, the desperate need to counter antibiotic resistance has brought this strategy of using viruses to attack bacteria back into vogue in the West.

BACTERIAL INFECTION: *The Magic Bullet*

The quest to identify and conquer bacteria that killed untold millions throughout human history is an unmatched story of achievement, and a struggle that continues today. Giants like Pasteur, Koch, Ehrlich, Fleming, Florey, Chain, and Heatley were brilliant and flawed individuals. Their sagas reflect stunning audacity and accomplishment, sometimes marred by arrogance, conflict, and petty grudges. Their competitive natures—the source of so much strife—undoubtedly also spurred them to greater heights. They bequeathed us the tools to defend ourselves against relentless, unseen enemies. May we have the good sense to use that gift well.

4

VIRAL INFECTION

Pandemic

In October 1952, a twelve-year-old boy named Arvid Schwartz tumbled off his bicycle while riding at his family's farm in northern Minnesota. This was unusual; there was nothing in his way and Arvid was an excellent rider. He stood up and noticed his legs felt a little weak, but this didn't stop him from brushing himself off and completing his chores later that afternoon. Strangely, the funny feeling in his legs persisted for a few more days. Then, he developed a headache, fever, and a feeling of malaise. The next morning, as he started to climb out of bed, he fell flat on his face.

To his horror, Arvid realized his legs were paralyzed.

He called out for help. His father carried him to a daybed downstairs and for the next three days his parents anxiously debated what to do. Should they watch and wait? Or fetch the doctor? Finally, a doctor was summoned. Unsure of the diagnosis, he recommended Arvid be taken to the University of Minnesota Hospital in Minneapolis. Arvid's father carried him to the car and laid him out in the backseat, propped up against a pile of pillows and blankets. It was a beautiful fall day, yet Arvid felt a premonition and took a long look around the farm. He had a feeling that he might not be back for a very long time.

At the hospital, a spinal tap revealed his parents' worst nightmare: poliomyelitis. Arvid was immediately separated from his parents and brought to a crowded contagion ward. The room was filled wall-to-wall with beds containing immobile children like him. For the next eight days, Arvid lay supine and feverish. Much of the time, he simply stared at the ceiling and cried.

"I had no idea what was going to happen to me," he later recalled of these first harrowing days. No one told Arvid whether he was expected to live or die. He didn't know if he would be there for a week, a month, or longer. Unable to move, he received enemas and had to be catheterized to relieve his bladder. When he wasn't crying, he tried to occupy his mind by imagining what he might be doing in the regular world, thinking, "Well, it's 8:30, we'd be in school now, and it's 9:30, we'd be having spelling or geography."

After the first eight days he was moved to another large ward—another sea of beds and what seemed like a multitude of paralyzed children. He got to see his parents but only very briefly. His mother wept. None of them could have guessed that his hospital stay would ultimately stretch to eight months, or that he would never walk unassisted again. Arvid's days were full of boredom, punctuated by stretching exercises and hot pack treatments with therapists. In time, the magnitude of his disability became clear. Muscle strength was rated on a one to ten scale, with ten being normal. Arvid's legs were consistently zero—no movement. His upper extremities were not as badly affected—only slight paralysis in one arm. Though his back was weak, his nurses helped him learn how to move with the aid of braces and crutches.

He counted himself lucky that, unlike so many other children, he was not confined to a coffin-like iron lung—that large and dreadful apparatus that entombed whole bodies, leaving only a child's pitifully small head poking out of one end. In these children, the poliovirus had damaged either the nerves that control the diaphragm or the part of the brainstem that maintains breathing. Without the assistance of the powerful motor that exerted alternate positive and negative pressure on their tiny bodies within the sealed confines of the iron lung, these children would not be able to inhale or exhale. Completely helpless, they were dependent on others for every aspect of daily living. Volunteers spoon-fed them. Through portals in the side of the machine, nurses washed them, changed their clothes, or scratched an itch. During power outages, the children risked suffocation until nurses or orderlies arrived to hand ventilate them.*

In retrospect, Arvid's experience would fall at the tail end of the polio pandemic, a scourge with a name practically no one had heard of before 1900 but that would terrify parents and transfix the nation for the next fifty years. Each summer, regions affected by polio closed swimming pools, movie theaters, and libraries. Parents kept their children indoors and forbade playing with other kids. No one knew how to combat this loathsome plague. It was a disease characterized by mystery as well as terror.

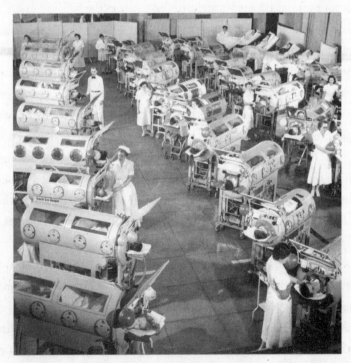

Hospital ward full of children in iron lungs

---

* Although most patients only needed iron lungs during the worst, most acute phase of their illness, some spent months and even years living almost entirely in them.

THE PARALYTIC DISEASE

Cases of poliomyelitis usually started innocuously: a mild headache, a slight cold. Parents sent their children early to bed, but then the pains would start—in the back, in the legs. After a sweat-drenched, sleepless night, a child might notice a slight weakness in a knee or an arm. Then, the next morning, perhaps the parents would be awakened by the sound of a small body crashing to the floor, or the high-pitched, terrified cry of their child now unable to move their neck or legs. At the bedside, doctors called to evaluate children with early symptoms might ask, "What's this I see here on your belly button?" A child who was unable to lift their head to look almost certainly had polio.

The mysterious disease attacked indiscriminately. Though associated with the young, it also struck adults. It affected rich and poor, urban and rural, and members of every race. Sometimes the family of an infected child would not become ill, but other times an entire household would contract polio. No one knew which town or city would next be stricken by an epidemic, nor how to combat it. Quarantines and strict sanitation measures that helped during outbreaks of cholera and typhoid seemed to have little effect. During a 1916 epidemic in New York City, 72,000 stray cats were killed and the streets were washed with four million gallons of water each day. Public health officials issued myriad directives, from avoiding restaurants and ice cream parlors, to closing off drinking fountains and forbidding public gatherings, to avoiding excessive tiredness, screening all windows, and obsessively killing insects. Neighbors avoided neighbors. Many wore gloves and refused to shake hands. Parents of children afflicted with polio burned their children's books, toys, and bedding—desperate attempts to protect against an invisible enemy.

What was not initially understood was that the emergence of polio in the twentieth century was an unintended consequence of improved sanitation in developed countries. Poliovirus is typically transmitted through fecal matter contaminating water sources. For most of human history, a lack of modern sanitation put people in close contact with contaminated water, which exposed them to low levels of the poliovirus regularly. Picture the nineteenth-century world of outhouses, lack of toilet paper, limited handwashing, and open sewers, and it's easy to see how every child encountered poliovirus as part of everyday living. A soiled hand that touched a mouth or nose, or a sip of contaminated water, gave poliovirus entry into the digestive system, where it multiplied in the small intestine and was shed in stools. For most, repeated, low-level exposure to the virus early in life—particularly while protective maternal antibodies still circulated in the bloodstream—allowed a child's immune system to generate its own antibodies against poliovirus. However, after developed countries improved sanitation infrastructure in urban areas, a broad swath of the population began to lack exposure to the virus in infancy. Now unprotected into their school-age years, these children never developed lymphocytes that would recognize and destroy poliovirus in the future. They were susceptible to the poliovirus in a way no prior generation had been.

Still, in 95 percent of children, the virus caused no symptoms or only very minor symptoms because the body's immune system repelled the intruder. In such cases, most parents did not even know their child had been exposed to poliovirus. But in 5 percent of children, poliovirus eluded the immune system to cause varying degrees of illness, from mild, flu-like symptoms to more serious signs including muscle pain, fatigue, and neck and back stiffness. This progressed to paralytic polio in about 1 percent of cases.

Paralysis resulted from the virus's predilection to migrate through the bloodstream to a part of the spinal cord called the anterior horn. There, poliovirus attacked gray matter motor neurons, which emanate from the spinal cord and carry signals to the body's muscles to command them to move.* Sometimes the damaged neurons recovered, and children with sudden paralysis regained some or all of their strength. Other times the damage was permanent. Up to 10 percent of patients with paralytic polio died. The death rate rose as high as 60 percent in a subset of patients with "bulbar polio"—the term given to cases in which the cranial nerves that control swallowing and

---

* The terms *polio*, *myelo*, and *itis* refer to gray matter, spinal cord, and inflammation, respectively.

the airway, or the medulla oblongata (a part of the brainstem that controls breathing), were injured.

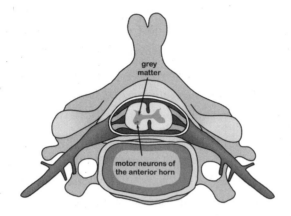

Cross section of the spinal cord. Poliovirus attacks the motor neurons of the anterior horn.

Doctors' turn-of-the-century ignorance of polio was so complete that there was lack of consensus regarding what the infectious agent actually was. Many believed polio was caused by bacteria. This was disproved in 1908 by Austrian physicians Karl Landsteiner* and Erwin Popper, who examined spinal cords of children who died from the disease and failed to find any signs of bacteria. If it wasn't bacterial, what was it? It had to be infectious because when they ground up the spinal tissue, filtered it, and injected it into monkeys' stomachs, the monkeys developed polio. Landsteiner and Popper suspected they were dealing with a virus, still a theoretical type of organism at the time.

Their conclusion was supported by American researchers Simon Flexner and Paul Lewis. In 1909, Flexner reported, "We failed utterly to discover bacteria, either in film preparations or in cultures . . . it would appear that the infecting agent of epidemic poliomyelitis belongs to the class of the minute

---

* Landsteiner later won a Nobel Prize for discovering the four different human blood types: A, B, AB, and O.

and filterable viruses that have not thus far been demonstrated with certainty under the microscope." Though scientists would become increasingly confident that viruses existed and caused disease, viruses themselves were too small to be seen with an optical microscope. It would not be until the 1930s that the advent of the electron microscope would make viruses visible for the first time.

Meanwhile, as scientists slowly uncovered polio's secrets in their laboratories, summer epidemics continued to provoke fear throughout the United States. A 1916 epidemic caused 27,000 cases of paralysis and 6,000 deaths. In New York City alone, there were 8,900 cases and 2,400 deaths. Eighty percent of victims were children younger than five, including many babies. Though polio killed far fewer than heart disease or cancer, its young and helpless victims made its tragic consequences more dreaded and heartrending than any other condition. A 1955 survey showed that Americans feared polio more than any other calamity except nuclear war.

But ironically, it was not the children's plight that proved to be the essential turning point in the war against polio. Instead, it was the determination of one man—a thirty-nine-year-old victim who refused to allow polio or paralysis to define him.

AT A BOY SCOUT JAMBOREE

On July 28, 1921, Franklin Delano Roosevelt boarded the steam yacht *Pocantico* in Manhattan for a forty-mile cruise up the Hudson River to Bear Mountain State Park. He was scheduled to make an appearance there at a Boy Scout Jamboree—for the 2,100 scouts in attendance, it was the highlight of the summer. Roosevelt loved the Scouts and felt all boys should spend time in the outdoors—especially boys from the urban jungle of New York City. He also welcomed the opportunity to take a break from his labors. He had spent the second half of 1920 campaigning as the Democratic vice presidential candidate on Governor James Cox's losing presidential ticket. After the election defeat to Warren

Harding and his running mate Calvin Coolidge, Roosevelt maintained a rigorous schedule. Now, the Jamboree was his last engagement before a desperately needed vacation. On the way to Bear Mountain State Park, he allowed himself to relax, telling stories and laughing with his friends onboard ship.

At the Jamboree, Roosevelt gave a speech, led a march, and joined the scouts in the mess hall for a "regular old-fashioned southern chicken dinner," according to one newspaper report. He enjoyed himself and shook countless hands. That evening he sailed back down to Manhattan for the night and the next day embarked on his much-anticipated holiday, first to spend a few days at his boyhood home in Hyde Park, New York, before traveling to join his family at their summer house on Campobello Island in New Brunswick, Canada.

On August 10, 1921, Roosevelt enjoyed a vigorous day with his family on Campobello. With his three eldest children, he sailed around the island, helped fight a mini-forest fire on a nearby isle, completed a two-mile hike, and went swimming in a pond. When he returned home after this action-packed day, he noticed a little soreness in his back at dinner. Telling his wife, Eleanor, not to worry—he was just overly tired—he went to bed early.

The next morning he woke up and felt worse.

He had a fever of 102 degrees, and there was an odd pain in his right knee, which felt very weak. He thought he just needed more rest, but by evening, the pain had worsened and spread to his back and neck. Then his left knee started to feel weak as well. He passed another night, sleeping fitfully.

The following morning, when he awoke, he was horrified to realize that he could not move either leg. He could not tighten the muscles of his abdomen or buttocks. He could not even hold a pencil.

Dr. Robert Lovett, a Harvard physician summoned from Boston, made the diagnosis and delivered the bad news. Roosevelt had polio. No one knew it then, but he would never walk again.

It's instructive to point out the ways in which someone like Roosevelt may have been uniquely poised to contract polio. He was the scion of a wealthy family who had lived a pampered and sheltered childhood. Instead of going to elementary school and gaining exposure to other children and new environments, he was educated by tutors who came to his home. Then he was sent to Groton, an exclusive boarding school, and finally, Harvard. Compared to the general public, Roosevelt probably lacked exposure to many everyday diseases during his youth, and he had certainly never spent time living in conditions considered even remotely unsanitary by contemporary standards.

Though it cannot be proven, historians believe that Roosevelt most likely contracted poliovirus sometime during his day visiting the Boy Scouts. Bear Mountain State Park was an extremely popular destination. Each summer approximately 75,000 tourists visited to hike the trails, camp, and swim in the lakes. This visitor burden overwhelmed the park's capacity to meet recommended sanitation guidelines and ensure clean drinking water. There were not enough portable toilets; outhouses were not purged and disinfected properly. It was common knowledge that visitors sometimes defecated in the woods or near bodies of water; therefore, it is quite possible that one of the water sources at the park was contaminated with poliovirus. It could have been in a glass of water or lemonade, or in the water used to wash dishes at the mess hall, or on the fingertips of a boy who shook Roosevelt's hand after failing to wash his own. However it happened, the thing that is certain is that poliovirus arrived at Roosevelt's mouth and entered his digestive tract.

Roosevelt would spend the next seven years working with therapists and searching for any treatment that might offer a chance to regain movement in his legs and the ability to walk. Through a friend, he discovered a resort in Warm Springs, Georgia, whose mineral-laden hot springs provided so much buoyancy that patients could almost imagine that they were walking normally again. This was an incredible feeling for a paralytic, and Roosevelt became convinced that the hot springs had a restorative effect that was allowing his legs to recover some function. He spent $200,000, approximately two-thirds of his fortune, to buy the resort, transforming it into a place where polio patients could take off their braces, crawl about on the lawns, wear shorts that exposed their atrophied legs without shame, and socialize comfortably—free of the stigma conferred by their affliction in the regular world.

Roosevelt also recognized the massive need for funds to care for polio victims' treatment and rehabilitation, and to pay for scientific research on the condition. With his law practice partner, Basil O'Connor, he founded an organization in 1927 that would revolutionize the link between charitable fundraising and scientific discovery—the Georgia Warm Springs Foundation, which transitioned to become the National Foundation for Infantile Paralysis in 1938. The National Foundation's "March of Dimes" campaigns raised millions of dollars annually to fund research and pay for patient care across the country. It also succeeded in making polio the most feared disease in America. In truth, polio was never the mass epidemic that the public was made to believe. Far more children died from accidents and cancer, for example. But by employing images of disabled children, often in wheelchairs or sometimes iron lungs, there was no more terrifying diagnosis in the minds of American parents, who willingly opened their pocketbooks to do their part in ending the scourge.

And so, it was the National Foundation, founded by Roosevelt and O'Connor, that redirected the donations of tens of millions of Americans into a multi-decade research effort to combat polio. It became clear that the best, and perhaps only, way forward was to develop a vaccine.

But for decades the challenges associated with such an endeavor appeared almost insurmountable.

THE MILKMAID

The first recorded human effort to purposely enhance the body's immune system through exposure to a pathogen was found in a Chinese book published in 1549, although the technique is believed to have been practiced in China since the tenth century. The Chinese understood that survivors of smallpox gained lifelong immunity to the disease. Practitioners obtained pustule samples from smallpox survivors, dried and crushed this material into powder, and blew the powder into uninfected patients' nostrils using a blowpipe. Another method was to rub the material into a superficial incision in a patient's hand or arm. The procedure usually precipitated only a mild case of the disease and conferred immunity based on the likelihood that a tiny, controlled exposure to the pathogen was far less likely to result in serious infection or death compared to natural infection associated with substantial inhalation of the virus. A report of the practice later reached the Royal Society in London via an employee of the East India Company who had been working in China. This practice was termed *variolation*, taken from the scientific name for smallpox: variola.

It was no coincidence that smallpox became the first infectious disease to be confronted in this way. Over three millennia, from the time of the ancient Egyptians to the twentieth century, it killed millions and ranks as one of the worst scourges in human history. The smallpox virus is highly contagious. In the eighteenth century alone, it killed 400,000 people worldwide annually, more than all other infectious diseases combined. When it struck a town, it was not unusual for almost the entire population to contract the disease and exhibit the high fever and blistering, pustular rash associated with it. Approximately 30 percent of those infected died. Those who survived were left with terribly disfiguring scars from healed pustules, often on their faces.

The first Westerner to adopt and promote variolation was a woman who knew firsthand the damage smallpox could do. Lady Mary Wortley Montagu, the wife of the British ambassador to the Ottoman Empire, was known for her beauty as a young woman, but after contracting smallpox in 1715, she lost all her eyelashes and her face became severely blemished by pockmarks. In Constantinople, she witnessed the practice of variolation in 1717 by two Arab physicians. She became so convinced of the practice's merits that she had her five-year-old son, Edward, variolated. Upon her return to England in 1721, her four-year-old daughter, Mary, was variolated by Dr. Charles Maitland, who had served as the embassy physician in Constantinople. Maitland subsequently tested the procedure at Newgate Prison on six condemned prisoners who were promised their freedom if they survived. All of them did.

Lady Montagu then convinced her friend Caroline, the Princess of Wales and future queen (as consort of George II), to variolate her two daughters in 1722. Although there was risk to the procedure—in about 2 percent of cases the patient contracted severe smallpox and died—the practice became fairly well adopted in England over the course of the eighteenth century.

The stage was now set for a new figure to alter the history of medicine by bucking tradition with an audacious experiment. Edward Jenner, born in Berkeley, England, in 1749, was thirteen years old when he began a seven-year apprenticeship with a country doctor named Daniel Ludlow. One day, Jenner overheard Ludlow discussing smallpox with a young milkmaid. The woman said, "I shall never have smallpox for I have had cowpox." Cowpox was a condition far less severe than smallpox that affected the udders of cows. Sometimes milkmaids contracted cowpox and developed pustules on their hands. The disease was never fatal, and after recovering from a mild illness of three or four days, the women were anecdotally believed to be protected from smallpox thereafter. Jenner did not do anything with this information at the time. He was only a young man. Cases of cowpox were not common, and outbreaks were rare. Jenner simply filed the information away in the back of his mind.

Then in the 1770s, after embarking on his own career, Jenner gradually collected, and eventually published, a case series of fifteen individuals who had previously had cowpox and did not contract smallpox, even though they had each been exposed to it through close contact with infected people (i.e., living in the same household with, or nursing, a smallpox patient). In May 1796, a cowpox outbreak occurred at a farm near Jenner's home. This gave Jenner the opportunity to devise a daring test. He drew some fluid from a pustule on the hand of a milkmaid named Sarah Nelms, and instilled it into two superficial, half-inch-long incisions in the arm of an eight-year-old boy named James Phipps. Pustules formed at the site, which later became healed scabs and left scars. Phipps had a mild fever and headache for a few days but then recovered. Jenner waited six weeks to make sure the boy had fully regained his health. Then he did something that would be considered entirely unethical today—he inoculated Phipps with live virus from the sores of a smallpox patient.

Phipps, a healthy boy, could have died to satisfy Jenner's curiosity and desire to validate a hunch.

But Phipps didn't die. The story had a happy ending. Jenner watched him closely for weeks and Phipps did not become ill at all. Jenner had shown that cowpox could be used to protect against smallpox, a practice later named *vaccination*, derived from the Latin word *vacca* for "cow." Jenner eventually repeated the experiment on over twenty people and became convinced of its effectiveness. In 1798, he published a book on the method. Vaccination against smallpox grew in popularity and spread across the Atlantic to the United States, where President Thomas Jefferson vaccinated his entire family and many neighbors from surrounding Virginia farms.* In France, Napoleon ordered that every French soldier be vaccinated.†

Jenner became famous throughout Europe and North America, but neither he, nor anyone else, understood why vaccination worked. He did not understand the human immune system. He did not know that smallpox was caused by a virus. He didn't know what a virus was.

VIRUSES

The term *virus* is derived from a Latin word meaning "poison." Viruses are very different from bacteria. They are the most common biological units on

---

* Smallpox brought by Europeans to North America killed tens of thousands of Native Americans. In 1807, a confederation of five tribes—the Mohawk, Onondaga, Seneca, Oneida, and Cayuga—sent Jenner a wampum belt as a token of gratitude for developing the method of smallpox vaccination.
† In 1979, the World Health Organization declared that smallpox had been fully eradicated worldwide through successful vaccination efforts, though frozen laboratory samples have been retained in the U.S. and Russia.

Earth and are, quite simply, a small fragment of DNA or RNA enclosed by a protein coat (called a capsid). Viruses are parasites that are not considered fully alive because they lack the ability to reproduce on their own. They are many times smaller than bacteria—too small to be seen with an optical microscope, which is the primary reason why their existence was not confirmed until the twentieth century. A virus's only goal is to infect a host cell and hijack that cell's protein-producing ribosomes to make copies of the virus itself. Viruses do this by latching onto cells and injecting their DNA or RNA into them. Messenger RNA produced from the DNA or RNA (by a process called *transcription*) is sent to ribosomes that read the messenger RNA in order to make viral proteins (by a process termed *translation*), resulting in exact replicas of the virus. An accumulation of new viruses usually overwhelms and destroys the host cell, which bursts apart and releases those viruses to seek out new hosts of their own.

The idea that nonbacterial microorganisms, later known as viruses, might exist originated with the findings of a German chemist named Adolf Mayer. In the 1880s, Mayer studied a mysterious disease that was ruining tobacco crops. Because it caused a characteristic discoloration of tobacco leaves, he called it "tobacco mosaic disease." When he mashed up diseased leaves and added a little water, he found that the sap-like fluid could be used to transfer the disease to healthy tobacco plants. When he examined the infected sap and plant specimens under a microscope, he saw nothing. He could not find any bacteria. He still believed, however, that an unseen bacteria must be the cause.

In 1892, a Russian scientist named Dmitri Ivanovsky passed infected tobacco leaf sap through a special, high-volume porcelain filter called a Chamberland filter, which was known to be fine enough to filter out bacteria. To his surprise, the liquid filtrate remained infectious—there was something smaller than bacteria in it that was causing the disease. He thought it might be a bacterial toxin. Then, in 1898, a Dutch botanist named Martinus Beijerinck proposed that this new infectious vector was not bacterial but instead a *contagium vivum fluidum*, or "contagious living fluid." He also called this a "virus," which he may have believed was liquid in nature.

Whereas the discovery of penicillin would initiate an era of mankind's dominance over bacterial diseases, the war against viruses would be prosecuted with far less success. Virology's pioneers did not know they were working in a field that would be called immunology. Though knowledge would increase, the prospect of a magic bullet to kill viruses never materialized, and twentieth-century virologists realized that the best, and often only, weapon in the fight against viruses would be prevention, through vaccines. Yet it was not easy to make virus vaccines. The key was to somehow weaken a virus enough that it was incapable of causing disease, but not so much that it would be unable to elicit a protective immune response in patients. This was a very fine line. How could one reliably attenuate a virus just the right amount?

In the 1930s, Max Theiler, a South African researcher, and his American colleague, Hugh Smith, were studying yellow fever at the Rockefeller Institute in New York City. They grew the causative virus in chick embryo cells and began to transfer it repeatedly though a succession of new tissue cultures. They found that with each passage through culture, the virus became weaker and caused less severe disease when injected into an animal. After more than 200 passages, the virus ceased to cause disease at all. They had developed a novel way to weaken a virus in a controlled manner that could be used to mass-produce a vaccine.*

Unlike Jenner, twentieth-century scientists like Theiler understood the basic underpinnings of the immune system, which is separated into two arms: the *innate* immune system, which we are born with, and the *acquired* immune system, which is influenced by pathogens we encounter throughout our lives. The innate, or "nonspecific," immune system is characterized by an army of constantly circulating cells and proteins that act as sentinels and first responders. If a foreign invader is detected, protein messengers called *cytokines*

---

* Their vaccine successfully prevented millions of U.S. soldiers from contracting yellow fever during service in tropical regions such as the South Pacific during World War II. Theiler won a Nobel Prize in 1951 for this work.

sound the alarm. A category of cells known as *phagocytes*, which includes cells with names like *macrophages*, *neutrophils*, *mast cells*, and *monocytes*, directly engage, engulf, and destroy enemy pathogens.

But it's the acquired, "specific" immune system that is pertinent to the study of vaccines. In this arm of the immune system, the "memory" capacity of *lymphocytes* (B cells and T cells) allows them to mount rapid and lethal responses to pathogens the body has encountered before. Such prior exposure gives lymphocytes the opportunity to quickly eradicate infectious agents before they can gain a foothold, by producing custom-tailored antibodies that recognize proteins (termed *antigens*) on the surface of the invaders and mark them for destruction by phagocytes. Lymphocytes, which are produced in the bone marrow, can stay in the bone marrow to become antibody-producing B cells, or they can travel to the thymus in the upper chest and develop into two types of T cells: CD4+ T cells that can be activated to become "helper" T cells that aid B cells in producing microbe-specific antibodies, or CD8+ T cells that can be activated to become "killer" T cells that recognize and eliminate virus-infected cells.

A successful attenuated-virus vaccine cannot be weakened so much that it loses its *antigenicity*—its ability to be recognized as a pathogen by the acquired immune system. An alternative approach could be to use a "killed"-virus vaccine, instead of a live, attenuated vaccine, but most virologists of the mid-twentieth century believed killed-virus vaccines to be inferior because their antigenicity was expected to be weaker.

The difference between these two approaches did not generate excessive controversy or debate in the first half of the twentieth century. But then, because of two opposing men, it suddenly became the most important thing in the virology universe. Ironically, these men who shared much in common—their dedication, brilliance, Jewish American heritage, and goal of curing a tragic disease—could never seem to agree that the other one might also be right. Their battle of ideas and ambition would play out on the grandest stage, in the midst of a worldwide struggle—the struggle to defeat polio.

THE RIVALRY

In the first half of the twentieth century, polio epidemics became annual summer events in communities throughout the United States. Each June, a handful of polio cases could prompt a city's schools and theaters to close. Playgrounds and beaches stood empty. Families stopped socializing with other families. If an outbreak occurred, newspapers broadcast headlines like "Polio Panic" and "Polio's Deadly Path." They printed daily death tolls and ran photos of polio wards full to bursting with row upon row of children encased in iron lungs. At mid-century, annual case numbers seemed to only march upward. In 1946, there were 25,000 polio cases in the United States; in 1949, 42,000 cases. In 1952, there were 58,000 cases.

And yet, even by the 1940s, scientists were little closer to understanding the poliovirus than they had been in the 1900s, when Landsteiner and Flexner were the first to isolate it using a monkey model. How the disease moved from person to person was a mystery and was clearly not by close contact alone. It was impossible to stem the spread by quarantining victims. Nor could polio be eradicated from the environment, though many tried by spraying their homes with DDT, both inside and out.

In 1947, a thirty-three-year-old researcher named Jonas Salk became the director of a new virology lab at the University of Pittsburgh. The position gave him the opportunity to direct his own research and build a team from scratch, something he longed for after years of working under the tutelage of others. Salk had a reputation for being soft-spoken, empathetic, hardworking, and ambitious. He had devoted much of his career to studying influenza; during the Second World War he'd served in Europe immunizing thousands of troops. In Pittsburgh, with generous funding from the National Foundation for Infantile Paralysis, Salk turned his attention to the poliovirus.

Prior to coming to the University of Pittsburgh, Salk had helped develop an effective killed-virus vaccine for influenza, and he believed a killed-virus vaccine would be the most realistic strategy for success with polio in the near term. He knew that most other virologists disagreed with him. They believed an attenuated-virus vaccine would be much better because the process of fully killing a virus was sure to eliminate most or all of its antigenicity and ability to arouse a protective immune response. And even if a killed virus did work, its effects were likely to be temporary, perhaps lasting only weeks or months, or a year at the most; this would necessitate multiple, ongoing vaccine inoculations over one's lifetime. In contrast, a live-virus vaccine would produce lifelong immunity by replicating a natural infection in the body—not try to trick the body into producing antibodies for only a limited period of time. Live-virus vaccines had been successfully created to combat yellow fever and smallpox.

But Salk thought a killed-virus vaccine would have advantages of its own. He believed it was the safer choice because a killed virus would never be able to cause disease. With live-virus vaccines, there was always the chance, albeit small, that inadequately weakened viruses might maintain their ability to produce infection. Killed-virus vaccines had worked not just for influenza, but also against rabies and the bacterial toxins of diphtheria, typhoid, and cholera. And practically all virologists acknowledged that a killed-virus vaccine would be much simpler to make than an attenuated-virus vaccine.

But before any vaccine could be contemplated, there were still many important questions that scientists had been unable to answer about poliovirus, questions as basic as: How did the virus enter the body and infect the spinal cord?

For decades, the prevailing belief was that poliovirus entered the nasal passages and traveled straight to the central nervous system, bypassing the bloodstream entirely. In fact, it was this very idea that had deterred many researchers from pursuing polio vaccine research because, if true, it meant a vaccine designed to produce antibodies in the bloodstream would do no good. But in 1941, another researcher named Albert Sabin helped debunk this idea. Sabin, who would become Salk's nemesis, performed autopsies that showed the virus was uncommonly found in the nasal cavity and was much more prevalent in the digestive tract. When other doctors succeeded in transmitting polio to chimpanzees by feeding it to them, instead of injecting it into their brains, the digestive tract became accepted as the mode of entry.

Another basic question was: How many different types of poliovirus existed?

No one knew for sure. By the late 1940s, researchers knew there were at least two types of poliovirus, but there could still be more. Any effective vaccine would have to protect against all of them. Many virus families consisted of multiple types; influenza was one that had so many different and frequently mutating types that an updated vaccine had to be given anew to patients each year. To answer the polio question, every possible poliovirus strain, worldwide, would have to be collected and examined to see if it matched one of the known types of poliovirus or might be a new type.

Salk volunteered to be a part of this gargantuan effort. His lab, and others, embarked on the multiyear task of testing virus samples from around the world. To describe this process as laborious would be a gross understatement. Stool samples, throat cultures, and spinal cord specimens were shipped to Salk's lab from near and far. Small portions of these samples were injected into the brains of monkeys. When these animals developed polio they were sacrificed so that their brains and spinal cords could be removed to isolate poliovirus. This virus was next injected into a new monkey. If that monkey was injected with a type 1 virus and survived, it gained immunity to type 1 viruses. Then, if that same monkey was injected with an unknown strain of virus and did not get sick, researchers could conclude that the unknown virus should also be classified as type 1. However, if the type 1–immune monkey got sick, then the unknown strain must be another type, such as type 2. Those monkeys that survived inoculation with a type 2 virus became immune to type 2, and thus became a vehicle upon which other type 2 viruses could be tested. And so on. It was painstaking and monotonous work. In 1949, a third type of poliovirus was identified. The typing program would last for three years, from 1949 to 1951. The effort, funded by the National Foundation, cost $1.19 million, and the lives of approximately 17,500 monkeys.

Ultimately, only three types of poliovirus were found.

VIRAL INFECTION: *Pandemic*

The next seemingly insurmountable challenge was to somehow grow enough poliovirus of each type to create a vaccine that could be mass-produced. Millions of doses would be required, and an inability to grow poliovirus in large quantities had always been a major obstacle. For decades, poliovirus had only been successfully grown in cultures made from the neural tissue of monkeys. However, monkey nervous tissue was known to cause a toxic reaction, and sometimes encephalitis and death, when injected into human subjects. No virus grown in neural tissue could be used in a safe vaccine, but for years, it was considered axiomatic that poliovirus could not replicate *in vitro* (in tissue culture) using any other medium.

Finally, a doctor named John Enders challenged this notion. Enders was the director of the Infectious Disease Research Laboratory at Boston Children's Hospital. In 1948, Enders and two assistants, Thomas Weller and Frederick Robbins, were working to grow varicella, also known as chicken pox, in cultures of fetal muscle and skin tissue. They happened to have some extra samples of poliovirus in the freezer and, on a hunch, decided to add this to the cultures, even though it was common knowledge that poliovirus could not grow in non-neural tissue. To their surprise and delight, poliovirus thrived in the muscle and skin tissue. Common knowledge was wrong. Now poliovirus could be grown in a petri dish or test tube using non-nervous monkey tissue—a medium that would not automatically be toxic to humans. Suddenly the prospect for a polio vaccine had taken a giant leap closer to reality.*

The stage was now set for an epic battle of ideas and wills. On the surface, it was a contest between two modes of thought that emerged from the setting of stuffy, academic conferences into public view—the feud between advocates of a killed-virus vaccine versus those who favored a live, attenuated-virus approach. Below the surface, it was a rivalry between two brilliant men, Jonas Salk and Albert Sabin, who would spend the rest of their lives waging their own personal Cold War—a war of words, science, and most important of all, legacy.

---

* Enders, Weller, and Robbins would share a Nobel Prize for this discovery in 1954.

Sabin was the more senior and experienced researcher. Eight years older than Salk, Sabin was the head of pediatric research at the Cincinnati Children's Hospital in Ohio. At academic meetings and conferences, he sometimes failed to veil his dim view of Salk as an overly ambitious, excessively striving neophyte who was very much junior to Sabin and the other luminaries in the field. The fact that Salk was doing the grunt work of typing viruses, considered too menial for virology's thought leaders, served to reinforce Sabin's imperious attitude. In one telling episode, at a meeting of leading researchers in 1948, Salk asked if the typing program might be more efficient if, instead of the time-consuming method of testing the infective capacity of unknown viruses, he could simply isolate the antibodies produced by each unknown virus and see which type of virus these antibodies neutralized.

It was a perfectly reasonable suggestion, one that would save a lot of time. But to Salk's chagrin, the first to respond was Sabin, whose reply practically dripped with condescension; he said, "Now, Dr. Salk, you should know better than to ask a question like that."

After a little further explanation by Salk failed to gain any support from the group, Salk was crestfallen. He recalled the experience was "like being kicked in the teeth."

"I could *feel* the resistance and the hostility and the disapproval," he remembered. "I never attended a single one of those meetings afterward without that same feeling."

In 1951, Salk and Sabin traveled to Europe on the same ship to attend a conference in Denmark. Salk later recounted Sabin's attitude toward him during the trip: "It became obvious to anyone who had not heard of it before that I was a nice young whippersnapper from Pittsburgh, going to Denmark to report on some drudgery I had performed. I might have failed abysmally, it seemed clear, if Albert had not been up in the flies, pulling the strings and setting the standards."

Sabin did not hide his opposition to Salk's idea for a killed-virus polio vaccine. In a 1949 letter to Basil O'Connor, the head of the National Foundation for Infantile Paralysis, Sabin wrote, "There is no valid reason for believing at this time that 'killed-virus' vaccines can be of any practical value in the control of a disease like poliomyelitis ... premature and irresponsible promises to the public, in my opinion, are not only unwarranted and unkind, but ultimately may also constitute a serious hazard to the important efforts of the NFIP."

Like many brilliant investigators, Sabin had a complex personality. He could be arrogant and condescending, but also highly ethical and single-minded in pursuit of a goal. Sabin acknowledged that a live-virus vaccine would take longer to develop than a killed-virus vaccine but believed that it would be superior in the long run. He felt there was still much to learn and understand about the poliovirus, and that the science could not be rushed.

Unfortunately for Sabin, Basil O'Connor was not a patient man. He had been Franklin Roosevelt's law firm partner and confidant, a man who had risen from modest beginnings to great wealth and influence. As the head of the largest, most influential medical charity organization in American history, O'Connor felt the urgency to find a polio vaccine acutely.

Jonas Salk was like-minded. Salk never needed to be reminded of the public exigency for a vaccine. His lab in Pittsburgh was in the basement of the Municipal Hospital, with wards of polio-stricken children occupying the floors above. He and his team would pass the wards each day on their way to the lab. A nurse recalled the ever-present tragedy of the place:

> There were sixteen or seventeen new admissions every day ... To leave the place you had to pass a certain number of rooms, and you'd hear a child crying for someone to read his mail to him or for a drink of water or why can't she move, and you couldn't be cruel enough to just pass by. It was an atmosphere of grief, terror and helpless rage. It was horrible. I remember a high school boy weeping because he was completely paralyzed and couldn't move a hand to kill himself. I remember paralyzed women giving birth to normal babies in iron lungs. I remember a little girl who lay motionless for days with her eyes closed, yet recovered, and I can remember how we all cried when she went home. And I can remember how the staff used to kid Dr. Salk—kidding in earnest—telling him to hurry up and do something.

Just like O'Connor, Salk felt that speed was of the essence, and perhaps also that sometimes in science, perfect is the enemy of good. He forged ahead with his own plans to develop and test a killed-virus vaccine, even though it went against the prevailing advice of his peers.

Salk and his team worked to replicate Enders's feat of growing poliovirus in non-neural tissue. They found the perfect medium in monkey kidney tissue. A single kidney could enable the growth of enough viral material to produce 6,000 vaccines. The method was meticulous and painstaking. Minced and filtered kidney tissue was seeded with live samples of the most virulent strains of each of the three types of poliovirus and carefully maintained in incubators. The next critical step was to neutralize the virus by killing it. Salk and his research team did this by immersing the cultures in formaldehyde. It was a delicate balance. Too high a concentration of formaldehyde might diminish the vaccine's ability to produce an immune response against the disease. But too little might inadequately kill the virus and leave unsafe, live vaccine that could actually cause polio, a horrendous thought. They tinkered with the protocol over and over. What were the best concentrations of formaldehyde and virus? At what temperature and acidity should these be combined? For how long? Safety was paramount. A sample of each batch of killed virus was tested by injection into monkeys that were watched for signs of polio. If a monkey became ill, the entire batch was thrown out. Even those monkeys that were not sick were later sacrificed so their nervous tissue could be examined for any signs of polio damage.

When Salk grew confident enough in his vaccine to test it on humans, he and his lab team first tried it on themselves. Salk also inoculated his wife and children. He then obtained approval to solicit volunteers at two institutions, the D. T. Watson Home for Crippled Children, a facility with many polio patients, and the Polk State School for the Retarded and Feeble-Minded. With consent from parents, testing began in June 1952. The basic strategy was to draw patients' blood before and after receiving the vaccine, looking for a resulting increase in the level of antibodies to poliovirus. Salk was on edge.

"When you inoculate children with a polio vaccine," he later admitted in an interview, "you don't sleep well for two or three months."

But to his relief, the vaccine appeared to work. No subject became ill. The vaccine produced a significant, protective antibody response against all three types of poliovirus.

In January 1953, Salk reported his study results at a conference of prominent virologists in Hershey, Pennsylvania. Many of the attendees thought a human vaccine was still years away and that a great deal of science remained to be discovered before that step should be taken. When Salk revealed his successful results in humans, the response was not favorable.

"It was a tense meeting," Salk remembered, "and I was by no means the tensest person there."

His peers expressed surprise, disbelief, and incredulity. Though unspoken, the room was probably infused with an undercurrent of envy. The effectiveness and safety of Salk's vaccine were questioned. His method of using a killed-virus approach was criticized. Albert Sabin was particularly harsh. Salk recalled, "It was almost as if he were trying to minimize what I had said. His interpretations made my work seem incredible, of no meaning or significance. We hadn't done this and we hadn't done that and this was premature and that was unsubstantiated. I remember asking him later, 'Why do you constantly emphasize the negative?' He answered that this was 'the scientific way of doing things.'"

Debate raged about whether Salk's trial results were enough to warrant the initiation of a large clinical trial. Most of the attendees agreed with Sabin and opposed the idea. But Harry Weaver, the National Foundation's director of research, considered Salk's study more than enough to go on. Impatient for a vaccine, two days later Weaver reported to the National Foundation's board of trustees that one of their scientists had created a polio vaccine that had been shown to be safe and effective in a small trial. He told them it was time to start thinking about a field trial. After someone leaked this information, the media picked up the story and a Pittsburgh reporter named John Troan singled out Salk; in the *Pittsburgh Press*, Troan wrote, "So far as anyone knows, there's only one scientist in this country who is working on exactly this kind of vaccine. He is Dr. Jonas E. Salk." Weaver tried to capitalize on this momentum by organizing a conference of influential government and medical thought leaders, along with the press, at the Waldorf Astoria hotel in New York City on February 26, 1953.

Now even Salk became concerned that things were moving too quickly. He wanted to conduct more research to perfect his vaccine and further confirm its safety. He worried that his peers would disapprove of the attention he was receiving, especially after *Time* magazine ran an article with a photo of "Researcher Salk," accompanied by the caption "Ready for the big attack."

Salk was right to be worried—his peers were aghast at the press attention. Increasingly, Salk appeared more opportunist than scientist, a man undeservedly named the leader in the fight against polio when this was far from true. It was unconscionable for a scientist to accept public accolades for an unproven vaccine, or to speak to the press about the results of a study before it had even been published in a peer-reviewed scientific journal. Sabin wrote to Salk, "Although it was nice to see your happy face in *TIME*, the stuff that went with it was awful." In a backhanded compliment, he cast the blame on gratuitous promotion by the National Foundation: "I knew you couldn't have had anything to do [with] it, for if you did they would have gotten the story straight." But there was nothing Sabin or anyone else could say or do to stop the groundswell of public pressure to produce a vaccine as quickly as possible. Both Salk and the National Foundation felt intense pressure to move forward, and newspaper headlines like "Polio Conquest Nearer" and "Hint Polio Vaccine Ready" didn't help to calm matters.

The truth about Salk's desire for attention was far more complex than his detractors would have liked to believe. In many instances, he shunned press attention. He disliked making time for reporters and hated allowing photographers to take pictures of him or his lab, mainly because such proceedings interfered with his work. But at the same time, he also took certain actions that he must have known could not fail to place him at the vanguard of the fight against polio in the public mind—at the expense of his virology colleagues.

One of these actions occurred on March 26, 1953, when, purportedly concerned that media enthusiasm and public optimism about a vaccine were grossly premature, Salk asked the National Foundation to support him in making a public address in order to convey the most accurate information. Basil O'Connor thought this was a great idea, and Salk went on national radio, with CBS, in a fifteen-minute program called "The Scientist Speaks for Himself." He reported his results but cautioned that far more work needed to be done and that a vaccine would not be available for a year or more. He also made one unequivocal statement that opened the floodgates of optimism: "In the studies that are being reported this week, it has also been shown that the amount of antibody induced by vaccination compares favorably with that which develops after natural infection." As a courtesy, Salk had sent Sabin a draft of the speech ahead of time. To Salk's annoyance, Sabin phoned him, incensed. "Told me I was misleading the public," Salk recalled. "Urged me not to do it. I was flabbergasted."

If the goal of Salk's radio address was to cool public expectations, he failed. Now the press began calling it the "Salk vaccine," and a few of Salk's staff even began referring to him derisively as "Jonas E. Christ" behind his back. Sabin became increasingly vocal in his opposition to Salk's approach. He testified before a congressional hearing that Salk's vaccine only promised weak protection and that its safety had yet to be confirmed by other scientists. Nevertheless, in November 1953, the National Foundation's Vaccine Advisory Board voted to embark on a large study of Salk's vaccine. It was a momentous decision. An experimental vaccine made by a thirty-nine-year-old scientist in a basement lab in Pittsburgh, one previously tested in only a couple hundred people, would now be given to hundreds of thousands of children across America.

The vaccine trial of 1954 was the largest medical study in history to date and gained worldwide attention. Seven million volunteers helped conduct the trial at 14,000 schools in forty-four states and Canada. In the U.S., 623,972 first, second, and third graders received three shots, each a month apart, of either vaccine or placebo. In addition to a placebo control group, another large control group consisted of observed children who underwent no procedure.

Not everyone supported the vaccine. Some parents refused to subject their children to an "experimental" treatment. On April 4, 1954, radio celebrity Walter Winchell publicly derided the vaccine in a monologue that included, "Attention everyone! In a few moments I will report on a new polio vaccine—it may be a killer!" Winchell emphasized the danger that the vaccine could actually contain inadequately killed virus, which could transmit polio to children.

Despite such critics, the trial proved extremely successful. Approximately 95 percent of children in the trial received all three shots—the dropout rate was very small. Basil O'Connor and the National Foundation had such confidence in the vaccine that they committed $9 million to pharmaceutical companies that would start producing the vaccine even before the results were known, in order to ensure a supply would be ready for distribution across the country right away.

The study results were scheduled to be announced on April 12, 1955, at the University of Michigan. The nation held its breath. By prior agreement, the 150 reporters present were to receive a press release and study report at 9:10 a.m., but they consented to postpone reporting the news until the scientist who had overseen the trial, Dr. Thomas Francis Jr., delivered his public address. When the report packets arrived on-site, pandemonium ensued. Frenzied reporters rushed the stacks of papers, forcing a university public relations officer to stand on a table and toss reports to the frantic sea of reporters below.

The first three words of the press release were, "The vaccine works."

Reporter John Troan recalled, "They brought the report in on dollies, and newsmen were jumping over each other and screaming, 'It works! It works! It works!' The whole place was bedlam."

The vaccine was safe and essentially 80 to 90 percent effective. In one arm of the trial, 200,745 children had been vaccinated and 201,229 children were injected with a placebo. Fifty-seven (28 per 100,000) children in the vaccinated group contracted polio, versus 142 (71 per 100,000) in the placebo group.

In the other arm of the trial that used observed controls, 221,988 children were vaccinated and 725,173 were observed. Fifty-six (25 per 100,000) vaccinated children contracted polio, versus 391 (54 per 100,000) in the observed group.

The total effort to develop the vaccine, from research grants to the completion of the field trial, cost the National Foundation $25,541,622.*

The news was immediately broadcast across the nation and around the world. Euphoria and spontaneous celebrations broke out in cities and towns. Church bells rang. Drivers honked their horns. In homes, offices, stores, and schools across America, people cheered and many wept. Newspaper headlines rejoiced: in the *Pittsburgh Press*, "POLIO IS CONQUERED," the *New York Post*, "POLIO ROUTED!" and in Hong Kong's *South China Morning Post*, "TRIUMPH OVER POLIO." A generation of Americans would remember what they were doing the moment they heard the news.

Jonas Salk became the most famous doctor in the world. After the announcement, he and his family were chauffeured in a limousine with police escort. In Pittsburgh, a delegation of city officials met him at the airport. His wife, Donna, overheard their five-year-old son tell a friend over the phone, "Hi Billy, I'm back from my vacation and I'm famous and so is my dad."

The press loved Salk. Soon, hundreds of awards, offers of honorary degrees, interview requests, and invitations to speak inundated him from all over the world. He received carloads of mail and thousands of telegrams. Schools and streets were named after him.

But there was a professional price to be paid for all the public adulation. Salk's scientific peers held a dim view of the hoopla, and of Salk's conduct. To them, he had been overly ambitious, always seeking the spotlight and too willing to accept sole credit for an achievement he did not accomplish alone. On top of that, he had not developed or invented any new scientific concept or technique. At best, they considered Salk's vaccine to be only a temporary solution until a longer-lasting attenuated-virus vaccine could be developed. Salk had taken the easy route, the road to fame and potential riches. Years later, Salk commented on being ostracized by his peers: "The worst tragedy that could have befallen me was my success. I knew right away that I was through—cast out."

---

* This amounted to approximately fifteen cents per American in 1955. In 2022 dollars, this total expense would translate to about $278 million.

In truth, it pained Salk to receive undeserved adulation from the masses that had chosen to elevate him to almost God-like status. He tried to get the press to stop calling it the "Salk vaccine." But there was little he, or anyone, could do to stem the overwhelming public adoration. The world needed a hero to thank, and it was a role partially thrust upon him. Plus, one of Salk's most unforgettable moments couldn't have been scripted better and did much to seal his reputation as a hero. In a live television broadcast with Edward R. Murrow's popular *See It Now* program on the day the national study results were released, Murrow asked Salk, "Who holds the patent on this vaccine?"

Salk responded, "Well, the people, I would say. There is no patent. Could you patent the sun?"

Jonas Salk

Salk would never earn a dime from the vaccine he pioneered. Later that evening, Murrow pulled Salk aside to warn him, "Young man, a great tragedy has just befallen you." When Salk asked him what he meant, Murrow said, "You've just lost your anonymity."

DISASTER

The National Foundation worked with a select group of pharmaceutical manufacturers to rush production and distribution of the vaccine. But before the public euphoria had even had a chance to abate, disaster struck. On April 24, 1955, a doctor in Idaho encountered a new case of polio in a first-grade girl who had just been vaccinated six days before. She died three days later. On April 25, an infant who had been vaccinated nine days before was admitted to a Chicago hospital with paralytic polio. On April 26, five more cases in vaccinated children were reported in California. Panic ensued at the National Foundation. What was happening?

All of the affected children had received vaccines manufactured by one firm: Cutter Laboratories of Berkeley, California. Administration of Cutter's vaccine was immediately halted, and their vaccine recalled, but 380,000 Cutter doses had already been given. An extensive investigation showed that Cutter was not properly adhering to the strict safety protocols established by Salk, and that live, inadequately killed virus had gotten into their vaccines. The botched Cutter vaccine would directly cause seventy-nine polio cases and precipitate additional cases in 105 family members and twenty other close contacts. Most of these patients became paralyzed, and eleven died.

To Albert Sabin and his like-minded peers, this was exactly what they had been warning could happen. How ironic that the very criticism leveled at attenuated-virus vaccines—that inadequately weakened viruses could cause disease—might prove the downfall of Salk's killed-virus vaccine. Sabin redoubled his attacks and testified before Congress that production of Salk's vaccine should stop. Salk became distraught over the Cutter incident. After being roundly criticized by peers at a meeting at the National Institutes of Health (NIH), he later admitted, "This was the first and only time in my life that I felt suicidal."

Cutter's production was shut down, but distribution of the vaccine from other manufacturers continued. Nationwide, cases of polio plummeted—from 58,000 cases in 1952, to 15,000 cases in 1956, to 7,000 cases in 1957. Salk's reputation survived the Cutter fiasco and soared to new heights, as polio no longer held America's children hostage each summer. Parents no longer felt fear. Swimming pools stayed open. Mothers let their kids out to play.

SABIN'S CHANCE

While Salk stood in the limelight, Sabin went to work. Never doubting his method was superior, he continued to pursue a live, attenuated-virus vaccine that promised several advantages over Salk's vaccine. It would be cheaper and easier to manufacture. It could be given orally, and in only one dose. The vaccine would take effect in the gut, where the poliovirus naturally establishes itself, to produce long-lasting, potentially lifelong, immunity. Targeting the digestive tract with an oral vaccine would also eliminate passive transfer of the poliovirus. A person immunized with Salk's vaccine might not get sick with polio, but the virus could still flourish in his or her gut and be shed through feces, potentially infecting others. Sabin's vaccine would turn this disadvantage into an advantage: if individuals were to ingest fecal matter from a patient protected by Sabin's vaccine, they would be ingesting attenuated virus within the feces that would essentially immunize them as well.

But perfecting a live-virus vaccine was difficult. One had to keep the virus strong enough to induce an immune response, but weak enough not to cause polio. Sabin weakened the poliovirus through multiple passages through monkey tissue. The endpoint occurred when a successfully attenuated virus did not cause paralysis in a chimpanzee after being injected into its spine. During the winter of 1954–1955, Sabin tested his vaccine in thirty inmates at a prison in Chillicothe, Ohio. Each prisoner received $25 and a little time off their sentence. The trial went well. All the prisoners survived and developed antibodies to poliovirus.

Now Sabin faced a dilemma. A full field test of his vaccine would require hundreds of thousands of children, none of whom could have previously received the Salk vaccine and developed protective antibodies. Unfortunately for him, the vast majority of children in America had already been vaccinated. Sabin's solution was to find a large, unvaccinated population outside the United States. He went to the Soviet Union.

Even at the height of the Cold War, Americans and Soviets were willing to share technology related to humanitarian purposes. Polio epidemics in the Soviet Union were a major problem, and the Soviets had had difficulty with administering the Salk vaccine, partly due to a shortage of glass syringes. Consequently, Sabin was invited to the U.S.S.R. in 1956. The Soviet government later approved the use of his vaccine and, in 1959, conducted a huge clinical trial, immunizing ten million children. Sabin's vaccine could be administered as a liquid, delivered orally with a dropper, or disguised as a piece of hard candy. The trial results proved Sabin's vaccine to be highly effective. Soon the Soviet government dictated that every citizen younger than twenty be vaccinated—77 million people.

Sabin would finally receive his long-awaited time in the sun. He continued to disparage Salk's vaccine, claiming that Salk's vaccine was only 60 to 70 percent effective (though the Public Health Service had deemed it 90 percent effective) and warning that "hundreds of children" would die if the Sabin vaccine was not made available immediately.

Meanwhile, Salk began to fight back. His vaccine had already proved longer-lasting than anyone expected and was responsible for reducing the incidence of polio in the U.S. by 92 percent in five years. At the end of 1961, government data would show this figure improving to a 97 percent reduction. But to Salk's dismay, the American Medical Association (AMA) advocated switching the national vaccination program from Salk's vaccine to Sabin's. Salk complained that they were switching to a vaccine that had not yet even been licensed for use in the United States. In a letter to the AMA, he wrote, "Not only is scientific justification lacking for the proposed new vaccination program but the *evidence* is notably lacking to support the *need* for 'change-over' from one form of vaccination procedure to another and for revaccination of the entire population." He believed his vaccine had the potential to eliminate polio completely if the government would just stick to the plan and continue its use. More than anything, Salk wanted his vaccine to be the one that eliminated polio, on its own.

In 1961, Sabin's vaccine was officially licensed for use in the U.S. and supplanted Salk's vaccine nationwide. Salk was crestfallen. In an interview decades later, his son Peter said, "Normally, my father tried to let these things go, but this one was so terribly painful, so personally insulting to him as a scientist, that he couldn't let it go. It is no exaggeration to say that it haunted him for the rest of his life."

Through the use of Sabin's vaccine, polio was almost completely eliminated from the developed world. From 1952 to 1981, the rate of paralytic poliomyelitis per 100,000 people dropped from 13.7 to 0.003. It seemed that Sabin had outlasted his rival to become the competition's final victor. Though the public never adored him the way it did Salk, Sabin reveled in the admiration of the scientific community, which showered him with awards. He received the National Medal of Science and was elected to the prestigious National Academy of Sciences, a body whose membership never voted to accept Salk. Sabin became the president of the Weizmann Institute of Science in Israel in 1969. He died in 1993, of heart failure.

Albert Sabin

Jonas Salk left the University of Pittsburgh in 1963 to found the Salk Institute for Biological Sciences near San Diego, a center for medical and scientific research whose work continues today. In the 1980s he worked on a vaccine for AIDS. Salk died in 1995, also of heart failure.

Yet even after their deaths, the battle between the Salk and Sabin vaccines continued. The Sabin vaccine almost fulfilled the dream of eradicating polio worldwide, but not quite. Unfortunately, one of the risks of a live-virus vaccine is the chance that it might cause polio in extremely rare cases. The risk was about two cases per million vaccine doses. By the mid-1990s, the only cases of polio worldwide were due to the Sabin vaccine; approximately a dozen occurred annually in the U.S. It became clear that usage of the Sabin vaccine alone would never fulfill the goal of fully eradicating polio. In 1996, the CDC instituted a new vaccine protocol for infants that combined two injections of Salk's vaccine followed by a dose of Sabin's oral vaccine—but even this did not eliminate polio completely. In 2000, the CDC switched fully to the use of the Salk vaccine alone. Since the Cutter fiasco, no Salk vaccine has ever caused a case of polio. So perhaps the final victory does belong to Salk.

In truth, there was plenty of victory to go around for both men. Because of them, American summers ceased to be shrouded by pervasive fear. Just as the competition between Louis Pasteur and Robert Koch spurred each to greater heights of achievement, the rivalry of Jonas Salk and Albert Sabin ultimately served to benefit mankind. Together, these two men, and the research teams they led, prevented hundreds of thousands of deaths and saved millions of children from paralysis.

PANDEMIC

In late December 2019, a thirty-four-year-old ophthalmologist named Li Wenliang was working at Wuhan Central Hospital in China. He became concerned about seven hospitalized patients who had been diagnosed with atypical pneumonia. These patients, all severely ill, had been quarantined.

When one patient's lab results came back positive for a SARS-type coronavirus, he and other doctors became alarmed. On December 30, at 5:43 p.m., he messaged a private WeChat group of physician colleagues: "7 confirmed cases of SARS were reported from Huanan Seafood Market. They are being isolated in the emergency department."

An hour later, a group member fearing government censorship responded, "Be careful, or else our chat group might be dismissed."

Li responded, "The latest news is, it has been confirmed that they are coronavirus infections, but the exact virus strain is being subtyped. Don't circulate the information outside of this group, tell your family and loved ones to take precautions."

Despite Li's personal wish to keep his messages private, screenshots of his posts went viral in China within hours. The next day, December 31, 2019, Chinese government officials in Wuhan confirmed that dozens of patients were being treated for a novel coronavirus.

On January 8, 2020, Li evaluated a woman with angle-closure glaucoma. Shortly thereafter, the patient, who was later found to be a storekeeper at the Huanan Seafood Market, developed a fever and tested positive for coronavirus. Li also became severely ill and was admitted to the intensive care unit in his own hospital. His initial WeChat posts had prompted criticism from the Chinese government, which accused Li of "rumor-mongering" and forced him to sign a statement admitting to making false statements that disrupted public order. At the end of January, Li was interviewed by the *New York Times*, which asked him what he planned to do after he recovered. Li responded, "I will join medical workers in fighting the epidemic. That's where my responsibilities lie."

Li Wenliang died on February 7, 2020. He left behind a son and a wife, who gave birth to their second child, a boy, in June 2020. As one of the first to sound the alarm regarding what would become a catastrophic worldwide pandemic, Li was acclaimed as a whistleblower. His death sparked an outpouring of public grief and anger at the government. At Wuhan Central Hospital, co-workers and citizens blew whistles in honor of his memory. A government investigation cleared him of wrongdoing, rescinded his statement of culpability, and offered an apology to his family. Li was officially honored as a "martyr," the highest civilian honor in China.

Meanwhile, the virus suspected of originating from an outdoor seafood and wild animal market had already spread beyond China with incredible speed. On January 13, 2020, a case was reported in Thailand. On January 20, the first detected American case occurred in Washington State, in a patient who had returned from Wuhan, China, five days earlier. On January 23, China shut down Wuhan, ceasing all public transit and prohibiting travel in and out of the city of eleven million. On January 30, the World Health Organization declared a global health emergency, and the next day the Trump administration restricted entry of travelers from China to the U.S. At that point, there were estimated to be approximately ten thousand cases in twelve countries.

On February 11, the disease caused by the novel coronavirus was officially named "Covid-19," a simple contraction of the words "coronavirus disease 2019." By mid-March, the virus had spread across the globe, rocking an unprepared world and straining every nation's resources. In the United States, concerts, parades, sporting events, and conferences were canceled. Mask-wearing became ubiquitous. Schools and universities shut their doors and transitioned to remote learning. By April, almost 10 million Americans had lost their jobs and 6.6 million filed for unemployment benefits (the previous record was 695,000, in 1982).

The U.S. government and every American had been caught flat-footed. Entire generations, with no memory of the 1918–1919 Spanish Flu pandemic that killed approximately fifty million worldwide, and scant memory of polio, had never contemplated the degree to which an infectious disease might one day disrupt their daily lives. Plagues of the recent past, including SARS, avian flu, and Ebola, were distant problems—diseases affecting foreigners across the seas, not Americans. The U.S. response to Covid-19 was marred by lack of testing, poor contact tracing, inadequate personal protective equipment, and shortages of medical supplies, particularly ventilators. American cases and deaths skyrocketed, leading the world's numbers. A nation that had forgotten the power of infectious disease was brought to its knees. Though several medications were tried in desperate attempts to combat the virus, none proved to be the panacea many prayed for.

The world's only hope would be a vaccine.

But the timeframe to develop a new vaccine, including preclinical studies, human trials, and regulatory approval, was typically between five and fifteen years. How many millions would die before that occurred?

Accelerating the creation of a vaccine to rapidly combat this extraordinary threat would require an equally extraordinary new approach, one that had never been tried before. Thanks to a handful of unsung heroes, that novel strategy was ready for the spotlight.

It harnessed the power of messenger RNA.

mRNA

On June 28, 1802, Francisco Javier de Balmis, a Spanish physician, received an order from his king. Viceroys from the New World were pleading with King Charles IV to help them combat smallpox epidemics, so Charles sent Balmis on a special mission—an expedition unlike any ever undertaken before.

Balmis was charged with bringing Edward Jenner's vaccination to the population of an entire continent. But how could he transport live cowpox virus across the Atlantic? He had no refrigeration or method of sterile storage. He could not fit herds of cattle on the expedition's small corvette, the *Maria Pita*. The king's demand seemed impossible.

Then Balmis thought of a way—using orphans.

He devised a plan to bring twenty-two orphans on the voyage. The boys, all between the ages of eight and ten, would serve as serial human incubators of the cowpox virus. Two children would be infected at the start; then Balmis would gradually transfer the virus to the other children, one after another, to keep the infection going like an Olympic torch relay until they could propagate it anew in South America.

Balmis's two-and-a-half-year mission brought the smallpox vaccination to present-day Cuba, Mexico, Venezuela, Colombia, Peru, and Bolivia. He refreshed his group of human incubators by obtaining twenty-five new orphans in Mexico and set out across the Pacific to the Philippines. He then sailed to Macau and proceeded west to complete a circumnavigation of the globe. The virus bequeathed by the original twenty-two orphans ultimately led to the successful vaccination of 1.5 million people.

This incredible humanitarian medical mission is unique in the annals of history, but today, it serves to demonstrate a crucial fact and limitation. From the very beginning—from Jenner to Pasteur to Sabin and beyond, every vaccine ever produced has been discovered, grown, transported, and manipulated in living cells. The development of Salk's polio vaccine cost the lives of tens of thousands of monkeys that were bred on farms for the sole purpose of one day serving as virologists' incubators. Each year, America's supply of flu vaccine requires the use of approximately 140 million eggs that serve as living incubators in which the virus is grown. Since the inception of virology as a discipline, living biological organisms and cells have been the only way to produce vaccines.

Until now.

In the early 2000s, anyone looking at the career of Katalin Karikó, a journeywoman biochemist at the University of Pennsylvania, would be largely unimpressed. Though her personal story was compelling—she had immigrated from Hungary in 1985 with the family's total savings of $1,246 sewn into their daughter's teddy bear—as a scientist she had failed far more often than she had ever succeeded. For decades, she labored in her chosen field of study, the science of messenger RNA (mRNA), but never managed to win enough grant funding to climb up from the bottom rung of the academic ladder. Without her own grant funding, she had little job security, and she never earned more than $60,000 per year.

Still, she stubbornly pursued the idea that mRNA could one day be used to direct cells to produce their own medicines. The fundamental function of every living cell is to produce proteins. The code that specifies which proteins are made is written in the chromosomal DNA inside each cell's nucleus. This DNA template is used to create a complementary strip of mRNA that exits the nucleus and travels to a ribosome, which reads the mRNA instructions to produce amino acid chains that fold to form proteins. The premise of Karikó's research was the idea that instilling synthetically produced mRNA into a cell could hijack that cell's production capacity and get its ribosomes to make any protein she desired.

In 1989, Karikó made a mini-breakthrough. Working with Dr. Elliot Barnathan, a cardiologist at the University of Pennsylvania, Karikó aimed to instruct cells to make a protein called a urokinase receptor. To detect whether they had succeeded, the researchers deployed a radioactively-labeled urokinase enzyme that would bind to the new protein.

It worked.

The cells were making the desired product; it was happening just as Karikó and Barnathan dreamed it would. The ability to direct nature's mechanisms to produce a protein at the whim of man was heady stuff. Karikó later told a *New York Times* reporter that, upon making this discovery, "I felt like a god."

But the years following this accomplishment were marked by frustration. Karikó endured repeated grant rejections and, in 1995, was essentially demoted from her official rank as an assistant research professor. Now, without formal academic standing at her university, Karikó had to face facts. She could do the expected thing, which was to give up—abandon academia to take a higher-paying job in the pharmaceutical industry or at a biotech company—but Karikó couldn't bring herself to quit. She loved research, and she was fascinated by mRNA. So she stayed at the university despite the fact that there was no guarantee her status would ever rise. She kept coming to work at 6 a.m., and on most weekends, sticking to her research projects even though she knew that few others believed they would bear fruit.

Then, in 1997, Dr. Barnathan left the university, leaving Karikó entirely without a position. She scrambled and landed in a neurosurgeon's lab; but, two years later, that doctor also left. At this point, a chance encounter around a photocopier changed Karikó's life and led to an important collaboration. She met Drew Weissman, an immunologist. "I am an RNA scientist—I can make anything with RNA," Karikó told him.

Weissman decided to take a chance on Karikó. In 2004, they made a breakthrough. For years, a major roadblock to using mRNA in live animals had been the fact that the immune system labeled synthetic mRNA as foreign and destroyed the mRNA before it could reach its target cells. Karikó and Weissman found that, if they added a molecule called pseudouridine to synthetic mRNA, they could trick the immune system into accepting it as a natural molecule. They used synthetic mRNA to induce monkeys' cells to upregulate production of erythropoietin (a protein that induces the body to make red blood cells); and, when this same approach made red blood cell counts in mice increase, they knew they had succeeded. They realized their method might be used to help the body generate any number of beneficial proteins: enzymes, hormones, naturally occurring pharmaceutical agents, and vaccines.

Two fledgling biotech companies noticed Karikó and Weissman's scientific reports: a company based in Cambridge, Massachusetts, called Moderna, and a German company called BioNTech. In the mid-2010s, these companies were working to developing vaccines against viruses like cytomegalovirus and Zika virus.

Then Covid-19 appeared. On January 10, 2020, Chinese scientists shared the novel coronavirus's (SARS-CoV-2) genetic sequence with the world. Within hours, BioNTech had developed an mRNA vaccine using the code. Moderna did the same within two days. To succeed, it was necessary for both companies to sort out what part of the code produced the coronavirus spike protein, a unique molecule studding the outside of the virus. If a vaccine could temporarily induce some cells to produce just the spike protein, this would be enough to gain immune recognition and allow the body to rapidly quell any future infection by actual Covid-19 coronavirus. Simply knowing the pertinent DNA code allowed researchers to assemble the same sequence using free nucleotides. This DNA was transcribed into mRNA, which was then encased in a lipid bubble that would protect it until it could enter a cell.

Just forty-two days after the genetic code was revealed, Moderna sent samples of their experimental vaccine to the NIH for testing. Less than nine months later, on November 8, 2020, the results of a BioNTech/Pfizer study (BioNTech had partnered with Pfizer) showed that their vaccine was very successful—90 percent effective against Covid-19. Moderna's vaccine trial soon showed similar success.

For the first time in history, creation of a vaccine did not depend on growth, attenuation, or transport in a cell culture, animal model, or unlucky orphan. Today, it is merely necessary to know the sequence of DNA nucleotides necessary to produce a desired product—a protein of choice. Because of this, drug making is now being likened to computer programming. Scientists assemble the instructions, which can be tweaked and adjusted at will, and send it to the body's cells, which then do all of the work. Unlike every attenuated vaccine ever made, including the Sabin vaccine, it is impossible for an mRNA-based Covid-19 vaccine to ever cause Covid-19 in a patient, because the protein produced is not the virus; it is merely a small piece of the viral envelope, a non-pathogenic coronavirus spike protein.

The potential utility of the mRNA technology pioneered by Karikó and Weissman is now recognized to be enormous and is difficult to overstate. Vaccines against many of the most dangerous viruses, including HIV, Ebola, H1N1, and Zika, could be only years away. It may become possible to direct cells to produce hormones, perhaps even insulin, on demand. For patients with heart disease, it might be feasible to induce cells to produce continuous levels of beneficial vasodilating agents like nitric oxide. The sequence of a patient's cancerous cells could be typed in order to create a personalized vaccine that primes the immune system to attack the cancer cells while leaving the body's normal cells alone.

The rapid development of Covid-19 vaccines in merely ten months was a true miracle of modern science. But before we become too fond of lauding our own cleverness, we would be wise to recognize how badly infectious diseases have humbled us. Though the incredible advances chronicled in these pages have demoted infectious diseases from the top rank of global killers, they remain responsible for two-thirds of child deaths worldwide, and one-quarter to one-third of all deaths in the developing world. Recent history has shown that roughly every decade, a novel infection becomes the cause of a regional epidemic, and that global pandemics occur about once a century.

The lessons of Covid-19 are sobering. The worldwide response was unquestionably late and extremely slow. In many respects, our pandemic response was not dissimilar to that of those living in the Middle Ages, when plagues prompted people to hide in their homes, quarantine the sick, cease all trade, and shun foreigners. The more ignorance and fear cloud our judgment, the more blunt and imprecise our tactics tend to be. A 1997 Hong Kong outbreak of an H5N1 avian influenza virus caused six deaths and prompted the killing of over 1.5 million chickens. A 2003 outbreak of an H7N7 virus in the Netherlands killed one person and resulted in the slaughter of almost thirty million chickens.

We take these dramatic measures because we lack the knowledge to do anything smarter. And so, as a grateful world emerges from the chaos of the Covid-19 pandemic, we should temper any temptation to congratulate ourselves and remember that the greatest benefits will come from the lessons of our failure.

Those lessons are myriad, and some of the most important have nothing to do with technology or even taking care of the sick. We could make great strides by simply improving basic sanitation and vaccination rates around the world. Straightforward measures such as ensuring adequate provisions for testing, contact tracing, and personal protective equipment will do much to rouse public confidence in governments and leaders during the next pandemic. Amassing specialized medical supplies and equipment, like ventilators, along with a readiness to rapidly transport these tools to the places they are needed, are not difficult tasks when planned out in advance. Future investments in healthcare infrastructure, testing technology, improved surveillance, and medical research are crucial.

We cannot fail to learn from what we have endured. And we should be supremely proud of the progress made against infectious diseases in the last 150 years. But we should also recognize that, no matter how many battles we win, we are fighting a war against infection that will never fully end.

5

CANCER

A Bewilderingly Complex Array

Eleven-year-old Einar Gustafson loved the Boston Braves. He followed them religiously, though he had never been lucky enough to attend a game. His family's potato farm near New Sweden, Maine, lay less than thirty miles from the Canadian border. At a distance of over four hundred miles, Boston was an entire day's drive away. Still, radio broadcasts made the Braves' games come alive, and Einar knew all about each player on the roster. In the late summer of 1947, he prayed the Braves might still catch the Brooklyn Dodgers and contend for the National League pennant.

Raised in the wilds of northern Maine, Einar was accustomed to the outdoors. Each weekday, he walked to a one-room schoolhouse to attend classes. He was slim and had an open, earnest, choirboy face. Though thoughtful and introverted, he was not reticent. Sometimes, he even came across as preternaturally confident and self-assured.

One day, Einar began to feel strange stomach cramps while walking to school. This discomfort gradually worsened until it had morphed into severe belly pain. His parents took him to the nearby town of Caribou, where the doctors were not sure what was wrong with him. A surgery was performed, which proved inconclusive. The family was then sent to a larger hospital in

Lewiston, a city 220 miles to the south, where a surgeon thought Einar might have an evolving case of appendicitis. Einar's worried parents consented to a second surgery. To the surgeon's surprise, Einar's appendix was not inflamed or infected. Instead, he found and recognized an entirely different problem—a tumor.

The tumor was removed and a diagnosis was finally made. When Einar's father heard the news, he wept, unnerving Einar because his stoic father rarely displayed emotion. Einar had cancer, a rare form of lymphoma. Removing the tumor would not cure Einar, but there was nothing more the doctors could do for him. Ninety percent of children with this diagnosis would die; most likely, Einar had only about six weeks to live.

Before eliminating all hope, the physicians in Lewiston told the Gustafsons that there might be one last place they could take Einar. There were doctors at the Children's Hospital in Boston that took in children with blood cancers like leukemia. Unfortunately, all these children died—leukemia was invariably fatal—but they might offer some experimental treatments. If anything further could be done for Einar's lymphoma, they would be the ones to know about it.

It was a Hail Mary, but even the slimmest chance of a treatment was better than no chance at all. The Gustafsons took their son to Boston. What happened to them there would dramatically change their lives, and the lives of thousands of children just like Einar.

OUR "WAR" ON CANCER

Each year, approximately 1.9 million Americans receive the bad news that they have cancer. Twenty-one percent of deaths in the United States—almost 600,000 annually—are due to cancer. By some estimates, it will eventually kill one out of every three of us. Though physicians have avidly combated cancer for well over a century, it was not considered a prime killer until the scourge of many infectious diseases had been surmounted, and average life expectancy had increased into the range at which many cancers emerge. By 1938, cancer had risen to become the second most prolific killer in the United States, following only heart disease. Cancer has retained the number two spot ever since; but, whereas deaths from heart disease steadily declined in the twentieth century, deaths from cancer rose, narrowing the gap considerably. With all the incredible technological advances we have witnessed in the last fifty years, from the Internet to the cloud, from the Human Genome Project to dramatic headway made in almost every other medical discipline, why has commensurate progress in oncology proved so elusive?

The answer is quite literally complex, as will become abundantly clear. But at the same time, oncology's ever-evolving story is peppered with incredible breakthroughs, some of which, by dint of luck, tenacity, or toil, beggar belief. The pioneers who sought to confront this deadly and devious foe exhibited unsurpassed courage and fortitude, because their patients almost always died. It is no small thing to fight battles one is sure to lose. It is not easy to subject patients to unproven therapies that are sure to make them sicker, or may even hasten their deaths, for benefits so modest that they are often measured in mere weeks or months of extended survival. Yet this is exactly what so many revolutionary oncologists did.

Cancer, in truth, is a term that describes more than a hundred different diseases. All generally stem from the same fundamental problem—unchecked and unwanted cell division—but cancer is remarkably heterogeneous. The prevalence, character, and prognosis of various malignancies are myriad. Practically every organ of the body can be afflicted. Some tumors remain dormant and self-contained, while others rapidly metastasize to far reaches of the body. Great variability can occur even within the same type of cancer, such as with leukemia, which can present in an acute, aggressive form that kills within weeks, or as a chronic, indolent disease that might remain stable for years.

This variability makes cancer a disease like no other. Is it any wonder that we spent much of the last century lunging forward, in fits and starts, and often blindly, against this capricious and shape-shifting foe? For decades, doctors dreamed of finding a magic bullet that might kill all cancers. This hapless belief belied their ignorance and elucidates why the earliest cancer breakthroughs seemed to stem not from rational, scientifically grounded inquiry, but from discoveries made via empirical, brute-force trial and error, or sometimes simple serendipity.

In the wake of the successful Apollo moon landings, President Richard Nixon famously declared an ambitious "war" on cancer in 1971. By many measures this effort was unsuccessful and even misguided. Two decades later, it was clear that we had not won the war on cancer, and most honest voices would admit that we were not even winning.

Yet we were not entirely losing either. The dedication of countless doctors, and the courage of innumerable patients, yielded advances that saved hundreds of thousands of lives. In fact, when one considers practitioners' woeful inadequacy at the beginning of this journey, the progress of the last fifty years begins to appear as something little short of miraculous.

TO CUT OR IRRADIATE

For millennia, the only possible treatment for cancerous tumors was surgical. Though anesthesia and antisepsis were lacking, sometimes tumors grew so large and disfiguring that patients willingly assumed the risk of excision at the hands of barber surgeons. Such operations were understandably rare—the excruciating pain of surgery alone rendered this a last resort. But the advent of anesthesia in the mid-nineteenth century made surgeons bolder. Among the boldest was undoubtedly the famed William Halsted, a brilliant surgical innovator at the Johns Hopkins Hospital in Baltimore. Halsted's indefatigable dedication to his work, aided by an inconvenient addiction to cocaine,*

---

* Halsted's cocaine addiction originated from self-experimentation during his investigation of the use of cocaine as a local anesthetic. At the time, the potential addictive properties of the drug were not fully understood.

helped him to develop many new and successful operations to remedy a wide range of conditions, including hernias, aneurysms, and diseases of the thyroid and gallbladder. He also established the first formal surgical residency training program in the United States.

Halsted's simple approach to cancer was to extirpate it. But there was a problem. Sometimes, even after a tumor had been entirely removed, cancer would spring up again months or years later around the margins of the excavation. Surgeons would then take patients back to the operating theater to remove the cancer again. If it recurred yet again, in the same area or farther afield, they would repeat the operation. Halsted's solution to this "whack-a-mole" predicament was to excise primary tumors with greater and greater margins of tissue around the cancer. He reasoned that cancer likely spread directly outward from the tumor, and therefore the best way to eradicate unseen cancer cells hiding in normal-appearing tissue was to excise as much of that tissue as reasonably possible.

In the case of breast cancer, this mentality led Halsted, in 1894, to pioneer what would become known as the "radical mastectomy," an extensive operation in which not only the breast, but also the underlying pectoralis (chest wall) muscles, and the lymph nodes of the neck and armpit, were removed. Though terribly disfiguring for patients, surgeons believed they were helping women by being as aggressive as possible—better to live disfigured than to die from cancer. One nineteenth-century English surgeon described a less aggressive approach to be "a mistaken kindness to the patient." Surgeons did succeed in saving many whose cancers remained localized and had not yet metastasized, unseen, to other parts of the body; but, in a significant percentage of cases, cancers recurred and death soon followed.

The origin of a second method of cancer treatment was birthed in 1895, when a German physicist named Wilhelm Röntgen made an accidental discovery. Röntgen was fascinated by the phenomenon of "cathode rays" generated by electric current inside a glass Crookes vacuum tube. He could detect these rays by the glow of a screen coated with barium platinocyanide, positioned very close to the tube (only a few centimeters away). On the evening of November 8, 1895, Röntgen decided to wrap his Crookes tube entirely with a black cardboard covering so that no light could escape. Without light interference, he thought he would be able to visualize the fluorescent glow much better. But when he began to darken the laboratory, he noticed the appearance of a dim, shimmering, greenish light in the room. It was coming from the barium screen, which stood at least six feet away because he had not yet moved it close to the Crookes tube.

This gave Röntgen a start. He knew that cathode rays could not travel more than a few centimeters, and certainly not a distance of six feet. Whatever was causing this glow across the room could not be cathode rays. He deduced there must be a different kind of invisible ray, something he termed an "X" ray because its properties were completely unknown. Röntgen found that X-rays were not blocked by two packs of cards, nor a thousand-page book, nor thick blocks of wood. While holding an object up to the screen, he made an amazing discovery—he could see the bones of his hands! He realized these rays could pass through human flesh but were blocked by bone.

News of Röntgen's discovery soon inspired the use of X-rays to treat cancer. The first to do so was a Chicago medical student named Émil Grubbé, who won a place in history as the first to use radiation to treat a local tumor in 1896. Grubbé fashioned his own cathode ray tubes and, upon working with them, noticed the skin of his left hand became swollen and painfully blistered after being in close proximity to the tubes. Realizing that X-rays damaged normal tissue, Grubbé tried to use X-rays to treat the breast tumor of an elderly woman who had suffered cancer relapse after a mastectomy. Grubbé treated the tumor for eighteen days in a row. It was a painful experience for the patient, but the tumor gradually began to shrink. Though the patient later died after metastases spread to her brain, this initial promising result emboldened Grubbé to treat other patients with local tumors. A novel method of treating cancer—radiation—had been born.

Meanwhile, in France, the discovery that similar invisible rays could also spontaneously arise from naturally occurring minerals like uranium inspired a husband-wife pair of scientists named Pierre and Marie Curie to seek out additional "radioactive" substances. In 1898, the Curies discovered that a composite mineral called pitchblende contained the elements polonium, named after Marie's home country of Poland, and radium, whose name was derived from the Latin word for "ray." To prove radium's existence, Marie endured four years of arduous work grinding, heating, dissolving, and filtering tons of pitchblende in a dilapidated wooden shed. She finally managed to extract one tenth of a gram of pure radium in 1902.

Marie Curie's triumph would unfortunately be marred by her ignorance of radiation's harmful effects. Working with pitchblende, the skin of Marie's hands became burned and blackened. Radiation exposure would eventually harm her bone marrow and cause her to develop aplastic anemia, the condition that killed her in 1934. Still, the Curies recognized radium's promise as a new way to treat cancer, one less risky and disfiguring than surgery. Doctors soon began using radium to treat superficial skin cancers. Radium pellets could be surgically inserted to combat deeper tumors.

And yet, despite these early steps forward, the diagnosis of cancer carried the same dreadful prognosis of certain death in most cases. Surgery and radiation only provided a glimmer of hope to individuals with local tumors. There was no defense against the invisible cancer cells that so often snuck away, undetected, to seed cancer in other, far-flung parts of the body.

Almost half a century would pass before a new kind of breakthrough would signal a different approach to treating cancer.

THE "SECOND PEARL HARBOR"

On December 2, 1943, the harbor of Bari, a city on Italy's southeastern coast, was full of Allied ships. Oil tankers and cargo ships filled with ammunition and supplies were so tightly packed that some ship hulls touched. Bari was in a British zone of occupation. Though the Italian government had quickly surrendered after the first Allied landings, their German counterparts seemed determined to hold Italy at all costs. To speed the unloading of so many ships, the usual blackout orders for Bari were suspended. The British, believing the Luftwaffe to be severely depleted, considered the odds of an enemy air attack to be remote. Anti-aircraft defenses were almost nonexistent—only one battery defended the city. That afternoon, British Air Marshal Sir Arthur Coningham revealed his dim view of the Germans by pompously quipping in a press conference: "I would regard it as a personal affront and insult if the Luftwaffe should attempt any significant action in this area."

That very evening at 7:25 p.m., an armada of 105 Junkers Ju 88 bombers descended on Bari in an overwhelming attack that lasted only twenty minutes. Rarely has a fleet of ships ever been caught so unprepared and undefended. German bombs set off multiple series of explosions amongst the closely anchored ships, laden with fuel and ammunition. In all, seventeen ships were sunk, and eight badly damaged. An oil pipeline located on a pier was hit, causing a torrent of fuel to stream into the harbor, where it ignited and fed a growing conflagration. The sky grew opaque with smoke as flames and oil coated the surface of the water below. Men leapt from burning ships into the frigid Adriatic waters; some were rescued, freezing cold and covered with oil. Luftwaffe bombs also fell in the city, leveling buildings and killing civilians. The total number of dead, counting both sailors and civilians, exceeded one thousand.

In the army, the attack became known as the "Second Pearl Harbor." The port of Bari would be out of action for three months. The Allies had suffered an unmitigated disaster. It would be hard to imagine matters growing even worse.

But then they did.

Even in the midst of the attack, men struggling in the water and civilians running for cover noticed something peculiar. It was a smell—the distinctive smell of garlic, or perhaps mustard. The aroma lay heavily across the harbor and no one knew what to make of it.

The reason was top secret. One American Liberty ship, the *John Harvey*, had held a secret cargo of 2,000 mustard gas bombs, each weighing sixty to seventy pounds. An explosion on the *John Harvey* sent the poison, aerosolized, into the air, and, in liquid form, spilling into the harbor. In the water, the toxic fluid mixed with viscous oil and coated hundreds of sailors swimming for their lives amidst the flames and explosions. Meanwhile, a poisonous cloud wafted over the city of Bari. Everywhere, the pungent odor of garlic permeated the air.

Because hospitals were inundated with the severely injured, those not as badly hurt were left waiting in soiled clothing or wrapped in blankets for up to a full day. If physicians had known these patients were coated in toxic fluid, they would have instructed the sailors to shed their clothes and shower immediately. In all, 628 military personnel were hospitalized with mustard poisoning, with symptoms ranging from extensive chemical burns and blisters, to severe conjunctivitis and temporary blindness. Within a month, eighty-three would die.

Driven by fear that enemy discovery of the facts might prompt a retaliatory German chemical attack, the Allies deliberately tried to cover up the mustard gas disaster. However, an army physician named Stewart Alexander correctly deduced that only chemical weapons could explain the constellation of patient findings he encountered. His suspicions were confirmed when an intact American bomb casing containing mustard toxin was retrieved from the bottom of the harbor. The 1925 Geneva Protocol, to which Germany, the U.S., and Britain were signatories, banned the use of poison gas in war, though not the possession of such gas. The American mustard gas had purportedly been brought to Europe as a precaution, to use in retaliation in case the enemy used it first.

In the days following the attack, Alexander noticed a peculiar and surprising finding among the surviving mustard gas victims: their white blood cell counts were extremely low. A normal white cell count usually ranged from 4,500 to 11,000 per cubic millimeter (microliter). Some victims in Bari had white cell counts lower than one hundred. Alexander realized that mustard gas harmed patients' bone marrow, where infection-fighting white blood cells were produced. This made him wonder if the chemical could be useful as a treatment for cancers related to white blood cell overproduction. He later recalled thinking at the time, "If mustard could do this, what could it do for a person with leukemia or lymphosarcoma?" Whereas mustard gas's effect on bone marrow had been largely ignored during World War I, and little studied afterward, the Bari incident had bestowed scientists with a very large set of human victims that could be carefully studied and followed.

Back in the U.S., Alexander's work was noticed by Colonel Cornelius Rhoads, the chief of the medical division of the Chemical Warfare Service. In 1944, Rhoads organized a classified clinical trial of 160 cancer patients who were treated with nitrogen mustard, a derivative of the chemical used in mustard gas. This study showed positive results in patients with lymphoma and, in 1949, the first cancer chemotherapy medication to be approved by the FDA was a form of nitrogen mustard named "Mustargen." It would be the vanguard of a class of chemotherapy drugs called *alkylating agents*, which are still used today.

Thanks largely to the catastrophic disaster at Bari, a new method of combating cancer had been launched. Surgery and radiation could strike at localized tumors but were insufficient to defeat cancers of the blood or cancer that had spread throughout the body. Now chemotherapy offered a tantalizing possibility—the ability to kill cancer wherever it lived.

A HOPELESS DISEASE

In the late 1940s, Sidney Farber was the chairman of pathology at Boston Children's Hospital, a position of prominence and prestige. He could have coasted on his laurels for the rest of his career. Instead, he did something almost unthinkable; he turned away from pathology to confront one of the world's most tragic and hopeless diseases—childhood leukemia.

Leukemia was a death sentence for children. It was a cancer of immature white blood cells—cells that multiplied so vigorously that they overwhelmed and wrecked the bone marrow by crowding out the normal production of healthy red blood cells, white blood cells, and platelets. Ninety-seven percent of childhood leukemias occurred acutely, characterized by rapid onset of fever, lethargy, pallor, and weight loss. The telltale blood draw revealed blood so full of abnormal white cells that the blood appeared like white sludge, or even pus. A lack of platelet production led to easy bleeding; a dearth of red blood cells caused anemia. Improper lymphocyte function made patients susceptible to infection. Death came incredibly quickly, often mere weeks after diagnosis.

The tragedy of watching children die from this terrible disease deterred many young doctors from ever contemplating a career in the depressing field of pediatric oncology. As a blood cancer, not a solid tumor, leukemia was untreatable because it could not be excised by a surgeon or irradiated with an X-ray. But Sidney Farber had an idea. His inspiration came from the work of an English doctor named Lucy Wills.

In the early 1930s, Wills was studying a nutritional anemia that was common among destitute, underfed mill workers in India. Wills found a nutrient that improved her patients' blood cell counts—Marmite, a food spread made from yeast extract. It was a sticky, brown paste with a strong, salty flavor. The most active ingredient in Marmite was folic acid, also known as folate. We now know that folic acid, which we normally ingest through eating fruits and vegetables, is an essential component of DNA's structure, playing a crucial role in cell division. The part of the body most susceptible to folic acid deficiency was the place where cell division occurred most actively—the bone marrow—hence the malnourished patients' anemias.

When Farber learned that Wills had corrected blood production in patients by giving them folic acid, he wondered: What if the same thing could help children with leukemia? In 1945, scientists from the Mount Sinai School of Medicine in New York reported that folic acid injections had made tumors in thirty-eight out of eighty-nine mice recede. It seemed as if folic acid might have the potential to reverse cancer. So Farber decided to try using it; he began a small trial in which he injected leukemia patients with folic acid.

This was a mistake.

To Farber's horror, his treatment not only failed to reverse leukemia, it actually worsened the disease and hastened children's deaths.

Farber was dismayed. His hospital colleagues were appalled. Farber had committed a terrible error. His treatment had accelerated the unchecked cell division characteristic of leukemia—it was like pouring fuel on a fire.

But Farber thought there was a lesson in his failure. If folic acid intensified the production of malignant leukemic cells, what if he could give patients an antagonist to folic acid? An anti-folate? What if he could replicate the anemia found in Wills's indigent Indian patients to the benefit of leukemic children?

An innovative physician and chemist named Yellapragada Subbarao supplied Farber with a folic acid antagonist molecule, a medicine that would later be called aminopterin. On December 28, 1947, Farber used it for the first time in a child, a two-year-old leukemic boy named Robert Sandler. The response was astonishing and far better than Farber anticipated. Sandler's white cell count, which had risen as high as 60,000 per microliter (normal value between 4,500 and 11,000), dropped dramatically, into the normal range within three days. The boy's color and appetite returned. Instead of being listless, he became alert and energetic. He began to look like a normal child again.

Unfortunately, this merciful reprieve proved ephemeral. Sandler's leukemia relapsed a few months later, and this time more anti-folates failed to help; he died in 1948. Still, Farber recognized he had achieved something no one had before—he had put a leukemic patient into *remission*, meaning reduction or disappearance of cancer cells, if only briefly. Farber soon began to treat additional children, but his methods prompted severe disapproval and even outrage among many of his physician colleagues. Anti-folate treatment had a toxic effect on all dividing cells, whether cancerous or normal. Rapidly growing cells found in hair follicles and the lining of the intestines also died, resulting in hair loss, diarrhea, mouth ulcers, nausea, and vomiting. Farber's experimental medicine was making children feel extremely sick, children that were going to die no matter what anyone did. Why was Farber torturing them instead of letting them die in peace?

But Farber persisted. In a 1948 article, he reported that ten of the first

CANCER: *A Bewilderingly Complex Array*

sixteen patients he treated benefited from temporary remissions. Some of the children had survived an additional six months, though all would eventually die. Casual observers may have understandably judged this a very modest achievement, but in the world of leukemia, it was a major accomplishment. Farber was willing to do undeniably harsh things in order to offer a modicum of hope to patients suffering from a dismal diagnosis. And there was no dearth of such patients. Desperate parents began to hear of Farber's work. Soon, they were traveling to Boston from near and far to bring him their dying, leukemic children, willing to try anything, eager for any chance Farber could provide.

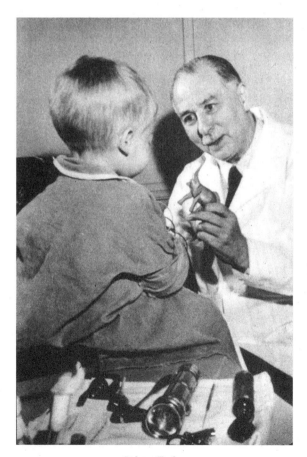

Sidney Farber

JIMMY

Ironically, Farber's nascent success in Boston, and the steady stream of families arriving at the Children's Hospital, seemed to highlight doctors' ignorance in the fight against cancer. Those who were previously easy to disregard as hopeless cases had now been granted a glimmer of hope. As children gained fleeting remissions of two or three months, it was only natural that the pressure to provide better results would soon follow. No one felt the weight of these expectations more than Sidney Farber, but his inadequacy to meet the moment was clear. Bluntly put, neither Farber nor any other doctor actually knew why cancer occurred. No one understood why malignant cells reproduced so endlessly or vigorously, nor what laws dictated their spread throughout the body. If future cancer treatments were to be discovered by methods beyond mere serendipity, doctors would first have to gain a far greater understanding of the basic science behind cancer. Attaining this knowledge would require a gargantuan, decades-long effort—one, Farber knew, that would require a massive amount of money.

As he contemplated how to raise the millions of dollars required to mount a serious research effort to defeat leukemia, Farber realized there was one successful organization that was already doing something very similar: the National Foundation for Infantile Paralysis, which was raising millions annually through their March of Dimes campaigns. Farber developed an ambitious vision to emulate the National Foundation by creating a charitable organization dedicated to curing childhood cancers. In collaboration with a charity called the Variety Club of New England, he helped establish the Children's Cancer Research Fund in 1948. The Fund's inaugural raffle event raised $45,456. This was a large amount, but not nearly enough to achieve Farber's aspiration to revolutionize cancer research.

It soon became clear that if Farber's cancer fund were to match the National Foundation's success, Farber would need to replicate its methods of effective marketing and public relations. The National Foundation inspired donations by producing posters of courageous kids in braces struggling to walk again. Farber needed something similar to tug at Americans' heartstrings. He needed to find

CANCER: *A Bewilderingly Complex Array*

a young patient who could represent all children with cancer. This exemplary child had to be friendly, fun, and spirited. He or she could not be too ill. Unfortunately, most of Farber's leukemic patients were languishing and near death. Photos of them would be depressing, not inspirational, and might only amplify the hopelessness of cancer. Farber thoughtfully considered each of the children residing on his inpatient ward. And then, he found the perfect one.

Einar Gustafson, a twelve-year-old boy from northern Maine, was healthier than most because he didn't have leukemia; instead, he had been diagnosed with an uncommon abdominal lymphoma. Farber had treated him with anti-folates, with good results. Einar looked and sounded like a model, all-American child. For Farber's purposes, he was perfect. To protect Einar's privacy, Farber thought he should give him a pseudonym.

He decided to use the name "Jimmy."

On May 22, 1948, Jimmy was introduced to the nation through the California-based, Saturday evening radio program *Truth or Consequences*. Before connecting to "Jimmy" in his Boston hospital room, the host, Ralph Edwards, furtively told his listeners, "Tonight we take you to a little fella named Jimmy. We're not going to give you his last name, because he's just like thousands of other young fellas and girls in private homes and hospitals all over the country. Jimmy is suffering from cancer, but he doesn't know he has it. He's a swell little guy, and although he can't figure out why he isn't out with the other kids, he does love his baseball and follows every move of his favorite team, the Boston Braves . . .

"Now, by the magic of radio, we're going to span the breadth of the United States and take you right up to the bedside of Jimmy . . . Up until now, Jimmy has not heard us. Now, we tune in a speaker in his room . . . give us Jimmy please. Hello, Jimmy?"

Now Jimmy's clear voice came over the airwaves.

JIMMY: Hi!

EDWARDS: Hi Jimmy! This is Ralph Edwards of the *Truth or Consequences* radio program. Well, I heard you like baseball, is that right?

JIMMY: Yeah, it's my favorite sport!

EDWARDS: It's your favorite sport! Who do you think is going to win the pennant this year?

JIMMY: The Boston Braves, I hope.

EDWARDS (CHUCKLING): Which one of the Boston Braves is your favorite player?

JIMMY: Johnny Sain.

EDWARDS: Johnny Sain, the pitcher? Yeah, he's won twenty games two years in a row, hasn't he? Who is the catcher?

JIMMY: Phil Masi.

EDWARDS: That's right... Have you ever met Phil Masi?

JIMMY: No...

PHIL MASI (ENTERING THE ROOM): Hi, Jimmy! My name is Phil Masi.

EDWARDS: What? Who's that, Jimmy?

JIMMY: Phil Masi!

EDWARDS: Where is he?

JIMMY: In my room!

EDWARDS (WITH RISING EXCITEMENT): Well, what do you know? Right there in your hospital room—Phil Masi from Berlin, Illinois. Who's the best home run hitter on the team, Jimmy?

JIMMY: Jeff Heath.

JEFF HEATH (WALKING IN): Thanks, Jimmy! I bet you can sock 'em, too!

EDWARDS: Who's that, Jimmy?

JIMMY (GASPING): Jeff... Heath!

EDWARDS (STUDIO AUDIENCE LAUGHING APPRECIATIVELY): Yeah! Right in your room, there?

JIMMY: Yes!

As the broadcast continued, more of Jimmy's heroes crowded into his room. Braves players Eddie Stanky, Johnny Sain, Warren Spahn, Bob Elliott, and Earl Torgeson entered, each with gifts including T-shirts, autographed photos, and even Torgeson's game bat. The Braves' manager, Billy Southworth, came on the air.

> SOUTHWORTH (TO JIMMY): What do you think of your ball club?
> JIMMY: They're good!
> EDWARDS: Jimmy, we saved Billy Southworth for the big surprise . . . Billy, what's the surprise for young Jimmy, there?
> SOUTHWORTH: Well, we play the Cubs tomorrow in a doubleheader at Braves Field in Boston, and we're calling it "Jimmy's Day"— dedicating the first game to you, Jimmy . . . We're going to win tomorrow's game for you, Jimmy! Aren't we, fellas?
> ALL BRAVES: Yeah!

Shortly thereafter, a piano was wheeled up to the door of the hospital room, and Jimmy and all the players sang a rousing rendition of "Take Me Out to the Ball Game." Jimmy's adorably off-key voice could be prominently heard. In total, seven minutes had elapsed on-air. Edwards bade Jimmy farewell and the connection to Boston was disengaged. Edwards next spoke directly to the audience, his tone somber and serious:

"Now folks, listen. Jimmy can't hear this, can he? Now look, really . . . Let's make Jimmy and thousands of other boys and girls happy who are suffering from cancer by aiding the research to help find a cure for cancer in children . . . this isn't a contest where you win anything, folks. This is our chance to help helpless little boys and girls such as Jimmy win a greater prize: the prize of life. Give from the heart for a cause so worthy it's impossible to describe."

The public reaction to the broadcast was overwhelming. That very evening, visitors came to the Children's Hospital lobby and stood in line to make their donations. Contributions poured in from around the country; Jimmy received heaps of letters and postcards addressed simply to "Jimmy, Boston, Massachusetts." In all, more than $231,000 was donated. Farber's research fund was soon renamed "The Jimmy Fund," and it would grow to surpass even the success of the National Foundation, continuing to the present day. By 1952, Boston's brand-new Jimmy Fund Building—with clinical space and research laboratories devoted solely to investigating and treating childhood cancer—was completed and ready to open its doors.*

Einar Gustafson photographed at his family's farm in 1948, wearing the uniform given to him by players of the Boston Braves

CHEMOTHERAPY

Through the 1950s and 1960s, more chemotherapeutic agents were discovered and developed. In 1949, a superior folic acid antagonist called methotrexate proved more effective and less toxic than aminopterin. It was also found to *cure* choriocarcinoma, a cancer originating from specialized cells of the placenta. In an era when many doctors thought of cancer as a single disease that might respond to a single magic-bullet cure—if only that cure could

---

* This institution was renamed the Sidney Farber Cancer Center in 1974, and later the Dana-Farber Cancer Institute in 1983, to recognize the philanthropy of industrialist Charles Dana.

be found—the search for new chemotherapeutic agents was on, both in the laboratory and in nature.

In 1951, two biochemists named George Hitchings and Gertrude Elion synthesized a new compound called 6-mercaptopurine that could inhibit cellular DNA function and therefore cell division. Soon, this new drug also proved successful at putting leukemia into remission. Other researchers scoured Earth's plant and animal kingdoms for compounds with anti-cancer properties. They searched mountaintops and seas, jungles and prairies, for substances to test. Even the dirt above the graves of cancer victims was tested for bacteria that might prove the basis for developing some helpful anti-cancer antibiotic.

There were rare successes. In 1952, Toronto doctor Clark Noble received a sample of the periwinkle plant (scientific name *Vinca rosea*) from Jamaica to investigate as an antidiabetic agent. Clark's brother, endocrinologist Robert Noble, injected extracts of the vinca plant into rats. To his surprise, instead of seeing any antidiabetic response, the animals died rapidly from infection. This was clearly not a useful diabetic drug; but, the acute susceptibility to infection made Robert suspect something had depressed the rats' immune systems. He confirmed that the plant significantly reduced white blood cell production in the bone marrow, and purification of the active ingredient yielded a new and effective cancer chemotherapeutic drug named vinblastine. This would be the first of a new class of chemotherapeutic agents termed *alkaloids*. Noble later cooperated with Eli Lilly & Company, which refined the molecule to create a version with slightly fewer side effects. This became vincristine, introduced in 1961. Vincristine worked by binding and inhibiting cells' microtubules, cytoskeletal proteins essential to the process of cell division.

This manner of serendipitously discovering anti-cancer treatments by noting the cell-killing properties of failed drug prospects would become a bit of a theme. In 1954, Sidney Farber obtained a sample of an unsuccessful antibiotic prospect from Selman Waksman, the Rutgers soil biologist who had won a Nobel Prize for co-discovering streptomycin. Waksman's drug, called actinomycin D, disrupted bacterial DNA but proved useless as an antibiotic because it killed normal human cells too, not just bacterial cells. This characteristic made Farber think actinomycin D might instead be useful to preferentially kill rapidly dividing cancer cells. When he treated mouse cancers with it, the tumors diminished and disappeared. When Farber's research team tested actinomycin D in humans, they found it unhelpful in leukemia; but, to their delight, it was highly effective against a pediatric kidney cancer called Wilms' tumor.

Unfortunately, all these successes carried a major drawback—terrible side effects. In truth, these "treatments" were poisons that killed all human cells, both normal and cancerous. Cancer cells died preferentially, due to their faster rate of multiplication, so the challenge with each patient was to give the drugs long enough to kill the cancer cells, but stop before too many healthy cells died as well. In addition to predictable complications like infection, bleeding, and anemia that stemmed from reduced normal blood cell production, each drug gained its own unique reputation for the harsh consequences of its use. Methotrexate damaged the liver and lungs; 6-mercaptopurine prompted unbearable nausea and vomiting. Vincristine precipitated low blood sodium, peripheral neuropathy, and constipation, while actinomycin D produced debilitating fatigue and skin necrosis if any of the drug extravasated outside a vein.

These awful side effects became cancer chemotherapy's limiting factor. But because higher and more prolonged doses of the drugs resulted in heightened killing of cancer cells, and therefore greater chance of remission, doctors aimed to give patients as much chemotherapy as possible, even to the brink of death. This morbid concept became commonly known as the "maximum tolerated dose," or MTD.

Many doctors lacked the stomach for playing this fatal game of chicken with cancer. It took a special kind of fortitude to poison patients and make them live out their last days in misery based on the slim hope of a remission that might only last weeks or a few months. Two oncologists who exhibited single-minded determination to combat cancer on these terms were almost identically named doctors at the National Cancer Institute, Emil Frei and Emil J. Freireich. Frei, who went by the nickname "Tom" to make identification at work easier, was calm and understated. In contrast, Freireich, who chose to go by "Jay," was loud, argumentative, and flamboyant. The two made an unlikely pair, but both became seized by the potential benefits of treating patients with multiple chemotherapeutic drugs at the same time. They believed that using drugs that attacked cancer cells by different mechanisms of action could reduce the incidence of drug resistance and have an additive effect on killing cancer cells. Since each drug caused different side effects—one might cause nausea while another liver damage, for example—they reasoned that the severity of any one type of side effect should not overwhelm the patient. If patients could endure an array of multiple side effects simultaneously, they would benefit from the greater cancer-killing capacity of two drugs—at least the theory went.

In short, it was hell.

And yet, it seemed to work. When Frei and Freireich started treating leukemic patients with two drugs, the children suffered greatly, but there was a definite increase in remission duration. The cancer cell-killing effect of some drug combinations proved to be not merely additive but synergistic. As a result, it did not take long for these daring physicians to advance to something many of their colleagues considered unthinkable. They started giving patients four drugs at a time. This protocol, started at the National Cancer Institute in 1961, was termed "VAMP," after the names of the four drugs administered: vincristine, amethopterin (the chemical name for methotrexate), 6-mercaptopurine, and prednisone.* Children were treated for two weeks, and then given a two-week break before another two weeks of treatment, and so on, for a total of six rounds of treatment.

Many of Frei and Freireich's colleagues were aghast. They considered VAMP to be not only unethical but unconscionably cruel. Even the great Sidney Farber adamantly opposed it, saying, "It's fine for rats and mice, but

---

* Prednisone is a steroid that is normally used as an anti-inflammatory agent. In cancer treatment, it was also shown to induce white blood cells to undergo premature *apoptosis*, the term used to describe a cell's death at the end of its normal lifespan.

such treatments would injure children terribly. These are all toxic drugs, and I will never allow such experimentation on my children." Farber used only one chemotherapy drug at a time and could not bear to make children suffer any more misery beyond that which one conferred alone.

Frei and Freireich were accused of using children as guinea pigs. At their lectures, they were roundly criticized and even jeered by righteously impassioned audience members, who blurted comments like, "This is a meat market! What a butcher shop!" Dr. Vincent DeVita, a future director of the National Cancer Institute who served as a trainee under Frei and Freireich, recalled the invective his mentors endured: "It was embarrassing and shocking to watch. Never had I seen doctors behave this way toward other doctors." DeVita and other young physicians began to doubt themselves. "Were we doing the right thing?" DeVita wondered. "Were we accessories to what other doctors were portraying as unethical behavior?" Before long, promising trainees were advised to avoid Frei and Freireich, deemed by many as radical nonconformists who had strayed far beyond the mainstream of oncology's established thought leaders.

For a while, it appeared the critics were probably right. Leukemic children receiving the toxic VAMP cocktail endured near total wipeout of normal bone marrow cells as well as cancer cells. As horrified parents watched, their helpless three-, four-, or five-year-old children lost their hair, became emaciated and listless, contracted terrible infections, bled spontaneously, and dropped into comas as their oncologists urged them to hang on and complete the weeks-long regimen of cellular poisons. Dr. David Nathan, a young physician at the National Cancer Institute, remembered watching the children's suffering and wanting to quit: "For me, this was a nightmare . . . This wasn't research. It was an execution. I was appalled by what I saw."

Then, something remarkable happened. As patient after patient was taken to the brink of death, 60 percent achieved remission of leukemia. This was much higher than what had been attained with conventional, single-drug treatment. Even better, as the children's bone marrow function began to slowly recover, these VAMP-induced remissions proved far longer-lasting than

CANCER: *A Bewilderingly Complex Array*

any that had ever been achieved before. Whereas Farber's single-drug regimens had produced remissions lasting only weeks or a few months, VAMP remissions lasted several months and even years. By 1975, the five-year survival rate for childhood leukemia had risen to 53 percent. For a disease that had been 100 percent fatal less than fifteen years before, this was an incredible achievement.

And yet, no one could pretend that this was a resounding victory over cancer. As time went by, survival rates for leukemia continued to improve,* but chemotherapy regimens failed to bring long-lasting remissions to many other cancers. A major problem was the common onset of "multidrug resistance." Echoing the ability of bacteria to evade antibiotics, cancer cells could also mutate and "learn" to survive the onslaught of chemotherapeutic drugs.† Despite successes against rarer cancers like leukemia, Hodgkin's lymphoma, Ewing's sarcoma, and testicular cancer, patients with more common afflictions like lung, breast, and colorectal cancer still held little hope. In 1963, the all-cancer five-year survival rate in the United States was approximately 37 percent.

Lacking better weapons, oncologists tinkered with various drug combination protocols for decades, varying how much of each drug to give, in what groupings, and for how long. The possible combinations grew so multitudinous that it became harder and harder to determine how effective the various regimens truly were, or if they were any better than what had been tried before. President Nixon's famous 1971 declaration of war on cancer via the National Cancer Act dramatically increased funding for cancer research ($100 million initially), but did not change the fact that, instead of discovering new knowledge about the molecular basis of cancer, "cancer research" largely reflected a

---

* Today acute childhood leukemia is curable in about 90 percent of cases.

† Here is one example of how a cancer cell might develop resistance. Methotrexate impedes DNA synthesis in cells by blocking a necessary enzyme called DHFR (dihydrofolate reductase). After exposure to this drug, cancer cells have shown the ability to upregulate DHFR's production, overcoming their susceptibility to the drug's actions.

strategy of empirical testing—trial-and-error sampling of tens of thousands of substances, both natural and chemical, to see if they killed cancer cells.*

Fifteen years after the war on cancer had been declared, an important 1986 study published in the *New England Journal of Medicine* revealed that cancer deaths were not decreasing in the United States; instead, they had increased by 8.7 percent between 1962 and 1982. The authors of this study plainly stated, "We are losing the war against cancer." Nobel laureate James Watson, who, with Francis Crick, had elucidated the double-helix structure of DNA, referred to cancer as "a black box that we're trying to influence with magic." In later years, as frustration grew that billions of research dollars were failing to improve death rates, a macabre joke sometimes uttered at oncology conferences questioned whether more lives might be saved if a bomb went off at the meeting, wiping out a sizable portion of the nation's oncologists.

A fifty-year odyssey of chemotherapeutic advances had scored some incredible successes. The chain of discoveries originating with the disaster at Bari and Sidney Farber's audacious vision had given hope to tens of thousands of previously hopeless patients; but, indiscriminate cell destruction via chemotherapy would never become the much-hoped-for magic bullet that might prove cancer's panacea.

To make further headway, many fundamental questions still had to be answered. Questions as basic as: Where did cancer come from? How did it thrive? What made it spread? For decades, the answers to these mysteries would generate more questions than solutions, and prove that cancer was a riddle far more complex than almost anyone could have guessed.

DEFINED BY HETEROGENEITY

As if unknowingly working together to complete a thousand-piece puzzle, investigators from across the globe would spend almost a century laboring

---

* Additional medicines discovered included paclitaxel (Taxol), derived from the yew tree; etoposide, from the fruit of the mayapple; and bleomycin, produced by a bacterium.

toward discoveries that might illuminate one small piece of the larger picture. In the 1930s, a Chicago urologist named Charles Huggins studied the prostate glands of dogs and found that removing dogs' testicles prompted shrinkage of their prostate glands. He deduced this resulted from reduced testosterone hormone and wondered if diminishing testosterone might have a beneficial impact on prostate cancer. Some of his dogs had prostate cancer; when he removed their testicles and eliminated testosterone production, their tumors shrank away.* Huggins had struck upon a remarkable new fact—some cancers could be influenced by hormones.

This information would be applied to the fight against another, more deadly form of cancer—breast cancer. Clues pointing to the dependence of breast tumors on hormones harkened back to 1896, when a Scottish surgeon named George Beatson made the remarkable observation that removing the ovaries of three women with breast cancer led to reduction of the size of their tumors. Decades later, scientists learned that ovaries are the source of estrogen, a hormone that maintains breast tissue by encouraging cell growth and division. In the 1960s, a chemist named Elwood Jensen used radioactive hormone labeling to identify the estrogen receptor lodged in the cell wall of many breast tissue cells. Interestingly, not all breast cells had the estrogen receptor.† Still, he wondered if reducing estrogen could be a way to treat breast cancer. That idea came to fruition in the form of tamoxifen, an estrogen antagonist shown in 1971 to effectively reduce both breast tumors and lung metastases.

Tamoxifen wasn't a cure—it soon became clear that the drug would only work on estrogen receptor-positive tumors—but for the first time, the public had access to an effective cancer drug that wasn't solely designed to poison cells indiscriminately. This was a novel instrument in oncologists' tool belts, a drug with a specific target, albeit limited to only one subset of one type of cancer.

---

* Today, this approach is known as "androgen deprivation therapy."

† About 60 to 80 percent of breast cancers are estrogen receptor-positive.

At the same time, another innovative doctor was also investigating a completely new approach to attacking cancer. In 1967, thirty-four-year-old Dr. Judah Folkman was the youngest-ever chief of surgery at Boston Children's Hospital. He had become obsessed with an unusual idea: that tumors emitted some sort of signal, possibly a growth factor, that drew blood vessels toward them. His theory, termed "angiogenesis," made him a laughingstock among oncologists for decades. Cancer research was devoted to chemotherapeutic drugs; practically no one thought blood vessels played any role at all. But Folkman persisted because, as a surgeon, he was often astonished by how heavily many tumors were invested with new blood vessels.

For years, Folkman had little evidence to support his claims. He continued to labor, blending up tumor tissue, hoping to isolate and purify a growth factor that tumors emitted to recruit blood vessels toward themselves. One doctor accused him of "purifying dirt." Another person at his hospital told him, "You're making a mockery of research here." Dr. Vincent DeVita recalled Folkman's dim reputation—at an annual meeting, "Judah made presentations almost every year. It was kind of a joke. Doctors in the audience actually made fun of him."

Finally, in 1983, Folkman proved his critics wrong. His lab isolated a molecule that made the vascular endothelial cells of blood vessels grow. They called it *fibroblast growth factor*, and published in *Science*. Soon, the tantalizing prospect of shrinking tumors by diminishing their blood supply sparked a race to develop angiogenesis inhibitors. One of the first to be discovered, *interferon alpha*, proved an effective treatment for hemangiomas. In 2004, the first FDA-approved anti-angiogenesis drug hit the market; it was Avastin (bevacizumab), a monoclonal antibody that blocked a protein called vascular endothelial growth factor (VEGF). Avastin is used to treat colon cancer, as well as malignancies of the lung, kidney, and brain. Though anti-angiogenesis could not cure cancer by itself, it armed oncologists with another important weapon in a widening war.

As more and more scientists devoted their careers to determining the true cause of cancer, their theories were somewhat muddled by the observation that cancer did not only arise spontaneously within an unlucky person's

CANCER: *A Bewilderingly Complex Array*

body but could also be caused by external factors, through exposure to carcinogens. The danger of carcinogens has been known for over two centuries. In 1775, a British surgeon named Percivall Pott noted that almost all cases of scrotal cancer he encountered occurred in young boys with the same occupation—chimney sweep. Pott discovered that soot residing chronically under the skin was a cause of scrotal cancer. He had identified a carcinogen. Other environmental carcinogens including radium, mesothelioma-causing asbestos, and tobacco became well known. Though these external factors may have befuddled doctors seeking a single, unifying cause of all cancers, they also presented a golden opportunity to reduce cancer deaths through prevention. Industry regulation and campaigns to curb smoking have undoubtedly saved far more lives than chemotherapy, radiation, or surgery ever have. In many respects, the best offense against cancer has been to play defense, through prevention and especially screening tests.

In the 1920s, a Greek cytologist named George Papanicolaou accepted a rather unusual assignment at Cornell University Medical College—to study the menstrual cycle of guinea pigs. This challenging task was made even more difficult by the fact that menstruating guinea pigs did not bleed visibly. Papanicolaou became adept at scraping off minute amounts of guinea pig cervical cells and viewing them under a microscope. He also spent decades studying human cervical cells and learned how the morphology of the cells changed at different points in the menstrual cycle, becoming so expert that he could actually tell what day in a menstrual cycle a woman was at, just by looking at the cells. In the course of this work, he inevitably came across some women with cervical cancer. In their cells, he identified characteristic cytological changes of the disease, including abnormalities in nuclei, cytoplasm, and organelles. And then it dawned on him—he had discovered a new way to test for cancer. In 1928, he published an article entitled "New Cancer Diagnosis," describing his technique, which was later called a "Pap smear."

For the next twenty years, practically no one took any notice.

The few pathologists who bothered to consider Papanicolaou's method did not think highly of his watery slide samples, which were not easy to evaluate. If one were truly worried about cervical cancer, a biopsy would be the more definitive way to discover it. Papanicolaou was compelled to take his test in a different direction. He had discovered that malignant cells of the cervix did not suddenly convert to cancer cells overnight. Instead, there were gradual, discrete changes in the cells that occurred before the cells became cancerous. Papanicolaou excitedly realized that these precancerous changes could make his technique useful as a screening test, allowing suspect patients to be identified and treated before they even developed the disease.

In 1952, the National Cancer Institute conducted a large-scale trial of Papanicolaou's Pap smears, proving their effectiveness. Women with precancerous changes could undergo a minor surgical procedure to remove the threatening lesions, making them far less likely to develop a malignancy. In the quest to defeat cancer, Papanicolaou's screening test, which is estimated to have saved the lives of more than six million women worldwide, revealed a new way of eliminating the enemy before it grew to full strength.

A doctor like Papanicolaou could take pride in creating a tool that helped identify cancer early, but his work had done nothing to explain why cervical cancer began in the first place. It would have shocked him to learn what we know now, that it is caused by the human papillomavirus (commonly known as HPV), which is transmitted via sexual intercourse. For those seeking to understand the origin of cancer, the knowledge that it could stem from carcinogens, viruses, sexually transmitted diseases, or arise spontaneously—and be influenced by hormones, blood vessel growth factors, and conceivably many other factors yet to be discovered—presented a bewildering panoply of possibilities that could only increase skepticism that cancer might ever be tamed.

The gripping idea that cancer might actually be an infectious disease, caused by viruses, was one that captivated both scientists and the general public in the twentieth century. In 1911, a thirty-one-year-old virologist named Peyton Rous did something that threw contemporary theories about cancer's cause into grave doubt. At the Rockefeller Institute in New York City, Rous was studying a rare type of sarcoma—a malignancy of connective tissues like muscle and tendon—found in chickens. He excised a tumor, ground it up in fluid, and put the mixture through a series of filters to remove all cells, bacteria, and contaminants, until the only thing left was the liquid filtrate that the cells had resided in. When he injected this cancer-cell-free fluid into a chicken, the animal developed cancer.

This was a surprise.

Here was evidence that the cause of cancer did not necessarily dwell within the cancer cell. The only organism small enough to pass through Rous's filters was a virus. He hypothesized that a virus caused cancer, an idea that contradicted every prevailing theory at the time. But Rous was correct. The virus that causes this type of cancer was later named the Rous sarcoma virus (RSV).

Those who recognized this began to wonder—was *all* cancer caused by viruses? Was cancer an infectious disease? It seemed entirely possible, since viruses were mysterious, poorly understood, and had not even been visualized yet—a fact that added to their mystique. Researchers began to look for viral causes of human cancers.

The search was fruitless until 1957, when an Irish surgeon named Denis Burkitt began to encounter children with terribly deforming jaw tumors while working in Uganda. Burkitt recognized this as an extremely aggressive form of lymphoma, one not previously studied in the West. By surveying doctors all over Africa, Burkitt discovered a pattern. All the cases appeared to occur in a defined geographic area, a band across sub-Saharan Africa. There were no cases in South Africa, and none in northern Africa. The pushpins covering Burkitt's map, each representing a case, covered the middle of the continent in a horizontal stripe. This inexplicable finding spurred Burkitt and two colleagues to embark on a ten-thousand-mile trek, driving a used station wagon to visit scores of hospitals and mission outposts in twelve different countries. Burkitt sought out anyone who had seen children with similar tumors, and discovered that the incidence of this unique lymphoma was related to a region's temperature and amount of rainfall. The cancer primarily appeared in regions where the temperature stayed above sixty degrees and annual rainfall was greater than twenty inches—in other words, warm and wet conditions.

This pattern of disease had been seen before—with transmissible conditions like sleeping sickness caused by the tsetse fly, and malaria spread by the *Anopheles* mosquito. Burkitt had discovered a cancer, today known as Burkitt's lymphoma, with a mode of spread that appeared typical of an infectious disease.

In 1963, Burkitt sent a series of tumor samples to a British virologist named Michael Anthony Epstein. Epstein, along with colleague Yvonne Barr and a pathologist named Bert Achong, examined these lymphoma cells using electron microscopy and identified a previously unknown virus: the Epstein-Barr virus.*

For the first time, a virus had been found to be the cause of a human cancer. *Life* magazine published a cover with the headline, "New Evidence That CANCER MAY BE INFECTIOUS." Peyton Rous, long ignored for his fringe idea, was awarded a Nobel Prize in 1966, more than fifty years after his initial discovery.† Additional cancer-causing viruses were later discovered, including the hepatitis B and C viruses that cause inflammation that leads to liver cancer, and the human papillomavirus that causes cervical cancer.

To make matters more complicated, in 1984, two Australian doctors, Barry Marshall and Robin Warren, discovered that a bacterium called *Helicobacter pylori* caused stomach inflammation that could lead to stomach cancer.‡ To prove bacteria was the cause of gastritis, and to fulfill Koch's postulates, Marshall took the audacious step of drinking a large dose of *Helicobacter pylori* mixed in meat broth. Within a few days, he became terribly ill, suffering nausea, vomiting, and fever. Endoscopic biopsies confirmed his stomach was teeming with *Helicobacter pylori*, and that he had developed gastritis and stomach ulcers, previously thought to originate from stress, not microbes. Anti-

---

* The Epstein-Barr virus is also commonly known as the cause of *infectious mononucleosis*, or "mono."

† Rous holds the record for the longest period from discovery (in his case 1911) to Nobel Prize (fifty-five years later). He shared the prize with Charles Huggins.

‡ Marshall and Warren won a Nobel Prize for their discovery in 2005.

biotic medicines helped Marshall make a full recovery, but now a bacterium had also been shown to be a cause of a cancer—a disease whose heterogeneity had become truly overwhelming. The search for cancer's cause and cure appeared bleaker than ever.

MOLECULAR UNDERPINNINGS

The startling realization that viruses could cause cancer ignited new investigations that would ultimately lead to the most compelling and unified theory of how so many disparate types of cancer might begin. In 1970, two virologists named Howard Temin, from the University of Wisconsin, and David Baltimore, from the Massachusetts Institute of Technology, independently discovered that certain RNA viruses, including the Rous sarcoma virus, used an enzyme called *reverse transcriptase* to convert RNA into a DNA copy.* This was the opposite of what normally happens in our cells. The usual direction of transcription is to make an RNA message from our DNA genome, which is then transported to a ribosome to instruct the production of a specified protein. Temin and Baltimore's discovery meant that an RNA virus with reverse transcriptase could insert its RNA genetic code—converted into a DNA copy—into a host cell's native DNA. This was how these particular RNA viruses caused cancer: by changing a cell's genome, with the result being endless cell division.

Using the Rous sarcoma virus to study this process made sense because the virus only contained four different genes. One of them, termed *src* (pronounced "sarc"), proved to be the culprit. This cancer-causing gene, called an *oncogene*, coded for a protein called a *kinase* that spurred cells to divide repeatedly; it was like flipping an "on" switch inside a cell to direct it to multiply. An even greater discovery was the fact that *src* was not a foreign gene. Research

---

* Temin and Baltimore shared a Nobel Prize in 1975 with Renato Dulbecco. Temin had worked with Rous sarcoma virus in Dulbecco's lab at the California Institute of Technology in the 1950s.

by two scientists, Harold Varmus and J. Michael Bishop at the University of California, San Francisco, revealed that the *src* gene was actually a normal part of the human genome and many animal genomes.* This revelation led to a new theory, that oncogenes were present but not necessarily activated in all of our cells. Such genes, termed *proto-oncogenes*, had the potential to turn into cancer-causing oncogenes if they mutated.

How could they mutate? They might do so spontaneously, from bad luck; or, they might be spurred to mutation by interacting with a carcinogen, a virus, or beams of radiation. This, at last, provided a common denominator to explain the cause of cancer: cancer originated from activating oncogenes. The viruses carrying *src* were not the cause of cancer. They were merely the method of transfer that brought meddlesome genes into normal cells and activated oncogenes that prompted unchecked cell division.

But there was more. Studies of an eye cancer called retinoblastoma revealed the importance of another type of gene, a tumor suppressor gene (also known as an *anti-oncogene*) that normally suppressed cell division. If a suppressor gene became inactive through mutation, its signal of restraint was lost and this could also push a cell to multiply. What emerged was the portrait of a human genome that was tightly regulated by both proto-oncogenes and tumor suppressor genes that normally kept cell division within the bounds of healthy growth. But if a change caused either type to malfunction—if an oncogene was turned on, or a suppressor gene turned off—the result was exponential, cancerous cell growth.

In the 1980s, at least a hundred oncogenes, with names like *ras*, *myc*, and *neu*, and tumor suppressor genes, such as *p53*,† *VHL*, and *APC*, were discovered. There were immediate victories to be gained by this remarkable increase in knowledge. For example, the discovery of two tumor suppressor genes, *BRCA1* and *BRCA2*, helped to identify many women susceptible to breast cancer. If sup-

---

* Bishop and Varmus won a Nobel Prize in 1989 for their discovery.

† *p53* is estimated to play a role in up to 50 percent of human cancers, though most cancers are polygenic (the result of multiple genetic mutations, not just one).

pression from either gene is lost through mutation, aggressive breast or ovarian cancer is more likely to develop (50 to 80 percent lifetime risk of cancer). This can prompt some women to consider prophylactic tamoxifen use, screening with more sensitive breast MRI instead of mammography, or even prophylactic mastectomy to completely eliminate the risk of future breast cancer.

An understanding of oncogenes and suppressor genes also allowed oncologists to aim at specific molecular targets and finally move beyond the indiscriminate, cell-killing strategies of conventional chemotherapy. In 1984, a new oncogene called *HER2* (which coded for a protein called *Human Epidermal Growth Factor Receptor 2*) was identified. This oncogene played a role in about 20 to 30 percent of breast cancers, a highly aggressive subset whose cells displayed an abnormally high abundance of HER2 proteins. An oncologist named Dennis Slamon and a Genentech scientist named Axel Ullrich spearheaded the effort to engineer a monoclonal antibody that could bind, and therefore block, cells' HER2 proteins. In 1998, this became a new drug for breast cancer called Herceptin (trastuzumab).

In this century, dozens of new cancer drugs aimed at specific molecular targets have been released. These include Gleevec (imatinib), for treating chronic myelogenous leukemia, Velcade (bortezomib) for multiple myeloma, Erbitux (cetuximab) for cancers of the head and neck, and Tarceva (erlotinib) for lung and pancreatic cancer. They are largely free of the horrible side effects common to their chemotherapeutic predecessors. A massive project called the Cancer Genome Atlas, undertaken between 2005 and 2018, aimed to sequence the total genome of more than ten thousand tumors representing over thirty of our most common and deadly cancers. Scientists around the world anticipated that this all-out effort would unlock cancer's underpinnings and lead to a multitude of new and effective drugs, possibly even portending the future introduction of "personalized" medicine.

Unfortunately, this is not what has happened. Despite the aforementioned novel pharmaceuticals, our increasing knowledge of cancer's molecular foundations has been more sobering than groundbreaking because we have learned that cancer is far more complex than we possibly imagined. While we now have a unifying explanation of cancer's cause, the truth is, the mutations that cause most cancers are so varied and diverse that it will take decades to exploit them for clinical benefit. Most cancers are caused by not one genetic abnormality but several. Pancreatic cancer has twelve. Several cancers display dozens or even hundreds of genetic variations, differing amongst tumors even within the same type. This means that two patients' breast cancers that are clinically identical can be characterized by completely different gene mutations. A 2006 study of breast and colon cancers identified 189 different gene mutations, with each tumor having an average of eleven. Even worse, sometimes not even the cells within a single tumor display genetic consistency. Neighboring cells in a single ovarian cancer tumor, for example, can contain many disparate genetic mutations, and none in common. This is termed *intratumoral* heterogeneity. The quaint idea that tumors originated from a single cell, cloned millions of times so that each cell of the tumor was identical, has proven grossly simplistic and naive.

Put simply, cancer has severely humbled us. In a 2002 *New England Journal of Medicine* article, scientists William Hahn and Robert Weinberg wrote, "The actual course of research on the molecular basis of cancer has been largely disappointing. Rather than revealing a small number of genetic and biochemical determinants operating within cancer cells, molecular analyses of human cancers have revealed a bewilderingly complex array of such factors." The more we learn about cancer, the more dizzying it proves to be. Cancers grow in a sophisticated ecosystem influenced by oncogenes, hormones, angiogenesis, inherited genetics, viruses, bacteria, and carcinogens. The advent of molecular targeting has not led to the subjugation of many cancers, and though this advance yielded several inspiring breakthroughs, chemotherapy, with all its drawbacks, continues to be the mainstay of treatment for most advanced malignancies today.

In short, we still need new weapons in the fight against cancer.

Thankfully, some of those weapons are now at hand—tools that hold great promise, despite stemming from an idea that, for almost a century, was ridiculed and considered little better than charlatanism.

CANCER: *A Bewilderingly Complex Array*

QUACKERY, AUDACITY, AND HOPE

In 1891, Dr. William Coley was a twenty-eight-year-old, newly minted surgeon practicing in New York City. As a graduate of Harvard Medical School, he was almost certain to achieve professional success by every conventional metric of the day. Instead, Coley's career trajectory would be anything but upward and smooth. One of his earliest cases, the plight of a pretty, vivacious seventeen-year-old girl named Elizabeth Dashiell, would haunt him for years and dramatically alter the course of his life.

Dashiell was a very close friend of John D. Rockefeller Jr., the son of Standard Oil's founder. Though the two were not romantically involved in a public way, their copious letter writing and habit of taking long, private carriage rides together hint that they were kindred spirits. Unfortunately, Dashiell developed a rare sarcoma on her hand. Coley operated on her four times, and even amputated her hand, but he was too late. The cancer, which Coley described as "one of the most malignant tumors I had ever seen," had already spread throughout her body, and she died.*

Coley was deeply affected by this young woman's death, and he became determined to study the rare disease that had killed her. As he searched his hospital's records for other patients who might have had the same type of cancer, he became intrigued by the case of a thirty-one-year-old German immigrant named Fred Stein, who had been admitted to the hospital seven years prior. Stein had presented with a large, fist-sized sarcoma tumor growing on his neck. Through the course of four surgeries over three years, his surgeon, William Bull, tried to fully remove the tumor but failed, and the cancer persisted to the point that Bull considered Stein's case "absolutely hopeless." Nothing more could be done for him.

Then, after his last surgery, Stein contracted a severe postoperative erysipelas infection of his surgical wound (caused by *Streptococcus pyogenes*). He

---

* The Rockefellers' close relationship with Dashiell spurred a dedication to cancer research that was part of their motivation to found the Rockefeller Institute for Medical Research in 1901.

developed a high fever and seemed ever closer to death. But something miraculous happened. In the midst of his infection, his residual tumor began to shrink. His fever would wax and wane, and whenever the fever peaked, his tumor seemed to soften and shrink a little more. After more than four months of careful monitoring, Stein's fevers stopped and his doctors realized his tumor was gone. There was no sign of residual cancer.

Stein had been cured . . . by what?

No one knew.

It appeared that he had somehow cured himself.

William Coley became absorbed by Stein's records, which reflected an inexplicable, spontaneous recovery from a terminal cancer. It seemed that the patient's high fever had spurred this cure—but Coley wondered if the cancer had later recurred. That was certainly possible. Was Stein still alive? Coley needed to find out, but to his disappointment, there was no record of Stein's current address. There was no way to find him.

In a display of remarkable tenacity, Coley started looking for Stein himself. He did this in his spare time, with no leads or certainty that Fred Stein was still alive. Coley roamed the Lower East Side tenements, going door to door, asking random passersby for information. He searched for weeks.

And then, incredibly, he found him.

When Coley came to Stein's apartment, he found Stein alive and healthy, still exhibiting a prominent neck scar from his prior surgeries. Coley brought Stein to William Bull, who confirmed this was the same man Bull had operated on and who had survived a terminal cancer diagnosis. Coley became convinced that Stein's infection and fever had been the key to overcoming his cancer—perhaps the bacteria directly attacked the tumor, he (mistakenly) thought.

Now Coley considered an audacious new stratagem: to find another sarcoma patient and *intentionally* infect him or her with *Streptococcus pyogenes* to induce erysipelas and reprise the same cure. In a later report, he wrote, "If erysipelas . . . could cure a case of undoubted sarcoma . . . it seemed fair to presume that the same benign action would be exerted in a similar case if erysipelas could be artificially produced." His choice of the word "benign" is curious, for he could not have failed to understand the risks of infecting patients with deadly bacteria. The physician's oath to "do no harm" notwithstanding, Coley believed that for desperate, terminally ill patients, his seemingly outrageous idea could be considered a rational option. He became "determined to try inoculations in the first suitable case."

It did not take long for Coley to find one. In May 1891, a thirty-five-year-old Italian man, known in the medical record only as "Signor Zola," sought his help. Zola had a terminal neck sarcoma. He agreed to let Coley inoculate him with bacteria from four different laboratories. To Coley's disappointment, these administrations failed to produce the desired high fever that Coley considered essential. Additional injections with more virulent bacteria also proved unhelpful. A frustrated Coley then asked a colleague who was traveling to Europe to visit the famous Robert Koch in Berlin, to see if the master of microbiology might be willing to supply Coley with his most potent sample of erysipelas-producing bacteria.

Koch complied. He sent Coley's friend home with a deadly strain that Coley promptly injected directly into Signor Zola's neck. Soon, Zola became intensely ill, with fever rising to 105 degrees. Though the "treatment" clearly risked Zola's life, Coley soon detected a definite reduction in the size of the tumor. Two weeks later, Zola had survived the infection and his tumor was completely gone.

William Coley had done it. He had cured a patient by inducing a fever, somehow galvanizing the body to purge the malignancy.

Coley devoted himself to furthering what he believed was a breathtaking new discovery. He tried the bacterial inoculations in a dozen cancer patients, but the results were mixed. Four patients displayed a positive response, but the deaths of two were hastened by his actions. Many physicians viewed Coley's approach with skepticism, and even derision, because they considered it unproven or because they could not replicate his reported successes themselves.

Still, Coley persisted. In 1899, he concocted a heat-killed mixture of bacterial organisms that he named "Coley's toxin," which became commercially produced by Parke, Davis & Company. Over decades, Coley treated hundreds of patients with his toxin, with variable results. Sometimes the tumors regressed, or symptoms improved, but other times patients contracted intractable infections that made them far worse. To many physicians, in comparison to more predictable treatments like radiation, and later chemotherapy, results with Coley's toxin seemed erratic and unreliable. Though he would endure decades of withering criticism and disparagement, Coley never stopped promoting his method. His product, which remained available for sale until 1952, was discredited by both the Food and Drug Administration and American Cancer Society in the 1960s. Decades later, a retrospective assessment claimed that, over a forty-year career, Coley treated more than a thousand cancer patients with bacterial injections, and that up to five hundred of these individuals experienced at least some degree of tumor regression.

William Coley, photographed in 1892

For a century after his initial investigations, Coley's work had teetered at the edge of acceptance and quackery, a liminal space on the fringe of oncology's consciousness. In contrast, today he is regarded as a something of a prophet, a man who audaciously, intentionally infected patients to, perhaps unknowingly, spur their immune systems to attack cancer. In the twentieth century, conventional wisdom held that the immune system ignored cancer cells because it could not recognize them as foreign. This premise seemed obvious, since malignant cells stem from normal cells.

But it is not necessarily true.

THE PROMISE OF IMMUNOTHERAPY

The subfield of oncology that William Coley initiated is now known as cancer immunotherapy. It is generally defined as any means by which the immune system is harnessed to treat cancer, and is made possible by the fact that cancer cells, like most cells, display unique proteins, or antigens, on their surfaces that are exposed to their outside environment and can be bound by other proteins, such as antibodies. Whereas cancer cells may mutate and develop the ability to evade chemotherapy drugs, they cannot elude the properly stimulated immune system, which, like them, can evolve and adjust to the threat of different adversaries. It is this key fact that makes immunotherapy such an alluring solution to cancer.

Yet it would not be until the mid-1970s that Coley's ideas would emerge from the shadows of ostracism, to be resurrected by Dr. Steven Rosenberg, the chief of surgery at the National Cancer Institute. In a story uncannily similar to that of Coley's, Rosenberg, as a young twenty-eight-year-old resident, had encountered a patient with a stunning, inexplicable recovery from cancer. The patient, James D'Angelo, was slated for gallbladder surgery with Rosenberg, but Rosenberg noted that eleven years before, in 1958, D'Angelo had suffered from terminal, metastatic stomach cancer that had unfathomably disappeared after a significant postoperative infection. Rosenberg performed D'Angelo's uncomplicated gallbladder operation and carefully inspected the liver and abdomen, which the records revealed had been riven with metastatic tumors in 1958. Rosenberg could find no trace of cancer.

Just like Coley, Rosenberg became gripped by this story of a man whose cancer had spontaneously disappeared. But there was a difference between these two doctors, separated in time by almost eighty years—unlike Coley, Rosenberg understood the immune system. Rosenberg believed the immune system possessed T cells that could recognize tumor cells as foreign and kill them, but that each tumor was so unique—so full of malignant cells displaying novel antigens—that there were rarely enough activated T cells to complete the job. In the battle between cancer and the immune system, cancer had an army billions strong, while the immune system had comparatively few T cell foot soldiers. This hypothesis led Rosenberg to consider a new method of treatment. His plan was to use a cytokine called IL-2—which promotes T cell growth—to supercharge a patient's T cell activity. He could give patients IL-2 directly; or, he could extract their T cells, use IL-2 to multiply them millions of times in a laboratory, and reinfuse these back into the body to provide a massive influx of reinforcements to turn the tide against cancer.

For four years, from 1980 to 1984, Rosenberg's ideas failed. He treated sixty-six patients. None experienced any benefit; all of them died. Then, in November 1984, he treated a Navy servicewoman named Linda Taylor who suffered from metastatic, terminal melanoma. Her medical file had been stamped "death imminent." Rosenberg grew billions of Taylor's T cells in a laboratory, infused these back into her bloodstream, and inoculated her with higher than typical doses of IL-2 repeatedly, hoping to put her immune system into overdrive.

It worked.

Over the course of four months, Taylor's tumors diminished, and then disappeared.

Rosenberg's success made him a brief medical celebrity. He was featured on the cover of *Newsweek* and became one of *People* magazine's "25 Most Intriguing People of 1985." But despite the hoopla, his success treating subsequent patients was mixed. He hadn't cured cancer, but he had confirmed an important principle, that T cells of the immune system could help defeat it. Though not all the time. And not, it seemed, without the help of infusing billions of reinforcement cells. If T cells could recognize some cancer cells as foreign, why didn't they do that all the time? Or at least more often? There was far more to learn and understand about how cancer eluded the immune system.

The work of a native Texan scientist named Jim Allison provided some of the answers. Allison carefully studied T cells and identified a crucial protein, the T cell receptor, that lies embedded in the cell membrane. Scientists naturally thought that an antigen binding to the T cell receptor would activate it and initiate the immune system's attack, but to their surprise, it wasn't that simple. Additional membrane proteins played important roles. One, called CD28, also needed to be activated in order to stimulate the cell. Another, CTLA-4, actually provided an *inactivation* signal to the cell—a constraint on the immune system. The body had evolved to feature a safety mechanism, what has now been termed a "checkpoint," to prevent the immune system from overreacting. A much more complicated view of the immune system was beginning to emerge. Activating a T cell to attack foreign antigens required the execution of multiple stimulatory steps, as well as the absence of inhibitory signals.

Then, experiments performed by Allison's lab team revealed a shocking new finding: that tumor cells in mice expressed a signal that stimulated CTLA-4, the inhibitory protein. Just as cancer cells altered microenvironments to their advantage by expressing VEGF to draw in blood vessels, they had also evolved to produce a signal that stimulated CTLA-4 to suppress the immune system's drive to attack it. Now, Allison realized that it might be possible to create an antibody to bind T cells' CTLA-4 receptors, blocking their normal, suppressive function, to unleash the full force of the immune system against cancer.

In 1996, that's exactly what he did. Allison's CTLA-4 inhibitor antibodies cured cancer in mice. After treatment, these animals' own immune systems became more active and attacked their tumors until they disappeared. It took many years for Allison to find a pharmaceutical company willing and able to test CTLA-4 inhibitors in human trials and bring them to market, but he eventually succeeded. In 2011, the first FDA-approved checkpoint inhibitor drug,

Yervoy (ipilimumab), was released to treat melanoma. Meanwhile, another checkpoint protein that muted the immune response had been discovered and named "PD-1."* In 2014, two new checkpoint inhibitors aimed at blocking PD-1, Opdivo (nivolumab) and Keytruda (pembrolizumab), were approved to treat melanoma and lung cancer. In 2015, ninety-one-year-old former president Jimmy Carter, suffering from terminal metastatic melanoma and tumors in his liver and brain, was treated with Keytruda after a course of radiation treatment. Within three months, all his tumors were gone. Carter was cured.

Even more success would soon follow.

Scientists have developed a way to engineer T cells with surface receptors that can specifically recognize the antigens displayed on the outside of patients' unique tumor cells. This is like creating a laser-guided, smart bomb to take out cancer cells with minimal collateral damage. Doing so requires inserting new genetic instructions into T cells (using nonpathogenic viruses as the delivery vehicle) to induce the T cells to produce a designer protein called a *chimeric antigen receptor* (or CAR). The receptors are "chimeric" because they are part antibody (the part that recognizes tumor antigens) and part T cell receptor protein. Their use is now called "CAR-T cell therapy."

Today, a patient's blood sample can be sent to a lab and, within two weeks, the lab can analyze the cancer cell's genome, identify a unique surface protein, and determine the DNA sequence required to make that protein. A virus vector carries the DNA into T cells, which then produce chimeric antigen receptors capable of targeting the cancer cells. These customized CAR-T cells, grown in the lab until they number in the billions, are then infused back into the patient where they seek and destroy the tumor. CAR-T cells have been termed a "living drug," because they multiply, remain in circulation for years, and attack tumor cells if they recur displaying the same antigen—potentially maintaining indefinite remission.

A clinical trial of CAR-T therapy struck a national chord in 2012 when

---

* In 2018, Jim Allison and Tasuku Honjo, the discoverer of PD-1, shared the Nobel Prize in Physiology or Medicine.

one of its first subjects, a terribly ill six-year-old girl with terminal leukemia named Emily Whitehead, received the treatment from Dr. Carl June at the University of Pennsylvania and was dramatically cured. Emily and her parents were interviewed on *Good Morning America* and her story was featured in a Ken Burns documentary on cancer. In 2017, CAR-T treatment medications Kymriah (tisagenlecleucel), FDA-approved to treat acute lymphoblastic leukemia, and Yescarta (axicabtagene ciloleucel), approved to treat diffuse large B-cell lymphoma, became available.

CAR-T therapy has made the once fantastical dream of a personalized cancer vaccine a reality. We are still in the early years of this immunotherapy revolution, and the treatment has not been shown to be effective against all cancers, but immunologists remain hard at work and are now extending their research beyond T cells to other leukocytes like macrophages. In addition, the advent of a new "liquid" screening test termed *circulating tumor DNA* (or CT-DNA) holds promise as a breakthrough in cancer detection. This exquisitely sensitive test detects tumor cell DNA that can be found free-floating in the bloodstream. It can be used to screen for both cancer and cancer recurrence after treatment; and since it sometimes captures a cancer's complete genome, it is conceivable that the information derived from it could guide CAR-T therapy even before there are enough cancer cells to form a tumor in the body.

JIMMY'S HOMECOMING

In 1997, Karen Cummings, a woman working at the Jimmy Fund's development office, opened an envelope that included a donation and a letter. The letter was from a woman named Phyllis Clauson, who claimed that Einar Gustafson, the original "Jimmy," was her sixty-one-year-old brother. Everyone at the Dana-Farber Cancer Institute had assumed Jimmy had died long before, as most children with cancer had. Cummings phoned Clauson to learn more. Soon thereafter, Einar Gustafson called Cummings, leaving a voice message that began, "This is Jimmy. Heard you were looking for me."

Cummings met with Gustafson and visited his home in New Sweden, Maine. There, Gustafson showed her the uniform the Boston Braves players had given him in the hospital. He talked about the fateful 1948 radio program that had successfully launched the Jimmy Fund, and recalled the long drives to Boston for treatments and follow-up visits. Gustafson had grown to six feet, five inches tall, married his high school sweetheart, had three daughters, and established a successful construction company. When the Braves left Boston for Milwaukee in 1953, he switched his allegiance to the Red Sox.

In 1998, Ralph Edwards, the former host of *Truth or Consequences*, met Gustafson in person at a game at Fenway Park, where the fiftieth anniversary of the broadcast was celebrated. When a *Boston Globe* reporter asked Gustafson how he anticipated feeling at the standing ovation the crowd was sure to give him, he replied, "I've had that before. They did that for me at the Braves game in '48, and I waved to the crowd." Gustafson was named an honorary chairman of the Jimmy Fund in 1999. He died of a stroke in 2001, at the age of sixty-five.

Since the war on cancer was declared in 1971, more than $100 billion has been spent on cancer research and prevention. Cancer deaths reached a peak in 1991 and have been declining ever since. From the 1990s to the 2010s, breast cancer deaths decreased 25 percent, colon cancer deaths decreased 45 percent, and prostate cancer mortality dropped 68 percent. The five-year survival rate for all cancers as a group is now approximately 67 percent.

Though the fight continues, the history of oncology reflects numerous incredible achievements. Cancers that were once invariably fatal but are now curable include leukemia, testicular cancer, and Hodgkin's lymphoma. Many more cancers have been rendered chronic, manageable illnesses that do not prevent patients from living full lives or only minimally affect their life spans. Prevention efforts, especially smoking cessation, save countless lives. Investigators have elucidated not only the molecular basis for cancer, as complex as that has proven to be, but also the means to deploy new weapons from another scientific discipline, immunology. In the coming decades, it is feasible to envision cancer screening by a simple blood test. If an abnormality is found, preventative treatments could be offered to prevent the formation of cancer at all.

The future is bright and, although we may never find a magic bullet that will cure all cancers, more and more people will be living longer and better lives despite their cancer diagnoses. Whereas many breakthroughs from the early decades of cancer research resulted from serendipity or trial-and-error methods, today's advances proceed in a far more intentional and strategic manner, giving us confidence that now, more than ever, we can say with conviction that the defeat of cancer will not be a matter of if but when.

6

TRAUMA

The Only Winner in War Is Medicine

June 8, 1880
Republican National Convention, Chicago, Illinois

Everyone expected Ulysses S. Grant to win.

The Civil War hero aspired to return to the presidency after a hiatus from office during the one-term Rutherford B. Hayes administration. But Grant failed. On the thirty-sixth ballot of the longest Republican national convention in history, a dark-horse candidate emerged to win the nomination. His name was James A. Garfield, a man who had not even sought the presidency at the outset of the convention. Born into poverty in Ohio in 1831, Garfield had risen far above his station to become an attorney, college president, Civil War general, and congressman. After defeating Democrat Winfield Hancock in the 1880 presidential election, Garfield became the twentieth president and was viewed as a unifying figure who might do much to heal the nation's wounds. At the outset of the Gilded Age, his up-from-nothing life story made him incredibly popular.

Taking office in March 1881, Garfield hoped to tackle one of the most difficult challenges in American politics—civil service reform. For decades, the spoils system had fueled a form of legal corruption wherein politicians rewarded their most loyal disciples with plum postings and sinecures. Garfield joined ranks with reformers who believed positions should be filled based on merit. To date, the malignancy of patronage had proven too entrenched to eradicate.

Then, like a cancer cell capable of mutating to defend itself, the spoils system managed to strike out against Garfield, the man who posed the greatest threat to its survival. It did this in the form of one man, Charles Guiteau, an educated but delusional narcissist and religious fanatic. Guiteau was an obsessive office seeker. At a time when anyone could call on the White House, he began showing up regularly, pressing Garfield's secretary with a request to be appointed Consul to Paris. After his requests were rejected numerous times, Guiteau began to despise the president and hatched a plan to assassinate Garfield. With Garfield out of the way, he felt certain the vice president, Chester A. Arthur, would be grateful to him for enabling Arthur's elevation and happily grant Guiteau any position he desired.

On July 2, 1881, President Garfield took a coach to the Baltimore and Potomac Railroad Station in Washington, D.C. As Garfield walked through the station without any police protection, Guiteau approached him from behind and fired a revolver from only six feet away.* The bullet passed through Garfield's right arm.

The president cried out, "My God! What is this?"

Guiteau immediately fired again and this bullet struck Garfield in the back. Garfield collapsed to the floor.

Chaos erupted. Guiteau tried to flee but was quickly apprehended. Garfield, conscious but in severe pain, was carried to an upstairs room of the station on a mattress. The first physician on the scene, Dr. Smith Townsend, used his unwashed finger to probe the wound in Garfield's back, which was four inches to the right of the spinal column. There was no exit wound. Garfield complained of back pain and pain in his right foot. A doctor named Robert Reyburn, who later became responsible for recording the president's full

---

* Guiteau was aware of the president's itinerary because it had been published in the newspapers.

clinical history, observed, "The President was deathly pale, almost pulseless, and apparently dying from internal hemorrhage." In all, ten physicians would examine Garfield at the train station, but the most experienced, fifty-five-year-old Dr. D. Willard Bliss, took charge.* Bliss had extensive experience caring for gunshot wounds during the Civil War, and knew Garfield personally, but he also had a reputation for arrogance. Bliss used a long, straight, non-sterile "Nelaton" probe to explore the president's wound and try to find the bullet. The probe had a porcelain ball at its tip. If the tip came in contact with a lead bullet, the bullet would leave a black mark. Bliss advanced the probe deeper and deeper into the president's back, moving it about, searching for the bullet in vain. The exploration was incredibly painful but the president did not cry out. Though one doctor protested Bliss's aggressive examination, Bliss persisted, next trying a flexible probe made of silver. In his own words, he "gently passed it downward and forward, and downward and backward in several directions, with a view of indicating the course of the ball." Unfortunately, in doing this, Bliss disrupted internal tissue layers and almost certainly created multiple false passages that would lead to adverse consequences later on.

Each doctor present believed Garfield's death was imminent. None of them thought he would last the night. Though not necessarily outside the standards of American medical treatment in 1881, practically everything about Garfield's care on that fateful day was incorrect. Carrying him on a mattress up a flight of stairs risked exacerbating spinal cord injury. The insertion of unwashed fingers into the wound was unwise. The exploration of the wound with rigid, unsterile probes was dreadful. These mistakes were especially unfortunate because, in 1881, Garfield's physicians knew about antiseptic practices. They had read about it in their medical journals for years and debated it at conferences. The shooting of Garfield would do much to reveal the cost of American physicians' arrogance, complacency, and resistance to new ideas.

And yet, in spite of his doctors' missteps, James Garfield was a determined

---

* Bliss's parents had felt so assured of his future in medicine that they gave him the first name "Doctor."

man. The strong, robust, forty-nine-year-old had never been known to fail at anything, and he was not about to start now. He had too much to live for.

A "YANKEE DODGE"

The effective management of traumatic injuries in modern times would hinge on the development of two distinct surgical advances: anesthesia and antisepsis. Prior to these, pain and poor odds of survival made surgeries rare; patients would not submit to the surgeon's knife except in dire circumstances. The gory nature of surgeons' work, and their general ineffectiveness, contributed to a centuries-old reputation for ineptitude. Surgery was considered little more than a basic trade, akin to blacksmithing or carpentry. The most common tradesmen to ply their hands at the human body were barbers, likely because they possessed the sharp implements required for bloodletting, tooth extractions, and, in wartime, amputations. In the nineteenth century, surgeons' reputations had still not significantly improved from medieval times.

Mid-nineteenth-century operating theaters were macabre. The most renowned surgeons operated at large university hospitals, where cases were performed in amphitheater-like settings. At the center of the operating "theater" lay a small circle of floor space with a worn, wooden table that was stained with the blood of numerous operations. Around this central stage rose rows of semicircular bleacher seating, rising to the rafters. Many domed theaters featured an oculus or skylight to cast light on the operating table at noon. When the sun wasn't shining, gas lights, or sometimes many candles, were brought in to illuminate the field. A thick layer of sawdust covered the floor to absorb copious amounts of blood. Surgeons did not routinely wash their hands. It didn't make sense to wash one's hands before an operation—if done at all this would logically occur after surgery. Nor were instruments cleaned; they were used repeatedly on patient after patient.

The "best" surgeons were those who simply operated fastest—for their patients, the torturous experience of an operation might at least be brief.

One of the world's most famous surgeons, Robert Liston of Scotland, earned his reputation primarily based on his speed. At the height of his career in the 1820s and 1830s, he could amputate a leg in less than thirty seconds. According to one medical historian, "the gleam of [Liston's] knife was followed so instantaneously by the sound of sawing as to make the two actions appear almost simultaneous." At the outset of cases, he sometimes self-importantly commanded his audience: "Now, gentlemen, time me!"

Liston's haste had consequences. During one operation, he worked so quickly that he inadvertently cut off a man's testicle during a leg amputation. Many of Liston's patients were terrified to fall under his care. One frightened man, suffering from a bladder stone, locked himself in a bathroom; Liston was forced to break down the door and haul him to the operating theater. Liston, more than anyone, understood the limitations pain imposed on surgery. To his credit, he would later leap at the chance to ameliorate it.

On March 30, 1842, in Jefferson, Georgia, Dr. Crawford Williamson Long became the first surgeon to employ anesthesia, by using ether to render a patient unconscious during an operation to remove a neck tumor. At the time, ether vapor, as well as nitrous oxide, were used by Long and others as recreational drugs. Long participated in so-called "ether frolics" and noticed that, in their altered, euphoric state, participants sometimes injured themselves without any sensation of pain or even recollection of the injury. He hypothesized that this unique effect might help patients endure their surgeries. At the conclusion of his historic, first operation with anesthesia, Long's disbelieving patient "seemed incredulous, until the tumour was shown him. He gave no evidence of suffering during the operation."*

Long did not publish a report about his use of ether, on this patient or in subsequent cases, until seven years later, in 1849. Meanwhile, in September 1846, a Boston dentist named William Morton used ether to aid the extraction of a patient's tooth. The success of this event was reported in the

---

* "National Doctor's Day" is celebrated on March 30th because this was the date of Long's first use of anesthesia.

press, which led Massachusetts General Hospital surgeon John Collins Warren to allow Morton to anesthetize one of his patients in the first-ever public demonstration of surgical anesthesia on October 16, 1846. At the conclusion of the successful operation the patient awoke and confirmed he had not felt any pain. At this, Warren declared, "Gentlemen, this is no humbug!" News of ether's miraculous properties spread across the globe.*

On December 21, 1846, at University College Hospital in London, Robert Liston consented to the use of ether on a patient undergoing leg amputation. To the audience that filled the operating theater, Liston announced, "We are going to try a Yankee dodge today, gentlemen, for making men insensible." With his patient unconscious, Liston proceeded to fully sever the leg in only twenty-five seconds. Within five minutes, he had finished tying off all the bleeding arteries and sewed a skin flap over the stump. The patient had not stirred. When the patient awoke, he became agitated, shouting, "When are you going to begin? Take me back, I can't have it done!" Only after seeing his amputated leg did he believe the surgery had already occurred. Liston, amazed at what had transpired, declared, "This Yankee dodge, gentlemen, beats mesmerism hollow!"†

The door to a new era in surgery had been flung open.

"THE ONLY MAN WHO HAS EVER STUCK A KNIFE INTO THE QUEEN!"

Sitting in the audience at Liston's historic operation was a student named Joseph Lister. No one knew it at the time, but this unassuming young man

---

* For decades, four men vied for credit as inventor of anesthesia: Crawford Long; William Morton; Charles T. Jackson, a physician/chemist who suggested the use of ether to Morton; and Horace Wells, a Hartford dentist who used nitrous oxide for dental procedures in 1844.
† Mesmerism was a method of hypnosis that had previously been tried to help patients endure surgery.

would be the hero to hurdle surgery's other daunting obstacle: infection. Though the advent of anesthesia gave surgeons more time to complete their operations, it did nothing to improve their rates of success, nor high mortality due to postoperative infection. In the mid-nineteenth century, as many as 80 percent of operations resulted in infection, and almost half of patients who underwent major surgeries died. Ironically, the development of anesthesia served to worsen surgical outcomes because, as operations could now proceed without pain, surgeons grew bolder and attempted far more operations of greater complexity and duration. Predictably, this increased the incidence of postoperative infections and death.

As hospitals' high infection rates contributed to their unwelcome reputation as "Houses of Death," most affluent people chose to have their operations in their own homes. In 1869, surgeon and obstetrician James Young Simpson wrote, "A man laid on the operating table in one of our surgical hospitals is exposed to more chances of death than was the English soldier on the field of Waterloo." Simpson had conducted an extensive survey of over 4,000 operations in Britain that revealed 40 percent of hospital amputations resulted in death, while only 10 percent of patients undergoing amputations outside of hospitals died. Postoperative infections were so mystifying and problematic that a Philadelphia hospital ceased performing operations from January to March, when outbreaks of erysipelas were thought to be most common. Anti-contagionists blamed such infections on miasma, or "bad air," that they hypothesized emanated from rotting flesh in wounds. When copious cleaning and strict quarantines of postoperative patients failed to stop the spread of infections in Britain, some hospitals were deemed so irreparably filthy or corrupted by miasma that they were intentionally destroyed so that new hospitals with purer air could be built in their place.

In the 1850s, Joseph Lister was a rapidly ascendant surgical star in Victorian England. He studied under the famed James Syme at the University of Edinburgh, and later became chief of surgery at the Glasgow Royal Infirmary. Throughout his early career, the problem of how wound infections spread to the bloodstream to cause sepsis and death perplexed Lister. An example of his personal case notes read: "11 P.M. Query. How does the poisonous matter get from the wound into the veins? Is it that the clot in the orifices of the cut veins suppurates, or is poisonous matter absorbed by minute veins & or carried into the venous trunks?"

Despite spending much time considering and theorizing about this problem, Lister could not deduce a solution. At the bedside of a patient with a broken arm one morning, he told his students: "It is a common observation that, when some injury is received without the skin being broken, the patient invariably recovers and that without any severe illness. On the other hand trouble of the gravest kind is always apt to follow, even in trivial injuries, when a wound of the skin is present. How is this? The man who is able to explain this problem will gain undying fame."

Lister's epiphany came in 1864.

He read about Louis Pasteur's work with fermentation and the discovery of microscopic organisms, and realized that the same microbes could be the source of wound infections. He set out to retest Pasteur's experiments and became convinced of the idea's veracity. Miasma theorists were correct that infections could be carried in the air, but it was not the air that was the culprit—it was minute organisms floating in the air. Furthermore, the causative microbes could be transmitted not just by air but also by contact, spread by physicians' fingers and instruments. Lister considered how he might employ antimicrobial substances to combat these organisms. At a medical conference, he would state, "When I read Pasteur's original paper I said to myself, 'Just as we may destroy lice in the head of a child who has pediculi, by poisonous applications which will not injure the scalp, so, I believe, we can use poisons on wounds to destroy bacteria without injuring the soft tissues of the patient.'"

These "poisons"—also known as *antiseptics*—were typically caustic solutions, or preparations of wine or quinine. They were not new. Though largely ineffective, they were commonly applied to infected wounds that appeared to demonstrate copious production of pus. A certain degree of pus was actually considered helpful and termed *laudable pus* because it was thought to represent part of the normal healing response to inflammation. But when pus production grew too abundant, such as in severe or chronic infections, doctors recognized this as a negative sign and tried to combat it with antiseptics, usually with little benefit at such a late stage.

But, Lister thought, what if antiseptics were applied as a preventative measure? To kill microbes before they could colonize a wound? Might that prevent them from flourishing and prove more useful than applying disinfectants after an infection had already taken hold?

Lister tested numerous types of caustic solutions. Some were too mild; others were too toxic and harmed normal tissue, or even promoted infection themselves by inhibiting the healing process.

Then Lister hit upon a top prospect.

He recalled that engineers in Carlisle had begun using carbolic acid to cleanse sewers and neutralize the fetid odors of pastureland fertilized with manure. It had been observed that cattle grazing this land had reduced incidence of infectious disease outbreaks. Lister surmised carbolic acid might be an excellent antiseptic.

In August 1865, in what would become one of the most famous cases in medical history, Lister encountered an eleven-year-old boy named James Greenlees with a compound leg fracture.* Greenlees had been crossing a Glasgow street when a large cart with metal-rimmed wheels ran into him and crushed his left leg. Greenlees's tibia was shattered; a jagged tip of bone protruded out of a laceration of the skin. Evaluating the boy three hours after the accident, Lister carefully considered what to do. Compound fractures usually required amputation because they were too difficult to repair and the break in the skin made infection almost inevitable—particularly in cases like Greenlees's where the wound had already become caked with dirt. The safest move would be prompt amputation. Not only was the leg likely unsalvageable, removing it might save the boy's life by reducing the risk of sepsis.

But at the same time, Lister knew the boy would forever be an invalid if

---

* Compound fractures denote when a broken bone has torn through the skin and is therefore exposed outside the body. In simple fractures, there is no break in the skin.

he lost a leg. Lister decided to put his faith in antisepsis and carbolic acid to the test. Instead of operating, he began to irrigate the wound with carbolic acid to clean it out and, hopefully, prevent the onset of infection. He set the bones, then dressed the wound with putty and placed a tin covering over it to prevent the carbolic acid from evaporating too quickly. Each day, Lister poured more carbolic acid over the dressing, allowing it to soak down into the laceration. After four days, he removed his initial dressing and inspected the wound. There was no sign of pus or infection, but the edges of the wound were red. Lister worried the carbolic acid was too caustic, so he diluted his mixture. To his relief, after several more days, the wound appeared to be healing cleanly. Six weeks later, James Greenlees's leg was fully mended and he walked out of the hospital.

The experiment was a triumph.

Lister began using carbolic acid, and eschewing amputation, whenever possible. Of the first ten compound fracture patients he treated, eight recovered without infection or loss of limb. Compared to past practice and results, this was miraculous.

Lister soon extended his use of carbolic acid antisepsis to the treatment of abscesses, superficial wounds, and surgical incisions. In such cases the incidence of infection declined dramatically; over a period of nine months in 1866-1867, not a single case of sepsis, gangrene, or erysipelas occurred in Lister's hospital wards. In an October 20, 1867, letter to his father, Lister wrote, "I now perform an operation for the removal of a tumour, etc., with a totally different feeling from what I used to have; in fact, surgery is becoming a different thing altogether." Before operations, all of Lister's instruments, every surgeon's hands, each suture, and postoperative dressings were dipped in carbolic acid. His success made him eager to disseminate his methods to others so that more lives could be saved.

Unfortunately, the reception to Lister's idea was not what he hoped. Many surgeons considered carbolic acid no different than many other antiseptics tried in the past, not recognizing that much of Lister's innovation was the practice of performing antisepsis preoperatively and in a meticulous way. The germ theory remained difficult for many physicians to swallow. They could not abide the idea that they, themselves, through their unwashed hands and instruments, played a role in their patients' infections and deaths. Many who purportedly tried carbolic acid failed to adhere to Lister's published practices, or followed them only half-heartedly. It did little good to douse a wound with antiseptic if the surgeon then used unsterile instruments to perform a surgery. And if the solution wasn't diluted properly, carbolic acid poisoning could result. A Scottish surgeon named Donald Campbell Black derided Lister's "carbolic acid mania," which Black believed was "based on false premises, and bolstered up by coincidences." One London surgeon mocked Lister by exclaiming, when someone entered his operating theater, "Shut the door quickly in case one of Mr. Lister's microbes comes in!" Ironically, Lister's system of antisepsis was adopted among surgeons in continental Europe long before those in Britain.*

In 1871, Lister gained an opportunity to do something that would prove more valuable than any scientific report in the effort to convince his colleagues, and the general public, that his antiseptic technique had merit: he operated on Queen Victoria. The queen had developed a painful abscess under her armpit. Though locked in constant battle with peers who were skeptical of antisepsis, Lister remained one of the most talented and renowned surgeons in the Empire. He was summoned to Balmoral Castle on September 4, 1871, and immediately saw that the queen's life was in danger. It was necessary to incise and drain the abscess. After anesthetizing the queen with chloroform, Lister carefully prepared the operative field with carbolic acid; and, employing the royal physician to spray carbolic acid into the air throughout the operation, he performed the surgery without complication. The queen survived. As a result, Lister's method gained far wider acceptance. Returning

---

* Not all of Lister's recommendations would prove useful. For example, he introduced a bulky, hand-pumped carbolic spray machine to aerosolize carbolic acid and spray it copiously about the operating room throughout an entire operation, thinking this would kill germs in the air. This was ineffective.

to the University of Edinburgh, where he had become the chief of surgery, he told his students, "Gentlemen, I am the only man who has ever stuck a knife into the Queen!"

In time, results would prove the value of Lister's methods. More and more surgeons across Britain saw that antisepsis significantly reduced infection rates. One assessment of published reports in the nineteenth century indicated that, before the adoption of Listerian principles, about 50 percent of amputation patients died; afterward, mortality dropped to less than 10 percent. In 1875, Lister embarked on a lecture tour of the continent. At multiple appearances in Germany he was hailed as a hero for his lifesaving breakthrough. In 1876, he was invited to speak at the International Medical Congress in Philadelphia, a meeting attended by approximately 450 physicians in the city simultaneously hosting America's Centennial Exhibition.

In the United States, Lister was surprised to find some American surgeons even more resistant to antisepsis than his British peers had ever been. Unbeknownst to him, Lister's invitation to Philadelphia by the prominent surgeon Samuel Gross had been made, in part, for the express purpose of discrediting antisepsis. Gross was a militant opponent of the germ theory of disease. In fact, many American hospitals, including Massachusetts General Hospital in Boston, had banned Lister's method, considering it unnecessary and even charlatanism. Too few American physicians had taken the intellectual leap to believe that infections were caused by tiny, invisible organisms. At the International Medical Congress, on September 4, 1876, Lister was forced to listen as multiple American physicians delivered diatribes attacking antisepsis. One after another disparaged the germ theory. Surgeon Frank Hamilton, who would later help care for President Garfield in 1881, slighted Lister by saying, "A large proportion of American surgeons seem not to have adopted the practice. Whether from a lack of confidence or for other reasons I cannot say." In a direct insult to their invited guest, Samuel Gross said, "Little, if any faith, is placed by any enlightened or experienced surgeon on this side of the Atlantic in the so-called treatment of Professor Lister."

Lister finally got his chance to deliver a three-hour lecture defending his methods but failed to gain many converts. He subsequently embarked on a cross-country tour, traveling by train to San Francisco and back, speaking at medical schools along the way to spread his antisepsis gospel.

To the detriment of countless American patients, many in his audiences remained unconvinced.

Joseph Lister, photographed circa 1855

A LANGUISHING PRESIDENT

After being shot at the Baltimore and Potomac train station in Washington, D.C., President James A. Garfield was conveyed to the White House in a horse-drawn ambulance. Dr. D. Willard Bliss, who had appointed himself in charge of the president's care, isolated Garfield in a second-floor bedroom where no one was allowed entry except with Bliss's permission. In July 1881, five years after Joseph Lister's visit to the United States, Bliss remained

TRAUMA: *The Only Winner in War Is Medicine*

strongly opposed to antiseptic methods. Covetous of the fame and honor associated with being the president's chief physician, he dismissed practically all of the doctors who had aided the president at the train station, and even barred Garfield's family physician, Jedediah Hyde Baxter, from seeing Garfield.

Although Garfield appeared near death immediately after the shooting, it soon became clear that he was not about to die. By the next day, his vital signs stabilized and, though still in significant pain and frequently vomiting, he was able to converse. Over the next several days, Bliss, who plied his patient with wine and daily injections of morphine, became optimistic that Garfield might recover. Bliss gave frequent interviews to the press and issued multiple bulletins on the president's status daily. In these, he was consistently optimistic. Just a few days after the assassination attempt, Bliss told a reporter, "I think that we have very little to fear . . . President Garfield has made a remarkable journey through this case . . . I think it almost certain that we shall pull him through." Incredibly, Bliss was also quoted as saying, with complete seriousness, "If I can't save him, no one can."

For three weeks, the president hung on, even showing signs of improvement with the return of his appetite and a request to smoke a cigar, which was denied. Then, on July 23, Garfield's condition abruptly changed. His wound spontaneously discharged a piece of bone and some fragments of clothing that had been carried into his body by the bullet. Garfield's temperature rose to 104 degrees. A "pus sac" was discovered in his back. By now, Bliss had sought the help of two prominent surgeons, David Hayes Agnew and Frank Hamilton, who incised the wound on July 24 and inserted a large drainage tube. Two days later, the wound was explored, drained, and dressed again; during this procedure bone fragments, one an inch long, were removed.

Infection began to slowly overwhelm Garfield's body. Bliss believed the infection stemmed from the bullet lodged in the president's abdomen, and the location of the bullet was considered of crucial importance. Bliss judged it to reside in the lower right quadrant of Garfield's back, though he had been unable to locate it after multiple probings. Tests were performed in which bullets were fired into the backs of cadavers to see if this might help determine where the bullet lay. Alexander Graham Bell, the famed inventor, tried to help by devising a rudimentary metal detector called an "induction balance," but two attempts to use it to find the bullet were unsuccessful.*

Meanwhile, Garfield continued to languish. Multiple, non-sterile drainage tubes were repeatedly inserted into, and removed from, his abdominal cavity over several days. Dr. Agnew incised and drained another abscess on August 8, at first believing he was dealing with an infection in a cavity formed by the path of the bullet, but that instead had resulted from the numerous, deep probings of Garfield's physicians. Suffering unrelenting nausea and vomiting, Garfield grew increasingly undernourished and dehydrated. His weight dropped from 210 pounds to 135 pounds. Bliss resorted to attempts to provide nourishment via enema, injecting mixtures of beef broth, milk, whiskey, egg yolk, and tincture of opium into the president's rectum, hoping some of the nutrients might be absorbed.

And yet, the president fought to live. Always retaining his good humor, often expressing profuse gratitude to those trying so hard to help him, Garfield exemplified the strength of character that had defined his life. He endured Agnew's operations without complaint. Some days, Garfield seemed to stabilize, and even expressed slight improvement in one of his symptoms, but overall, there was gradual, inexorable decline. By August 25, Garfield was a mere shadow of his former self. Inflammation of the parotid gland rendered half his face paralyzed. Infection became so widespread that pus drained freely out of his ear. He sometimes lost consciousness and began to have hallucinations.

On August 30, a probe inserted into the original wound was "easily passed downwards to the extent of twelve inches" in the direction of the right groin. If Bliss realized this cavity was not due to the bullet's path, he did not

---

* Röntgen's discovery of X-rays would not occur for another fourteen years.

permit the knowledge to alter his opinion of the bullet's location. As the nation continued to pray and anxiously awaited each updated bulletin reporting the president's condition, Garfield managed to hang on through the end of August and into September, though he and everyone else were beginning to realize it would take a miracle for him to survive.

For days, Garfield had grown increasingly insistent that he be allowed to leave the stifling confines of the White House. On September 5, he was transported by train to Elberon, New Jersey, where 2,000 people had worked through the night to install 3,200 feet of track from the train depot to the door of the seaside home he would occupy. Two weeks later, on September 19—seventy-nine days after being shot—Garfield succumbed to overwhelming sepsis and died.

An autopsy revealed that Bliss had been terribly wrong about the location of the bullet. Instead of lying on the president's right side, it had ricocheted leftward off one of Garfield's ribs, perforated a lumbar vertebrae, and come to rest on his left side, behind the pancreas, where the bullet was found encapsulated by fibrotic tissue. Multiple pus-filled cavities were found in the abdomen. The immediate cause of death appeared to be the rupture of a splenic artery aneurysm. It is possible that the bullet came in tangential contact with the splenic artery and damaged it, or the rupture could have been unrelated to the bullet and due to the president's dire overall medical condition, deranged by sepsis, malnutrition, dehydration, and multiple organ failure.

Medical historians have studied Garfield's case for over a century. Some have proposed alternative, contributory causes of death, including heart attack, pneumonia, and cholecystitis. Whatever the ultimate cause—and there could have been several acting in concert at the end—what is clear is that Garfield's overwhelming infection was a decisive factor in his demise. The consensus of most medical historians is that Garfield's bullet wound did not kill him, but that infection promoted by his physicians' repeated, unsterile examinations did. As experienced military surgeons, Bliss and his colleagues knew that men could live with bullets lodged in their bodies—many Civil

War veterans did. Once it was clear that Garfield's vital signs were stable, the search for the bullet should not have become the obsession that it did. In retrospect, it would have been better to simply cover the wound, manage pain, maintain hydration, and avoid any internal intervention.

Not surprisingly, Willard Bliss was roundly criticized for his handling of the president's care. His disdain for sterile technique, imperious manner, and overly optimistic public bulletins seemed to exemplify all that was wrong and backward about American medicine and its practitioners. One physician, the publisher of a popular periodical called *Walsh's Retrospect*, wrote that Bliss had done "more to cast distrust upon American surgery than any time heretofore known to our medical history." It became popular to quip that the lesson from the entire sad affair was simply that "ignorance is Bliss."

The idea that Garfield's doctors caused his death became so widespread that even the assassin, Charles Guiteau—who pleaded not guilty by reason of insanity—argued that his gunshot did not kill the president; medical malpractice did. At the outset of his trial, Guiteau wrote in a statement, "General Garfield died from malpractice . . . according to his own physicians, he was not fatally shot. The doctors who mistreated him ought to bear the odium of his death, and not his assailant. They ought to be indicted for murdering James A. Garfield, and not me." It took a jury less than an hour to find Guiteau guilty on January 26, 1882. He was executed by hanging on June 30, two days shy of the one-year anniversary of Garfield's shooting.

Though Garfield's six-month presidency was brief, the consequences of his assassination were significant. Chester A. Arthur, previously mocked by many as a political tool, found the courage to pursue the civil service reforms Garfield had endorsed, culminating in the Pendleton Civil Service Reform Act of 1883. Medically speaking, the debacle of Garfield's care helped American physicians finally recognize the importance of antiseptic technique, a key step that would spark the dawning of American surgical innovation, led by pioneers like William Halsted. Though Americans were late to accept Lister's methods, once adopted, the stage became set for the United States to become a focus of surgical advances in the twentieth century.

A BATTLE BETWEEN BLOOD SUPPLY AND BEAUTY

The twentieth century is replete with the stories of medical advances made by long-suffering, maverick doctors who, through intelligence, serendipity, and perseverance, made astonishing progress against humankind's deadliest scourges, including heart disease, diabetes, infectious disease, and cancer. But the discipline of trauma surgery is different. In this arena progress did not predominantly come from years of dedicated research, timely epiphanies, better organization, or brilliant nonconformists. Only one thing precipitated the most dramatic advances in trauma treatment.

War.

Two catastrophic world wars produced a quantity and degree of trauma that dwarfed that of prior human conflicts. But not every kind of injury was novel. As in prior wars, soldiers suffered gunshot wounds, stabbings, hemorrhaging, and shattered bones. Such casualties benefited from decades of slow but undeniable improvements in triage, surgical technique, speed of care, and infection control—all advances that would still have eventually been made in civilian medicine, albeit at a slower pace. What the world wars truly bequeathed to doctors was the opportunity to confront new, horrific types of injuries they had never encountered before on a large scale. The history of this time vividly validates the adage: "The only winner in war is medicine."

The predominance of trench warfare in World War I was largely responsible for that conflict's gruesome new medical challenge: terribly disfiguring facial injuries to tens of thousands on both sides. Soldiers who might peer over the top of a parapet, or risk a quick glance into no-man's-land, exposed only their faces to enemy gunfire and ferocious artillery bombardments that filled the air with razor-sharp metal fragments that could tear flesh and bone away in an instant. Men with fractured jaws or missing cheekbones crowded casualty stations. Others had no mouths, noses, or eyes. Sometimes, entire faces were shot away or burned beyond recognition. An examination of 48,000 British Army casualty records showed that 16 percent suffered face, head, and neck wounds. Surgeons had little to no experience treating such injuries. And yet, these mutilated men were sometimes otherwise fully alive and without further injury. Those who survived were transported to hospitals in England and could expect to live long lives. It was their disfigured faces that would, oftentimes, prove far more limiting than the loss of a limb or the ability to walk. This was the state of affairs when a surgeon from New Zealand took it upon himself to observe, learn, innovate, and help create an entirely new surgical specialty.

In 1914, at the outset of World War I, thirty-two-year-old Harold Gillies was an otolaryngologist—a surgeon specializing in the ear, nose, and throat—in London. Originally from the small town of Dunedin, on the southeastern coast of New Zealand's South Island, Gillies had studied medicine at Cambridge and was on his way to building a successful private practice. After Germany invaded France, he volunteered to join the Red Cross and Royal Army Medical Corps. He was soon sent to France as a general surgeon and assigned to oversee a French American dentist named Charles Valadier.

As a mere dentist, Valadier was not licensed to perform surgery without a qualified surgeon like Gillies overseeing him, but it quickly became clear that the army desperately needed Valadier's skills working on wounds of the lower face. Already, soldiers with terrible facial wounds had begun to arrive at aid stations in great numbers. Valadier tried his best to patch them up, experimenting with autologous tissue transplants and bone grafting. Gillies had never seen anything like it, and he was captivated.

Convinced of the immense need for new methods of handling facial trauma, Gillies traveled to Paris to observe a French surgeon named Hippolyte Morestin, who had gained a reputation for performing reconstructive surgeries. Amazed, Gillies later wrote, "I stood spellbound as he removed half of a face distorted with a horrible cancer and then deftly turned a neck flap to restore not only the cheek but the side of the nose and lip . . . it was the most thrilling thing I had ever seen. I fell in love with the work on the spot." Gillies became convinced that the British Army needed someone like Morestin and decided it should be himself. In January 1916, he returned to England and gained approval to form a 200-bed unit for facial trauma at the Cambridge

Military Hospital, in Aldershot. He began to recruit surgeons and dentists from across the U.K. To spread the word about their unit, he bought a stack of luggage tags, labeled them with his hospital's address, and shipped them to casualty stations in France. Within a few weeks, wounded soldiers began to arrive wearing Gillies's luggage tags fastened to their uniforms.

On July 1, 1916, one of the largest and bloodiest battles in world history commenced—the Battle of the Somme. This battle would last until November and result in over one million casualties, including more than three hundred thousand deaths. In the first day alone, the British Army suffered 60,000 casualties, including 21,000 killed. Harold Gillies's hospital soon received 2,000 dreadfully disfigured men. In a letter to a colleague, Gillies later wrote, "There were wounds far worse than anything we had met before. Men without half their faces; men burned and maimed to the condition of animals." No British surgeon had ever confronted anything like this.

Gillies got to work. He learned it was unwise to attempt primary closure of large facial wounds—stretching surrounding skin too much to cover sizeable gaps led to terrible distortion, scarring, and contractures. Where noses, cheeks, mouths, and eyelids used to be, Gillies confronted gaping holes. New tissue and skin would be required to fill in and close such cavities.

But how?

The answer was to transport it in from elsewhere.

Gillies employed old—and experimented with new—methods of moving tissue in and around the face using skin flaps and fragments of cartilage and bone from the ribs, tibia, or pelvic crest. The limiting feature of any skin flap was its blood supply. It did no good to move a piece of skin without maintaining an adequate source of blood to keep it alive. It was impossible to cut a large patch of skin from part of the body and simply sew it to the face—the graft's absence of nourishing blood supply would cause it to die like a piece of sod drying out in the sun before it had the chance to establish roots. Therefore, Gillies had to cut skin flaps in a way that maintained constant blood perfusion. He did this by always keeping a part of the flap connected to the body. If he cut three sides of a rectangle, and kept the fourth side attached to the body like a hinge, the entire patch would remain alive and nourished with blood.

Gillies's simplest skin flap was an advancement flap, in which three sides would be cut, and the skin stretched over a slightly greater area.* Another method was a rotation flap, where, instead of being stretched forward in one direction, it was stretched in an arc. A transposition flap was more complex, involving a section of skin—still attached at one end to maintain blood supply—being moved to an entirely new position to cover a defect. This left an open wound bed at the donor site that was ideally amenable to direct closure with sutures, or could be covered by a simpler advancement or rotation flap.

To reconstruct a nose, for example, Gillies could transpose a skin flap down from the forehead and cover the forehead donor site using an advancement flap. In cases where the nose was completely missing, he might start by implanting a small piece of shaped cartilage under the forehead skin, in the triangular form of a nose, and wait several weeks for the graft to become fully integrated with the surrounding tissues. Meanwhile, the patient would appear to have a prominent, nose-like bump protruding from his forehead. Later, this new "nose" would be part of a flap that was rotated down into the proper position. Subsequent surgeries might be performed to trim excess tissue, form the nostrils, and shape the nose for better cosmesis.

Since no two injuries were alike, Gillies and his colleagues often had to improvise. In a patient who had lost his lips in a shell burst, Gillies's associate, Henry Pickerill, brought down a forehead flap that included part of the hairline, so that the hair would continue growing on the upper lip as a mustache to hide the extensive scarring underneath. Because there were clear limitations to how far a skin flap could be stretched or moved, quite often, treatment of severe injuries required multiple—sometimes over twenty—carefully planned surgeries over months or years.

---

* Today, many plastic surgeons follow a general guideline that a skin flap should not be longer than three times the width of its base—the edge that remains connected to the body.

TRAUMA: *The Only Winner in War Is Medicine*

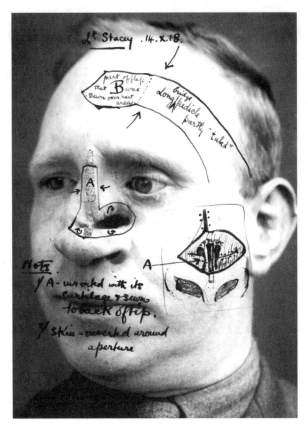

1918 photograph of a wounded serviceman, with facial drawings showing Harold Gillies's plan to stage repair of trauma

The influx of Somme casualties convinced Gillies and others that a larger facility was required to meet the army's needs. In August 1917, an entire hospital dedicated to facial and reconstructive surgery was opened, the Queen's Hospital at Sidcup, in Kent. Between 1917 and 1921, 11,752 surgeries on 8,749 patients would take place at Queen's Hospital. There, Gillies introduced a multidisciplinary approach. His team included physicians, anesthesiologists, radiologists, dentists, mask-makers, photographers, and even an artist—Henry Tonks, a former surgeon who switched careers and painted many of Gillies's patients before and after their surgeries. The horrific injuries of one man, twenty-year-old able seaman Willie Vicarage, would prompt

Gillies to develop one of his most important contributions to surgery—the tubed pedicle flap.

Vicarage had been badly burned from an explosion in the Battle of Jutland, the largest naval battle of the war. Examining Vicarage, Gillies noted, "This poor sailor was rendered hideously repulsive and well-nigh incapacitated by terrible burns . . . In addition to the total facial burns . . . the ears and neck were burnt; and the hands were contracted into frightful deformities." Because Vicarage's entire face and neck were severely damaged, there was no easy place from which to draw skin to reconstruct the face. Gillies's only option was to try a skin flap from farther away—the upper chest. Leaving one edge of the flap attached at the chest to maintain blood supply, he planned to attach the opposite end to the patient's face. As he raised the flap, he noticed that its sides had a natural tendency to curl inward. This sparked an epiphany. What if he joined the sides of the flap to create a tube? That would keep the flap's raw underside healthy, moist, and protected from the outside air. Once the skin at the face had taken root, he could then cut the flap free from the chest, unroll the tube, and use the skin at the face to reconstruct the nose.

This is exactly what he did. In other patients, he brought skin from the arm to the face. The tubed pedicle flap opened up dramatic new possibilities in reconstructive surgery, greatly augmenting the size of viable skin flaps and the distance they could travel. Without fully realizing it, Gillies was inventing a brand-new surgical specialty, and today it is fitting that he is remembered as the modern-day "father of plastic surgery." One of Gillies's most famous aphorisms was the observation that "plastic surgery is a constant battle between blood supply and beauty."

After the war, Gillies opened a private practice, wrote an important textbook on plastic surgery, and advanced the techniques he had developed by lecturing abroad. His wartime contributions were recognized with a knighthood in 1930. He could not have guessed that another, even larger conflict would soon consume Europe, one that would severely test the limits of the discipline he had pioneered.

TRIAL BY FIRE

August 31, 1940
Southeastern England, RAF Kenley

Thirty-one-year-old Royal Air Force pilot Tom Gleave was worried. Assigned to the 253 Squadron's emergency reserve group, he was currently grounded while most of his comrades engaged Luftwaffe bombers crossing the Channel. Now Gleave anxiously counted the Hurricanes returning, watching for friends. Several planes were badly damaged—the group had gotten into a serious clash with the Germans. To Gleave's dismay, squadron leader Harold Morley Starr did not return. Starr was a jolly fellow—well liked and a gifted pilot who never failed to put duty first. Gleave's unease grew. He hoped Starr had somehow put down somewhere safe. Gleave sought out the base commanding officer, who delivered shocking news.

Starr had bailed out of his plane and was descending by parachute when a German fighter opened fire on him and killed him.

Gleave was devastated. And furious. He recalled, "I went back to the dispersal point and broke the news to the Squadron. Nothing can describe their feelings . . . Tempers were raised to white heat . . . Their determination to smash the Hun now knew no bounds."

Now Tom Gleave was squadron leader. He deserved the position; he was already an ace—to date, he'd been credited with shooting down five Messerschmitt Me 109s, and one Junkers Ju 88. Less than two hours later, the order to scramble rang through the base. A large group of enemy bombers was heading toward Biggin Hill airfield. Gleave sprinted to his Hawker Hurricane, rumbled out across the tarmac, and lifted into the sky. As the squadron raced toward Biggin Hill, the sun shone brightly, the sky a clear, pale blue.

Suddenly, Gleave saw the enemy, an armada of Junkers Ju 88 bombers flying high above him. He climbed and closed the distance rapidly. In his first pass he singled out the fifth bomber from the front and let loose with all eight of his wing-mounted Colt-Browning machine guns. To avoid colliding, he dove away, then rose again to attack the third Junkers in line, hitting the bomber's port engine, which began to smoke. He next set his sights on the lead bomber, which was about to begin its bombing run, but at that moment, Gleave "heard a metallic click above the roar of my engine." An enemy bullet had struck the fuel tank in the root of his starboard wing. Gleave felt a sudden flash of heat and looked down. "A long spout of flame was issuing from the hollow starboard wing-root, curling up along the port side of the cockpit and then across towards my right shoulder," he later recalled. He tried to rock, then slow his plane, hoping this might somehow extinguish the fire, but the flames only grew. The cockpit was "like the centre of a blow-lamp nozzle . . . the skin was already rising off my right wrist and hand, and my left hand was starting to blister."

He had to get out.

Gleave undid his harness and ripped off his helmet. Now the heat was unbearable, the pain excruciating. Practically his whole body was engulfed in flame. He tried to lift himself out of his seat but lacked the strength. Frantic, his only chance was to open his canopy and then roll the plane, allowing himself to drop out. He slid the canopy back.

Then, in a blinding flash, his plane seemed to explode.

Gleave was somehow thrown clear. He found himself falling through space, tumbling head over heels. Managing to pull his rip cord, his parachute opened, and he drifted to the ground.

Landing in an open field, he staggered to his feet. Assessing his injuries for the first time, he saw his pants had completely burned away except for small fragments that had been covered by his parachute harness. In his memoir, Gleave recalled, "The skin on my right leg, from the top of the thigh to just above the ankle, had lifted and draped my leg like outsize plus-fours . . . Above each ankle I had a bracelet of unburnt skin: my socks, which were always wrinkled, had refused to burn properly . . . the skin from my wrists and hands hung down like paper bags."

The farmhand who found Gleave must have been horrified at the sight of him. The skin of Gleave's head, neck, and arms was severely burned and sloughing off. His nose was practically gone, his eyes mere slits. The farmhand put Gleave on his back and carried him to the farmhouse, where the farmer's wife put Gleave on the bed in their best bedroom. All Gleave could do was apologize profusely for muddying their beautiful, clean sheets.

When burns exceed 30 percent of our total skin area, they are potentially fatal. More than 50 percent of Tom Gleave's body had been burned. He was one of hundreds of RAF pilots in the Battle of Britain, and thousands during the war, who suffered grievous burns from cockpit fires—a condition so common that it became known as "airman's burn." Gleave was first taken to Orpington General Hospital, where orderlies used a wheelbarrow to bring him into the building. Doctors covered his charred skin with gentian violet and tannic acid, chemicals thought to prevent infection and promote tissue growth, but that actually hindered the healing process in extensive burns like Gleave's. Also unhelpful were the dry gauze bandages used to wrap his skin; these clung to his wounds, pulling off more skin whenever they were removed and replaced.

When Gleave's wife saw him for the first time, she had the composure to say, "What on earth have you been doing with yourself, darling?"

He answered, "I had a row with a German."

Gleave endured weeks of pain from his burns, which became infected. In late October 1940, he was transferred to a specialized burn unit at Queen Victoria Hospital in East Grinstead. There he met the "Maestro."

WALTZING

The "Maestro" was a New Zealand surgeon named Archibald McIndoe, who happened to be a distant cousin of Harold Gillies. McIndoe was passionate, headstrong, quick-tempered, and brilliant. After training in abdominal surgery at the Mayo Clinic in Minnesota, McIndoe moved to Britain in 1930. Under Gillies's influence, he gravitated toward the nascent field of plastic surgery. After the outbreak of the Second World War in September 1939,

McIndoe was put in charge of the Queen Victoria Hospital in East Grinstead, about forty miles south of London, which became the center of care for RAF casualties.

Unlike Gillies's experience in World War I, when facial trauma from bullet and shrapnel wounds constituted a new challenge, McIndoe found himself confronting a different problem—severe burns. In England during the war, about 4,500 Allied airmen survived aircraft fires in crash landings or by bailing out midair. Eighty percent of these men suffered burns of the face and hands. During the Battle of Britain, most of McIndoe's patients were fighter pilots, but later in the war his hospital was filled with wounded bomber crewmen.

At the outset of hostilities, both Allied and Axis physicians were largely ignorant of effective burn care treatment. McIndoe introduced several innovations. To clean and disinfect the men's wounds, he discarded the old tannic acid treatment and instead put men in warm saline baths because burned pilots who had gone down in the salt water of the English Channel were observed to have fewer infections and better skin condition than those who had parachuted over land. In the salt baths men were able to freely flex their joints, which helped reduce contractures and scarring, and bandages simply floated off wounded skin atraumatically. After the baths, the men dried off by standing naked in front of large heat lamps because toweling off would abrade the skin. Their wounds were dusted with sulfonamides; and, instead of dry gauze, their dressings were coated with Vaseline so they wouldn't stick and tear away fragile, healing skin.

McIndoe next approached the daunting task of reconstructing the men's horribly burned faces. This was a gigantic challenge. Unlike Gillies, who often had access to areas of normal skin at nearby parts of the face, neck, or chest, many of McIndoe's pilots had burns affecting over half their body surface area. How could skin be grafted to the face if there was no viable skin within reach?

Here, McIndoe took Gillies's development of the tubed pedicle much further than had ever been routinely done before. For a pilot whose closest healthy skin (that was also pliable enough to be used for the face) might be at the thigh, McIndoe could raise a skin flap from the thigh and first attach it to the arm. Using Gillies's basic method, he would cut three edges of a large skin flap (keeping one edge hinged like a trap door to maintain the flap's blood supply), wrap it in a tube so the underside was protected, and sew the free end to a pocket made by an incision in the arm. During the healing period of several weeks, the patient's leg was thus tethered to his arm in this unnatural state. After the blood supply to the flap at the arm had been established, McIndoe could sever the flap at the thigh and swing that end up to the face. This technique of marching a tubular flap of skin up the body, end-over-end, was known as "waltzing."

At any particular time, several men at the Queen Victoria Hospital could be seen in contorted positions, their heads down and bent at an angle, nose attached to arm by a tube of skin. Finally, after the connection at the face established adequate blood supply, the flap could be cut at the arm and used to help reconstruct the face.

When Tom Gleave arrived at McIndoe's Ward Three at Queen Victoria Hospital, he went into the saline bath and, for the first time in weeks, felt pain-free. He watched his bandages float away and reveled in the feeling of moving his arms and legs. His burns were re-dressed using Vaseline-coated bandages that were sprinkled with fluid every two hours with a watering can to keep his skin moist. Early in Gleave's hospitalization, the Minister of Aircraft Production, Sir Stafford Cripps, visited the ward. Upon meeting Gleave and viewing his horribly wounded face, which was partially devoid of skin and displayed exposed bone, Cripps fainted and collapsed to the floor. Such visitors would never forget the unnatural sights in McIndoe's section of the hospital.

"One day," Gleave remembered, "the Maestro toured the ward and I met him for the first time. He stood over me looking at me through those horn-rimmed spectacles with a clear candid gaze I have never seen in the eyes of any other man. It gave one immediate strength. You had an instinctive feeling that this man was not going to fool you. He weighed up my face, and then said the first facial grafts I needed were eyelids, top and bottom . . . Your eyes will be covered for a week. You won't like it, but it's worth it." Gleave added, "It was just the way he said it. I believed it. I felt like a shipwrecked man who has been floating around in the sea for hours when a lifebelt suddenly comes by."

In January 1941, weeks after Gleave's successful eyelid reconstruction, McIndoe came to him again. This time, he declared, "You need a new nose. What about it?"

Gleave replied, "I'll do whatever you suggest."

The same day, all the hair on the front half of Gleave's head was shaved so that McIndoe could harvest enough tissue to perform a forehead flap and reconstruct Gleave's nose. In all, between 1940 and 1953, Gleave would endure at least ten operations. Along the way, he was able to return to the RAF, not as a pilot but as a commander with his feet firmly planted on the ground.

Meanwhile, McIndoe's efforts in the operating room were matched by another innovative aspect of his treatment that was completely nonsurgical: the realization that his patients' psychological health was as important as their physical health. Many men, understandably distressed at their horrible wounds and appearance, found it difficult to go on and many contemplated suicide. McIndoe once explained to a friend, "Imagine how they feel . . . On Friday night they are dancing in a nightclub with a beautiful girl and by Saturday afternoon they are a burned cinder . . . One minute has changed him from a Don Juan into an object of pity—and it's too much to bear."

As a result, McIndoe introduced many morale-boosting adjustments. Flowers and live music were common at the ward. McIndoe purposely assembled a group of extroverted staff and lively young nurses who would buoy the men's spirits. Rank among his patients was ignored. The lowest-ranking private was treated and managed the same as any officer. This equal standing, and a steady supply of free beer, improved camaraderie among the patients, who eventually established a unique club for themselves called the "Guinea Pig Club." The guinea pigs took great pleasure in perfecting their gallows humor and playing practical jokes. By the end of the war, there were 649 members of the Guinea Pig Club, and Tom Gleave was elected "First Pig," its leader. McIndoe, who was crowned the club's first president, described it as "the most exclusive Club in the world, but the entrance fee is something most men would not care to pay and the conditions of membership are arduous in the extreme."*

Members of the Guinea Pig Club

McIndoe also relied on the compassion of East Grinstead's residents to help his patients readjust to civilian life. When patients ventured into town, citizens took great pains to accommodate them. Restaurants and pubs removed all mirrors. Families invited the airmen into their homes for tea, and the town soon became known as "the town that never stared." It wasn't unusual to see men with long, tubed pedicles hanging from their faces at a bar, at the movie theater, or playing a game of soccer in public. In time, East Grinstead became famous for the way it cared for the RAF's wounded warriors. A November 1943 *Reader's Digest* article highlighted the town's unique role: "The first time you see one of these boys the blood goes out of your face and your stomach rocks . . . But the good people of East Grinstead stop these chaps in the street and chat with them . . . not even the children stare at them. One obvious shudder might undo weeks of excruciating work at the QVH. So in East Grinstead the most ghastly burned boy is the most welcome. His

---

* After McIndoe's death in 1960, HRH Prince Philip, Duke of Edinburgh, served as president of the Guinea Pig Club. The club's last reunion took place in 2007.

face is the job of the hospital, but his will to live is a job that is in the hands of the townsfolk."

Archibald McIndoe was later knighted by the Queen. In 1958, two years before his death at the age of fifty-nine, he acknowledged how far the field of plastic surgery had come in a speech to the Royal College of Surgeons: "We have now arrived at the time when . . . we can within a reasonable time create order out of chaos and make a face which does not excite pity or horror. By doing so we can restore a lost soul to normal living." McIndoe became the only civilian ever to have the honor of burial at the RAF church of St. Clement Danes in London.

PUSHING THE ENVELOPE

In the decades after the Second World War, surgeons' capacity to improve on McIndoe's wartime advances continued apace. In McIndoe's day, the idea of taking large flaps of skin completely off the body and transplanting them to other, distant locations was science fiction. He might have been able to transplant very small pieces of skin, to reconstruct a patient's eyelids, for example, because tiny pieces of transplanted skin might quickly re-establish enough blood flow to survive; but the so-called "free flap" was impossible. There was no way a surgeon could dream of reconnecting all the myriad blood vessels underneath a large rectangle of skin at a recipient site before the skin flap died from ischemia.

Then surgeons realized they didn't have to. In the 1970s and 1980s, an Australian surgeon named Ian Taylor used radiopaque dye to study the subcutaneous vasculature of over 2,000 cadavers and realized that the pattern of blood vessels supplying the skin was neither homogeneous nor random. Instead, there were key, "source" arteries that consistently perfused certain regions of the skin. From these larger trunk vessels spread tributaries that endlessly arborized as their size diminished to the level of microscopic capillaries. "The body is a three-dimensional jigsaw puzzle made up of composite blocks of tissue supplied by named source arteries," Taylor wrote in his landmark article. He realized that keeping an entire piece of transplanted skin alive required only the prompt identification and reconnection of the critical source artery and vein at the graft site. Find the right artery and vein, and it was possible to keep a large skin and tissue flap alive by connecting only two blood vessels to their counterparts. These units of tissue supplied by consistent patterns of vasculature came to be called *angiosomes*.

But even these crucial trunk vessels were often very fine, sometimes almost as thin as angel-hair pasta. How could surgeons possibly reconnect them?

The ability to join the ends of two flimsy, delicate blood vessels (of any size) has always been considered a challenge. A successful connection would not only have to be watertight, but also done in a way that preserves completely smooth blood flow, lest undue turbulence occur and promote formation of blood clots. The surgeon who discovered a method to accomplish this would change the future of surgery forever. His name was Alexis Carrel.

In 1894, Carrel was a medical student in Lyon, France. That year, France's president, Sadi Carnot, was visiting the city when he was stabbed in the abdomen by an Italian anarchist at a public appearance. The dagger severed Carnot's portal vein, a major vessel below the liver, and he bled to death on the operating table. This event had a lasting impact on Carrel, who recognized that if surgeons could find a way to reconnect blood vessels, many lives might be saved.

Carrel, whose family owned a lace factory, prepared himself for the challenge of joining blood vessels by first learning precise stitching skills from a master of embroidery. He became so expert at using tiny needles and fine silk thread that he could reportedly fit 500 identical stitches in a small piece of cigarette paper. He soon moved on to experiments sewing blood vessels together in dogs.

Carrel's breakthrough technique, termed *triangulation*, came from his inspiration to first join the ends of two vessels using three stay sutures evenly spaced around the circumference of the connection. Leaving long threads of suture attached at these three points, an assistant could gently exert traction on the strings and draw the round shape of the vessels into a straight line between the stay sutures. In doing so, Carrel changed the circular, cross-sectional shape of the vessels into a triangle. Now the joined edges of tissue were positioned in a flat, straight line, making it easy to sew the vessels together, from point to point, in three sections. Carrel's simple and ingenious method overcame the difficulty and awkwardness of suturing two flaccid blood vessels along a curve.*

The advent of the operating microscope greatly enhanced surgeons' ability to anastomose smaller and smaller blood vessels in the second half of the twentieth century. This allowed vascular, cardiothoracic, and plastic surgeons to accomplish feats their predecessors could have only dreamed of. Taylor's angiosome concept soon bequeathed surgeons the ability to transfer large free flaps all over the body, leading to incredible advances in reconstructive surgery. Perhaps inevitably, such progress would eventually lead to an astonishing achievement: the full face transplant.

In March 2011, at the Brigham and Women's Hospital in Boston, plastic surgeon Bohdan Pomahač led a team that performed the first full face transplant in the United States. The recipient was Dallas Wiens, a man from Texas who, in 2008, had been up in a cherry picker helping to paint the side of his church. When Wiens accidentally maneuvered into a nearby high-voltage power line, the wire touched his forehead and he was electrocuted. Wiens's face was instantly obliterated.

After twenty-two surgeries and months of healing, the front of Wiens's head was covered by a featureless, uniform mask of skin that had been transplanted from his back. There was no face at all. His eyes, nose, and lips were completely gone; his mouth was a mere slit through which he could drink

---

* Carrel won a Nobel Prize in 1912. He later made pioneering contributions in the field of organ transplantation; however, his legacy was marred by his strong endorsement of eugenics. During World War II, he ran a research institute in Vichy France and was accused of being a Nazi sympathizer and collaborator. He died in 1944, before a trial could occur.

through a straw. Without lips, it was difficult to keep food in his mouth. Though permanently blind, he could still hear others' shocked gasps at the sight of him, or perhaps worse, the abrupt silence when he entered a room. Yet through these trials, Wiens retained an incredibly positive attitude. His optimistic personality made him a uniquely appropriate candidate for an experimental and potentially transformational surgery. After meeting Dr. Pomahač, Wiens ultimately decided to pursue a face transplant for the sake of his four-year-old daughter, Scarlette. He wanted her to see him with a face. In an interview, he said, "I could not bear the thought of her growing up and being asked questions, 'Why does your daddy look different?' And dealing with that all of her childhood."

How does a physician approach a donor's family members to discuss transferring their loved one's face to another human being? In those challenging moments, when the donor is brain dead, the physician may explain how the face could dramatically improve the life of another person. The family will learn that when the face is transplanted to the recipient, the recipient will look nothing like the donor because the face will change as it melds to the recipient's unique underlying tissue and bone structure.

Another major ethical consideration is whether, for a non-life-threatening condition, recipients should be sentenced to using immunosuppressive drugs, necessary to prevent graft rejection, for the rest of their lives, when such drugs are likely to shorten their lifespans.

When Wiens received word that a donor face had become available, he quickly traveled to Boston where his surgery took seventeen hours and involved over thirty team members.* It was important that the donor face be transplanted as soon as possible, yet the completion of Pomahač's harvest had been postponed until the donor's heart, urgently needed by a recipient patient, was removed first. Ideally, a transplanted face will not be without blood flow for more than four hours, which is exactly how long it took Wiens's

---

* Wiens's surgery was funded by the U.S. military, which hoped it might help lead to advances that would benefit future wounded service members.

new face to be harvested, transferred to Brigham and Women's Hospital, and brought into the operating room where surgeons began to carefully reconnect the arteries, veins, and nerves that would improve his ability to breathe, eat, and speak, as well as help him regain more normal sensation and taste.

Wiens's surgery was a resounding success. Pioneers like Harold Gillies and Archibald McIndoe would have found this achievement truly astounding. The face Wiens received looked far more human than the featureless tissue that had been previously present, and he recovered significant nerve function which allowed him to smile, frown, and feel his daughter's kisses again. He even regained his sense of smell. Today, wearing sunglasses, he can walk into a restaurant without strangers noticing or staring. When asked in an interview what message he would want to give to the anonymous donor's family, Wiens said, "I wouldn't even know where to start. What they did has given me a new life."

A DIFFERENT CENTURY, A DIFFERENT PRESIDENT

On March 30, 1981, almost exactly one hundred years after James Garfield's shooting, President Ronald Reagan exited the Hilton Hotel in Washington, D.C., at 2:25 p.m., after delivering a speech to the AFL-CIO federation of labor unions. As he was in the process of walking the thirty feet from a hotel door to his limousine, a reporter shouted, "Mr. President," and Reagan turned, lifting his arm to wave to the crowd gathered behind the rope lines, about fifteen feet away. At this moment, John Hinckley, a twenty-five-year-old college dropout with a history of mental illness, rapidly fired six shots in Reagan's direction with a .22 caliber revolver.

Every shot missed. The first struck press secretary James Brady in the head. The second hit police officer Thomas Delahanty in the neck. The third struck a building across the street. The fourth hit Secret Service agent Tim McCarthy in the abdomen. The fifth and sixth bounced off the side of the limousine. The sixth shot, after ricocheting off the vehicle, struck Reagan in the left torso just as Secret Service agent Jerry Parr was shoving him through the rear door. Inside the limousine, neither Reagan nor Parr realized the president had been shot. Reagan, feeling great pain in his chest, believed Parr had broken a rib by landing heavily on top of him in the back seat. When Reagan began to cough up blood, he said, "You not only broke a rib, I think the rib punctured my lung."

Parr directed the limousine to nearby George Washington University Hospital. In the emergency department, Reagan initially insisted on walking on his own, but soon experienced chest pain and severe difficulty breathing. He fell to one knee and had to be assisted into the trauma bay. The president's appearance was ashen. There was blood in his mouth and his systolic blood pressure was 80 mmHg (normal level 120 mmHg). When his clothes were fully removed, it was finally discovered that he had been shot—a 1.5-centimeter entry wound was found below the left armpit between the fourth and fifth ribs. No exit wound was seen. No breath sounds were heard over the left lung.

A chest tube inserted to drain fluid from around the lung yielded an initial 1.2 liters of pooled blood, with steady flow thereafter at a rate of 200 to 300 milliliters every fifteen minutes. Reagan received intravenous fluid and antibiotics. A chest X-ray revealed the bullet was located in the left lung near the border of the heart. As more blood was drawn out of the chest, breath sounds became audible. Reagan remained conscious but in severe pain. When his wife, Nancy, arrived at the hospital, he managed to quip, "Honey, I forgot to duck."*

Quickly, a decision was made to perform emergency surgery on the president. Before the operation was about to begin, Reagan lifted his oxygen mask and joked to one of his surgeons, "I hope you're a Republican." The surgeon, a professed Democrat, replied, "Today, Mr. President, we're all Republicans."

---

* Reagan was repeating the line made famous by Jack Dempsey, who stated it after losing the world heavyweight boxing championship to Gene Tunney in 1926.

A left thoracotomy provided access to Reagan's chest. Half a liter of undrained blood was removed. With the help of another, intraoperative X-ray, the bullet was localized and extracted through a small incision in the lung. It was discovered that the bullet had ricocheted off the top of a rib and traveled through the entire width of the left lung, coming to rest just one inch from the heart.

The operation took two hours and forty minutes to complete. Blood transfusions played an important role in the president's care—from the shooting to the completion of surgery, he lost more than half his total blood volume. Later that evening, awake but intubated in the intensive care unit, the ever-humorous Reagan scribbled a message to his staff that read, "Am I dead?"

Reagan was discharged from the hospital twelve days later. His full recovery, at the age of seventy, was made possible by numerous modern medical advances beyond the imagining of James Garfield's physicians. Reagan was the first president in history to survive being shot while in office.

John Hinckley had made the assassination attempt in the hope that the act would impress actress Jodie Foster, a woman he had stalked and been infatuated with for years. In June 1982, Hinckley was found not guilty by reason of insanity. He called the shooting "the greatest love offering in the history of the world." In 2016, he was released from an institutional psychiatric care facility to live with his mother, with required court supervision and periodic psychiatric evaluations. In June 2022, all restrictions imposed on Hinckley were lifted and he was granted full and unconditional freedom at the age of sixty-seven.

Press secretary James Brady was permanently disabled from his brain injury and died in 2014. Secret Service agent Tim McCarthy recovered from his wound; police officer Thomas Delahanty suffered permanent nerve damage in his left arm, a disability that forced his early retirement.

A century of medical progress made Reagan's wound, far worse than Garfield's, a mere footnote at the outset of a long and consequential presidential administration. No longer were surgeons blind to the inner workings of the body, nor ignorant of the location of foreign material. Operations were no longer games of Russian roulette, with countless patients dying from infection.

"FLYING AMBULANCES"

The modern emergency room care that President Reagan received reflected technological advances and societal values, including principles of triage and ethics, bequeathed to us from a unique source: military medicine. Napoleon's chief surgeon, Dominique Jean Larrey, is given the most credit for promoting swift treatment of battlefield wounds and establishing new practices and doctrines that are orthodox today. In the 1790s, Larrey was dismayed to witness the lack of care given to wounded soldiers, who were generally left on the field until a battle was over—often more than twenty-four hours after being injured. He observed that such casualties were far less likely to survive than men who had been treated sooner.

In 1793, Larrey began equipping aid wagons and assigned them to each infantry unit, ready to speed injured soldiers to dressing stations in the rear. These wagons, known as "flying ambulances," became standard in the French Army. Larrey also instituted a basic method of triage, with casualties separated into three groups: those who could be quickly attended to and returned to battle, those with severe wounds but who had a chance to live, and those who were almost certain to die. He also established an important code of ethics, dictating that the most gravely injured patients would be treated first, regardless of rank, and enemy soldiers would be treated as well.*

Today, dramatic advances in military medicine have greatly increased wounded soldiers' odds of survival, and battlefield treatments have often

---

* Larrey served in the French Army for over twenty years, taking part in twenty-five campaigns and sixty battles. After the colossal Battle of Borodino, during the 1812 Russian campaign, Larrey performed more than 200 amputations in twenty-four hours—about one amputation every eight minutes—working without ceasing, and by candlelight at night.

served as the incubator in which new standards of care for civilian medicine have been tested and proved. In the twentieth century, injured soldiers benefited most from development of aseptic surgery, antibiotics, intravenous fluids, blood transfusions, rapid helicopter transport, and mobile army surgical hospitals (MASH units). Twenty-first-century wars in Iraq and Afghanistan saw forward-deployed surgical units tasked with stabilizing severely wounded soldiers through "damage control surgery" meant to buy time until transport could be arranged to more advanced medical centers. Army helicopter crews included critical care nurses trained in advanced life support. Whereas in the Vietnam War injured troops were kept in Vietnam until stable enough to fly, in Afghanistan the Air Force evacuated soldiers more promptly using huge C-17 cargo planes outfitted as "flying ICUs," with highly trained medical staff to treat critically injured patients en route to bases in Germany or the U.S. The in-flight mortality of such transported patients was incredibly low, only 0.25 percent.

Because of these efforts and resources, it was not unusual for a soldier wounded in a remote part of Afghanistan to be transported to Walter Reed Army Medical Center in the U.S. for surgery within forty-eight to seventy-two hours. Many thousands of service members alive today would have died in past wars, and military necessity has contributed to advances, not only in surgical treatment but also in a wide range of important arenas, including mental health care, prosthetic limb development, telemedicine, and tissue regeneration.

Technology, more than any other factor, promises to revolutionize what we view as the limits of trauma care today. Already, stem cells are being used to grow artificial organs that replace injured ones. The first successful bioartificial transplant took place in 2008, when a portion of trachea grown from the recipient's stem cells was implanted and became properly vascularized and incorporated by the body. Producing such organs in a lab often requires scaffolds that provide an organ's basic shape and contain tiny, weblike spaces where stem cells can reside and grow. These scaffolds may be made of biodegradable protein material, like collagen, elastin, or glycosaminoglycan; or, they can be composed of non-biodegradable, biocompatible material like silicone or 3D-printed plastic made to the exact size and shape of the recipient's original organ. Such methods have been used to grow artificial skin, bone, ears, bladders, blood vessels, and even functioning heart tissue. Advances in the field of tissue engineering raise hopes for a future where artificial organs, including livers, kidneys, hearts, and pancreases, might be grown in a lab using a patient's own cells, alleviating the challenges of immune rejection and organ scarcity.

We will always live in a world where traumatic injuries threaten health. Someday, when diseases like heart disease, diabetes, and cancer are cured or become chronically managed illnesses that no longer shorten lifespans, surgeons will still be operating on wounded soldiers, gunshot victims, motor vehicle accident survivors, and children injured on the playground. These patients will benefit from a century and a half of progress in surgery, anesthesia, antisepsis, diagnostic imaging, and—when the damage is too severe or organs fail—transplants, tissue engineering, and regenerative methods that effectively replace what was broken. The long and dramatic story of trauma surgery's evolution highlights the brilliance of both innovative physicians and their courageous patients. Together, they represent a shining example of the triumph of the human spirit, and their sacrifices continue to serve us all.

7

CHILDBIRTH

The Mysterious Killer

November 3, 1817
Claremont House, near Esher, England

At last the day had arrived. Twenty-one-year-old Princess Charlotte of Wales was in labor, two weeks after her predicted due date. For months, it seemed the entire Empire had been holding its collective breath, sharing that unique blend of excitement, anticipation, and anxiety known to all parents. The stakes were high. As the only legitimate granddaughter of King George III, Charlotte was second in line to the British throne and the only eligible heir of her generation.[*] This fact alone made Britons pay close attention, but there was also something else that made Charlotte unique.

She was loved.

Though she was far from perfect, the people loved her. In her adolescence, Charlotte was intemperate, headstrong, and unconventional; but, compared

---

[*] Because Charlotte's parents, Prince George IV and Princess Caroline, loathed each other and lived separate lives, there was no chance the couple would conceive a male heir that would supplant Charlotte's position.

to the rest of her debauched, indulgent family, she was relatable and popular. The public liked her independence, famously displayed when she defied her father's order to wed the future king of the Netherlands, partly because she could not abide the idea of ever leaving England. They respected her for marrying for love, and liked that she appeared blissfully happy with her husband, Prince Leopold. More than anything, they prayed that she would one day become a mature and responsible queen, far more like Queen Elizabeth than her doddering grandfather, King George III, who had famously lost the American colonies, or her hedonistic father, Prince George IV, who had served as regent since 1811 due to George III's descent into madness—possibly caused by porphyria or bipolar disorder.

Thus far, Charlotte's pregnancy had been healthy and uneventful. In an attempt to keep the baby small, which would ease delivery, she was put on a diet and intermittently bled, a practice performed as a panacea for many ailments. As her due date came and went, the public's prayers multiplied. Charlotte tried to keep busy and exercise a little each day, but it was hard not to feel nervous about the upcoming delivery.

At Claremont House, Charlotte's contractions began at 7 p.m. According to Sir Richard Croft, Princess Charlotte's accoucheur (male midwife), by 11 p.m. the cervix was "the size of a halfpenny" (about 2.5 cm). Though Croft hoped for a smooth and relatively quick delivery, the first stage of labor—the period from onset of contractions to full cervical dilation of ten centimeters—progressed slowly and somewhat erratically. Charlotte's contractions, which occurred approximately every eight to ten minutes, were noticeably weak. By 11 a.m. the next day, the cervix had only dilated to the size of "a crownpiece" (about 3.5 cm).

Croft began to worry. He could not deny that the princess's labor was not going smoothly. It might become necessary to use forceps to aid the delivery, but although he was one of the most highly regarded birth attendants in England, this was not a tool he felt comfortable wielding. Like most accoucheurs of his generation, he much preferred to let nature take its course and disliked forceps, which had fallen out of favor for the harm they could do in untrained hands.

Still, just in case forceps became necessary, Croft sent for backup. He summoned Dr. John Sims, whom Croft knew would be skilled at employing forceps if that became advisable. Finally, at 9 p.m., the cervix was fully dilated after twenty-six hours of labor. Croft must have felt great relief. Things were going all right after all. Charlotte remained calm and in good spirits. She sometimes got up and took small walks around her room. Prince Leopold was present to lend emotional support.

The second stage of labor, a period of intensifying contractions that denote a baby's transit through the birth canal and delivery, typically lasts between twenty minutes and two hours. Charlotte's second stage did not fit this mold. It progressed slowly, characterized by weak, irregular contractions. As hour after hour passed with little to no apparent progress, Croft grew increasingly alarmed. Yet he never requested Dr. Sims to come into the room and apply forceps. Near noon the following day, now November 5, after more than fifteen hours in the second stage, a small amount of dark green meconium, fetal feces, emanated from the uterus. This was a concerning sign of fetal distress, or possibly even fetal demise. It also raised the frightful specter of infection.

At 9 p.m., after laboring for fifty hours in total, Princess Charlotte delivered a stillborn, nine-pound boy. Croft and Sims immediately tried to resuscitate the baby, by rubbing it vigorously and blowing into its nose, to no avail. Charlotte took the news stoically. Quite possibly, she was too exhausted to do anything more than nod. Prince Leopold was bereft. He left his wife and went to his bedroom, where he took an opiate to help him fall asleep.

Croft now steadied himself for the third stage of labor, delivery of the placenta. Though overall, Charlotte appeared healthy and fine, he began to realize something remained amiss. Her uterus was still contracting irregularly, and he could not feel the placenta descending as it normally should. To his dismay, he realized the princess's uterus was undergoing hourglass contractions, with the center of the uterus contracting more strongly than its top or bottom. Even worse, the placenta was in the uterus's upper half. At that location, it was less likely to descend.

Then, thirty minutes after the birth, the princess began to bleed.

CHILDBIRTH: *The Mysterious Killer*

As blood oozed from the birth canal, Croft faced a full-blown crisis and was forced to make a difficult decision. He was conservative and did not like to intervene with maneuvers or instruments, but a bleeding uterus, whether from a retained placenta or weak postpartum contractions, could rapidly turn into a life-threatening disaster. Now both Croft and Sims agreed that they had no choice but to attempt manual removal of the placenta.

Croft managed to insert his fingers into the uterus and peel the placenta from the uterine wall. Doing so made Charlotte cry out in pain. Croft left the placenta in the vagina to be gradually expunged by the body, a practice commonly thought to spur greater uterine contractions by physiologically helping the body recognize the birth was complete. Twenty-five minutes later, the placenta had still not come out on its own, so Croft decided to remove it manually. When he did so, more bleeding ensued.

There was little he could do about it. But then, before Croft gave in to despair, the bleeding mercifully slowed, and then stopped.

By now, Croft was undoubtedly exhausted from stress and lack of rest. Amazingly, the princess still looked reassuringly normal, given the circumstances. Her heart rate was steady. She did not complain of pain, and she even managed to eat a little food. Despite all that had transpired, it looked like Charlotte would be all right, though the loss of the baby weighed heavily on all.

Then, around midnight, something changed. Charlotte complained of feeling ill and hearing a singing noise in her head. She tried to sleep but grew restless and irritable. She vomited. Croft gave her some laudanum, hoping to calm her nerves. Within an hour—now the early morning of November 6—Charlotte's pulse began to race and then became irregular. She complained of severe abdominal pain. Croft and Sims began to administer anything they could think of that might have a positive effect. They gave her wine, food, and opiates. They warmed her with hot water bottles.

Charlotte's condition worsened. She became delirious and began to ramble. She complained of feeling uneasiness in her chest and had difficulty breathing. Her pulse became weak and erratic.

At 2:30 a.m., she died.

THE MYSTERIOUS KILLER

Childbirth is not a disease, but it does kill. It would be difficult to consider any other medical calamity its equal in terms of tragedy or heartbreak. The unexpected loss of one, or sometimes two, lives, just at the moment when new life is imminent, seems too much to bear. And yet, throughout human history, so many women and their families endured exactly that. For millennia, childbirth was *the* major cause of maternal and fetal mortality. In developing nations, it remains a prevalent killer. Worldwide, about 300,000 women die from causes related to pregnancy and childbirth each year—over 800 per day—with approximately two-thirds of deaths occurring in Africa, and one-fifth in southern Asia. Moreover, the risk of injury and death in childbirth is about one hundred times higher for the fetus than the mother.

In the developed world, maternal death in childbirth has only become rare in the last seventy years. Prior to that, women had no choice but to associate childbirth with the very real possibility of dying or being permanently injured. The historical record attests to this grim reality. In seventeenth- and eighteenth-century America, the rate of maternal death per pregnancy was about 1 to 1.5 percent. Since most women birthed multiple children, the average lifetime risk likely approached 4 percent or more. A seventeenth-century American advice book instructed women to pray, repent, and prepare for the possibility of death before their due dates. Pioneer women completed wills before giving birth. Nineteenth-century mothers filled diaries and letters with recollections of excruciatingly painful labor and delivery.

When birth didn't kill, it commonly maimed. From perineal tears, to prolapsed uteruses, to incontinence, sexual dysfunction, and chronic pain, many debilitating injuries could last a lifetime and have devastating impacts on women's lives. Lacking birth control, it was not uncommon for mothers to endure ten or more pregnancies in their lifetimes, causing many to live in an almost perpetual state of protracted dread. This terrible fear was an equalizer of the rich and poor, of the educated and uneducated.

Why was childbirth so deadly? And how did that change?

CHILDBIRTH: *The Mysterious Killer*

The answers to these questions define what is perhaps the most tragic story in medical history. Quite often, the answer to the first was physicians themselves, and a truthful answer to the second cannot fail to reveal centuries of missteps, errors, and avoidable deaths.

It seems paradoxical that evolution has not conceived a better way for the most intelligent organisms on the planet to perpetuate their species. From a mechanistic viewpoint, the challenge of human birth is simple—it is difficult to pass a too-large human head through a too-small opening, the birth canal. Compared to the anatomy of other mammals, and particularly primates, the human pelvis is narrow. This characteristic allows us to walk upright with ease, but it also makes childbirth problematic. A woman's birth canal is so constricted that a baby's head can only traverse it smoothly by rotating in transit to match the head's greatest diameter with the pelvis's greatest diameter, which is side-to-side at the top of the pelvis and front-to-back at the bottom. This delicate pirouette ideally ends with the baby facing backward at delivery. It would be far easier to deliver offspring through a wider pelvis that allows babies to drop out with ease, but the price of this convenience would be wider hips and a bow-legged walk—like the gait of a chimpanzee.

The second half of this problematic equation is the baby's enormous head. Large brains, encased in large skulls, seem to be the cost of claiming the title of Earth's smartest animals. Anatomical head size may be *the* determining factor in the length of human gestation. Beyond nine months, the growing head simply becomes too large to pass through the birth canal. In modern times, improved maternal diet has made fetal heads even bigger than in centuries past, exacerbating the problem. A study of births in a prominent Dublin hospital showed the average birth weight of newborns in 2000 was seven pounds, ten ounces—about a full pound greater than the average weight in 1950.

The anatomic constraints that determine gestational limits lead to one very inconvenient consequence: human babies are born very underdeveloped. At birth, we are helpless. Unlike the newborn calf that emerges from the womb, staggers to its feet, and begins to walk around, or newborn whale that immediately starts swimming, we arrive as useless beings unable to even lift our heads. From the standpoint of human development, it would be better to be born months later, when we are larger, stronger, and perhaps able to crawl. But we can't, because our brainy heads are too big and our mothers' pelvises too small. These are the loathsome facts that make childbirth difficult, and sometimes deadly.

For centuries, the physiology of pregnancy, labor, and delivery were mysteries, especially to men. Some believed babies grew fully formed in the womb like mini-adults who actively struggled to fight their way out of the uterus—one way of explaining how membranes ruptured. Others thought the man's ejaculate caused the baby to be formed from the woman's menstrual blood, or that semen produced the full human form by itself and the woman was merely a receptacle for growing it. Another theory held that eggs from the right ovary became boys, and eggs from the left ovary became girls.

For most of human history, older, experienced women helped new mothers give birth. In Europe, these helpers became known as midwives, and the vast majority of them had no formal training. Midwives were often uneducated, illiterate women who helped their daughters, friends, and neighbors. Their role sometimes went far beyond obstetrics—some delivered animals, provided pediatric care, and performed abortions. They were called upon to confirm virginity, assess fertility, or determine the identity of a baby's father. For centuries, midwives enjoyed a monopoly on their field because men were not allowed to attend births. In 1522, curiosity compelled one German physician to sneak into a woman's delivery room dressed as a woman. He was discovered, arrested, and burned at the stake. In a medieval world governed more by ritual and religion than knowledge or science, childbirth seemed as miraculous as it was mysterious.

In time, men did begin to insert themselves into the birthing room, but this process remained slow due to religious and social customs. Since it was considered improper for men to view female genitalia, accoucheurs, or "male midwives," as they were known, often attended births blindfolded and tried to assist birth with their hands solely by feel. Another practice was to tie one end of a giant sheet around the man's neck like a bib, and carry the sheet across the mother's body in order to entirely block the accoucheur's view of the vagina. Further tactics to eliminate any suggestion of impropriety included keeping the mother mostly clothed during labor, having the male midwife keep his eyes averted at all times, or sometimes the opposite—making the practitioner lock eyes with the mother so she could be sure his eyes did not wander.

A 1711 woodblock print displaying a male midwife blindly delivering a baby under a bedsheet intended to preserve the modesty of the mother

In England, a unique family of successful accoucheurs invented a device that gave them an advantage over their competitors—obstetric forceps. Most credit Peter Chamberlen "the elder," who was born in 1560, with inventing this tool, although it is possible his younger brother of the same name, Peter Chamberlen "the younger," did so, or contributed to its design. Their tool, likely invented in the 1580s or 1590s, was initially comprised of two long-handled spoons, fixed at a central joint, that could be placed in the birth canal on either side of the fetus's head to help draw it out. The Chamberlens became adept at using forceps to aid in cases of arrested labor, when the baby's size or position made it difficult to transit the birth canal. In such cases, the typical practice was to kill the fetus using hooks, screws, or compressive forceps to impale or crush the fetus's cranium and drag it out, often in pieces, in an attempt to save the mother. The Chamberlens' forceps had the potential to save countless maternal and fetal lives.

But it didn't, because they did something shrewd.

They kept it a secret. For generations.

To keep competitors from learning about the instrument that helped them extract babies, the Chamberlens made sure never to let anyone outside of their family view it. It was transported by carriage, in an ornate and heavy wooden chest that required two men to lift and carry into a birthing room; this gave the false impression that their tool might be a large and complex machine. Inside the room, all family members were forced to leave and the mother herself was blindfolded. A drawn sheet, ostensibly to preserve the mother's modesty, served to further conceal the Chamberlens' secret device from view. To disguise sounds that might provide clues to their tool's character, or perhaps to simply further their mystique, the Chamberlens banged bells with hammers and chains inside the room to create distracting noises.

Chamberlen forceps

CHILDBIRTH: *The Mysterious Killer*

The Chamberlens became wealthy and famous. Peter the elder attended the deliveries of King James I and King Charles I's wives. The secret of the forceps remained in the family for over a hundred years, passed down from Peter the younger to his son, also named Peter, and subsequently to Hugh Chamberlen, a son in the next generation. In 1670, Hugh Chamberlen attempted to profit from the family secret by traveling to France, aiming to sell it to a famous accoucheur named François Mauriceau, or the French government. Neither took him up on the offer, and Hugh is believed to have later sold his family's intellectual property to a Dutchman named Roger van Roonhuysen in 1693, whose family also kept it a secret for decades.

Eventually, the forceps concept became known in the 1700s, and many practitioners began to produce and experiment with them. In the 1740s, a British doctor named William Smellie improved the forceps design by adding greater curvature to the blades. He also utilized an "English lock," which enhanced the ability to insert the blades individually and then fix them at a central joint. Smellie would famously attend births wearing a dress, presumably to reduce maternal anxiety, or possibly to hide his instruments under his clothing.

The use of forceps was not a panacea for complicated childbirth. In skilled hands, it could be a valuable tool, but in unskilled hands, forceps were dangerous. Many babies suffered grievous head and facial trauma, or death, from forceps use. It was also easy for forceps to damage the mother's vagina, cervix, and uterus. Still, its use proliferated, largely because the instrument helped male midwives steal business from female midwives who did not use them. At times driven by hubris and impatience, accoucheurs used forceps early and often. This frequently proved a disservice to mothers. Obstetrical training remained woefully inadequate. Even well into the nineteenth century, it was not unusual for medical graduates to have never witnessed a live birth. Any forceps training usually employed mannequins (sometimes fitted with an actual pelvis taken from a cadaver), dead fetuses, or rag dolls. A taboo against viewing women's genitalia meant that forceps were often applied without looking, increasing the chance of injury.

By the second half of the eighteenth century, bad outcomes due to forceps deliveries had besmirched the reputations of many obstetricians and the nascent field as a whole. Doctors were criticized for resorting to forceps too readily, when normal labor might still progress if practitioners remained patient and waited. A backlash ensued, and forceps use fell out of favor. New doctors were not trained to use them and were taught to allow nature to take its course in almost all cases. This was the conventional wisdom at the time of Princess Charlotte's pregnancy in 1817. Sir Richard Croft, and most other well-regarded accoucheurs of that era, almost always preferred observation to intervention.

A TRIPLE OBSTETRIC TRAGEDY

World history is full of examples of maternal death in childbirth that altered royal lineage, influenced the rise and fall of dynasties, and changed the course of nations.* But the death of Princess Charlotte of Wales in November 1817 is undoubtedly one of the most consequential; its effects would reverberate throughout the nineteenth and twentieth centuries.

It is probable that Princess Charlotte died from hemorrhaging. In addition to the blood lost during the third stage of labor, an autopsy revealed a large blood clot in her uterus. Pulmonary embolism, a clot blocking blood supply to the lung, is another possible culprit.

Charlotte's death stunned the British Empire. In the words of British statesman and Member of Parliament Henry Brougham, the country was struck, "as if by an earthquake at dead of night . . . It really was as if every household throughout Great Britain had lost a favourite child." Public mourning went on for weeks. Men and women wore black armbands. Theaters, businesses, and courts closed. Public events were canceled and shipping came to a halt. In more recent times, only the death of Diana, Princess of Wales, in 1997, provides some indication of how widespread the grief and

---

* Other famous victims of death due to childbirth include Henry VIII's wives Jane Seymour and Catherine Parr.

CHILDBIRTH: *The Mysterious Killer*

sense of heartbreak must have been; many have called Charlotte the original "people's princess." Prince Leopold was bereft. In one crushing moment, he had lost his beloved wife, son and heir, royal title as consort, and much of his fortune because he was not independently wealthy at the time. Though he later became King of Belgium, he never regained the blissful happiness of the one and a half years he was married to Charlotte.

Immediate criticism of the birth fell to Sir Richard Croft, though the royal family was quick to commend his care to the public. Croft was condemned for not employing forceps during the long second stage of labor when it was clear the delivery was not progressing steadily. If he had done so, perhaps the baby would have lived and Charlotte's life been spared. From the standpoint of obstetric history, Charlotte's death prompted renewed interest in the use of forceps, a movement that grew and persisted into the twentieth century.

Croft himself was devastated and unable to forgive himself for the princess's death. Three months later, he attended the delivery of another mother whose labor began to stall and became reminiscent of Charlotte's ill-fated childbirth. The similarity was too much to bear. Even before the woman delivered, Croft went into another room and shot himself in the head. His suicide has prompted this heartbreaking episode in British history to be termed a "triple obstetric tragedy," due to deaths of baby, mother, and accoucheur.

Charlotte's death meant that there was no eligible heir in her generation. The crown now seemed likely to pass to the Duke of Brunswick, a thirteen-year-old, feeble-minded boy who was the grandson of King George III's sister. Alarmed, the royal family realized the Hanoverian dynasty was now in great jeopardy. A more suitable heir was required. Several of Charlotte's uncles promptly dismissed their mistresses and sought proper marriages. Within a year, three of them had married. George III's fourth son, Edward, Duke of Kent, married Princess Victoria of Saxe-Coburg-Saalfeld, who happened to be Prince Leopold's sister. A year later, in 1819, they had a daughter. Her name was Victoria.

Ascending to the throne in 1837, Queen Victoria would reign for almost sixty-four years and have an entire era named after her. Had Charlotte lived, Victoria would likely never have been born. Historians have written countless books chronicling the lives of Victoria and her offspring—children and grandchildren who populated the thrones of Europe and, in the twentieth century, set in motion events that would lead to the First World War.

A PROPHET IN HIS OWN COUNTRY

As the nineteenth century progressed, childbirth remained risky. The most common cause of death was not due to breech presentation, obstructed labor, or even hemorrhaging. Instead, it was an infection of the uterus during, or shortly after, childbirth. This killer was called *puerperal*, or childbed, fever. Infection would arise, seemingly spontaneously, a few days after uncomplicated deliveries. Victims could then spiral downward with astonishing speed. As historian Irvine Loudon described it, "A woman could be delivered on Monday, happy and well with her newborn baby on Tuesday, feverish and ill by Wednesday evening, delirious and in agony with peritonitis on Thursday, and dead on Friday or Saturday." This was a torturous death. When the infection spread to the central nervous system, meningitis resulted in mental instability, hysterics, convulsions, and unconsciousness.

In the decades before the discoveries of Pasteur and Koch, no one knew why these women were dying. Some believed infection stemmed from rotting breast milk inside the mother. Others blamed constipation or an overanxious personality. Those who held miasma responsible made attempts to improve sanitation and institute quarantines in vain. Other failed tactics included newly painting hospital walls, replacing beds and linens, and fumigating wards with smoke or a form of chlorine gas. Doctors with a run of childbed fever deaths might burn their clothes. Others subjected patients to myriad treatments, including douches, laxatives, enemas, quinine, and shaving pubic hair. Some believed the best remedy was to ventilate women's genitals with fresh, clean air—so they made women lie in beds placed outside on the roofs of hospitals.

Then, in 1847, an eccentric and passionate twenty-eight-year-old Hungarian doctor named Ignaz Semmelweis came to a remarkable realization. He discovered that doctors, himself included, were responsible for killing these mothers.

Semmelweis held a staff position at the Vienna Maternity Hospital, the largest maternity hospital in the world—a massive institution that cared for about 8,000 patients per year (by comparison, the largest maternity hospitals in Boston and London only cared for about 200 to 300 per year). This Vienna hospital, which was tasked with providing care for the city's poor and unmarried women, became a highly regarded center of obstetrical training due to its surfeit of laboring mothers for trainees to practice upon. The maternity wards were divided into two clinics. The first clinic was staffed by medical students and doctors like Semmelweis. The second clinic was attended by midwives. Mothers were admitted to each clinic on alternate days.

At the time of Semmelweis's arrival in 1846, there was already a well-known and perplexing phenomenon at the hospital—the death rate from puerperal fever was about 11 percent in the doctors' clinic, but less than 3 percent in the midwives' clinic. Some months, the death rate in the first clinic rose to over 18 percent—almost one in five women attended by the physicians died. Since this difference between the clinics was common knowledge, mothers strongly favored admission to the midwives' clinic. Sometimes women would beg not to be admitted to the doctors' clinic. Because the changeover in clinic admissions occurred at 4 p.m. each afternoon, many patients would postpone their arrival to ensure admittance to the midwives' clinic—resulting in some giving birth outside on the street, in carriages, on the steps of the hospital, or in corridors.

Semmelweis observed all of this with increasing alarm. He saw mothers' fear when they arrived at his clinic: "That they were afraid of the First Division there was abundant evidence. Many heart-rending scenes occurred when patients found out that they had entered the First Division by mistake. They knelt down, wrung their hands and begged that they might be discharged . . . for they believed that the doctor's interference was always the precursor of death." He did not know why mothers were more than three times more likely to die in the doctors' clinic. He was amazed to find that even mothers who delivered on the street had far less likelihood of getting childbed fever than patients in his ward.

"To me," he wrote, "it appeared logical that patients who experienced street births would become ill at least as frequently as those who delivered in the clinic . . . What protected those who delivered outside the clinic from these destructive unknown endemic influences?"

Semmelweis began to collect data about the two clinics to see if he could divine what was causing the difference in mortality. Patients in both wards ate the same food, slept in the same linens, and experienced the same temperature and climate. He thought overcrowding might lead to more cases of fever, but the midwives' clinic was always the more crowded of the two. Miasma couldn't be the reason; the clinics shared an anteroom and all the patients were breathing the same air. Semmelweis examined patients' religious practices and even asked the priest who delivered last rites to change his typical walking path through the wards, but he found nothing that explained the differing outcomes. He was forced to conclude that there was only one major difference between the clinics: one was staffed by doctors and medical students, and the other by midwives.

But how could that explain why so many more died under doctors' care?

After a year of inquiry, Semmelweis finally found his answer.

In 1847, Semmelweis's friend and colleague Jakob Kolletschka was accidentally cut on the finger by the scalpel of a medical student during an autopsy. The cut became infected, and Kolletschka developed sepsis and died. Heartbroken, Semmelweis reviewed his friend's autopsy report and realized the pathology affecting Kolletschka's internal organs was virtually identical to those of women who died from puerperal fever—here was the same peritonitis, pericarditis, pleurisy, and meningitis that Semmelweis had seen in numerous dead mothers. He began to suspect that his friend had been infected with the same evil pathogen that plagued pregnant women. Kolletschka had been cut during an autopsy—could dangerous particles from the dead body have gotten into his wound and killed him?

As this thought crossed Semmelweis's mind, he realized, with mounting horror, that one thing that medical students and doctors from the first clinic did routinely was perform autopsies in the hospital. In fact, an enormous number of autopsies were being done because government rules dictated that every patient who died in the hospital must have one. Doctors and students were going back and forth between autopsies and deliveries multiple times a day, and no one was washing their hands.

Semmelweis hypothesized that something from the cadavers was being transferred, via doctors' unwashed fingers, into mothers' birth canals. He called these unseen poisons "cadaverous particles," and his theory explained why midwives, who did not attend autopsies and performed vaginal exams far less frequently than physicians, had much lower rates of puerperal fever in their clinic.* Stricken, Semmelweis realized that he, himself, had caused women's deaths, writing, "Only God knows the number of patients who went prematurely to their graves because of me. I have examined corpses to an extent equaled by few other obstetricians."

With religious zealotry, Semmelweis made immediate changes to combat these cadaverous particles. He made every doctor wash his hands in chlorinated lime, a bleach-like solution, before and after performing autopsies. The beneficial results came quickly. The mortality rate in the first clinic soon dropped to less than 2 percent. This was an astonishing achievement. Semmelweis told everyone who would listen about his breakthrough, certain it would save countless lives in maternity wards across Europe.

But it didn't.

Unfortunately, most doctors considered Semmelweis's theory to be ludicrous. It was offensive to physicians, as educated gentlemen, that anyone might regard their hands as unclean. It seemed impossible that invisible things found in cadavers could cause childbed fever, and it was cumbersome and irritating to wash one's hands multiple times a day in caustic solutions. More likely, critics thought, the disease stemmed from many different causes,

---

* Postpartum women were particularly vulnerable to infection. Perineal lacerations, and uterine linings raw from childbirth and placental detachment, became susceptible locales for microbial growth and quick infiltration into the bloodstream. In many respects, infection of the birth canal after childbirth was not dissimilar to a wound infection.

such as fecal contamination during birth, bad air from sewage or poor ventilation, or "bad milk" from the mother—a notion termed "milk metastasis theory." One doctor blamed puerperal fever on tight women's petticoats that caused a backup of feces to poison the bloodstream. Most seemed content to accept maternal deaths as dictated by the hand of God and something impossible for man to influence.

Though some friends and colleagues supported Semmelweis and even helped disseminate his theory at medical conferences and in articles, many more obstetricians in Vienna and other European capitals ridiculed, or more often ignored, his gospel. Semmelweis became increasingly frustrated. After witnessing the dramatic reduction in childbed fever in his clinic, he considered it unconscionable that others did not adopt the same practices. His obsession with an unpopular theory began to grate on many of his colleagues, particularly older physicians like his immediate superior at his hospital. Semmelweis often became argumentative when others disagreed with him, and he became an increasingly ostracized figure. In 1849, his hospital reappointment was denied.

He moved on to a humble position at a community hospital in Budapest, where his hand disinfection practices again produced an amazing reduction in mortality, dropping deaths from puerperal fever to less than 1 percent. In 1855, he gained a more prestigious position as a professor of obstetrics at the University of Pest, but his ongoing inability to convince more colleagues of the merits of his methods incensed him. Though he long delayed writing in medical journals to defend his ideas, in 1858, he finally did publish an article; and, in 1861, he followed it up with a book called *The Etiology, Concept, and Prophylaxis of Childbed Fever*. Unfortunately, much of this voluminous treatise was poorly written, rambling, and pedantic. A significant portion amounted to a petulant diatribe against his enemies and detractors. Few read it. After some who did wrote negative critical reviews, Semmelweis began to fire off hostile open letters addressed to leading obstetricians throughout Europe, accusing them of malpractice and even going so far as to call them killers of young mothers. He described other doctors as "ignoramuses," and penned declarations like, "I denounce you before God as a murderer."

CHILDBIRTH: *The Mysterious Killer*

Semmelweis's wife began to notice his behavior growing increasingly erratic. Sometimes belligerent and hyperactive, other times morose and languid, his comportment seemed progressively more detached from reality. Plagued by anxiety, Semmelweis became depressed, drank excessively, and frequented prostitutes. He sometimes talked to himself or imaginary persons, and wandered the streets at night. His conduct grew so alarming that his family believed he was becoming insane. In 1865, they arranged for him to be committed to an asylum in Vienna.* Semmelweis was duped into believing he was simply going to visit the institution, at the request of a trusted colleague, but once there he was forcibly restrained, put into a straitjacket, and locked away. When Semmelweis resisted, he was beaten by guards. He soon developed an infection from a hand wound, probably a result of his beating. The infection led to sepsis, and two weeks later, on August 13, 1865, he died.

Ignaz Semmelweis

---

* Medical historians have speculated that Semmelweis's behavior may have stemmed from Alzheimer's disease, bipolar disorder, syphilis, or mental exhaustion after years of defending his work.

In the long and tragic saga of obstetric history, Semmelweis's story ranks among the most heartbreaking—for the way he was treated, and because he was right. The failure to recognize the effectiveness of his methods resulted in decades of enduring ignorance and hundreds of thousands of maternal deaths worldwide, stretching well into the twentieth century. In Semmelweis's absence, the maternal mortality rate at his hospital at the University of Pest promptly increased six times, from 1 to 6 percent.

Today Semmelweis is regarded as a prophet who, though never properly acknowledged in his lifetime, proved prescient and correct. He is honored around the world as the "savior of mothers," and the "Semmelweis reflex" is a term that refers to the rejection of new scientific ideas that come into conflict with traditional, established thought and practice. His theory would not be validated until Louis Pasteur's discovery of microbes, but even then, another two generations of obstetricians failed to adhere to strict hygienic practices, and the scourge of puerperal fever would continue until the advent of sulfonamide antibiotics in the 1930s. Too many doctors failed to recognize or admit the truth: that they, themselves, were the greatest danger to pregnant women. A great number of their patients would have been better off without any medical assistance at all.

BETSEY, ANARCHA, AND LUCY

Ignaz Semmelweis made his contribution to women's health before the world was ready for it, but not every nineteenth-century innovator was ignored. One physician whose discoveries were amply lauded was Dr. J. Marion Sims, of Alabama. Sims was Semmelweis's mirror image. Whereas Semmelweis was ridiculed in life and revered after death, Sims earned great fame and fortune during his lifetime, but today is disparaged and vilified by many. He is one of the most controversial figures in medical history.

Why?

Because the discoveries he made, which have benefited women ever since, came from experimenting on enslaved Black Americans.

CHILDBIRTH: *The Mysterious Killer*

In 1845, Sims was a respected surgeon practicing in Montgomery, Alabama. It was not uncommon for slave owners to bring their slaves to him for treatment, often for gynecological issues because enslaved women were valued for their reproductive potential. A common and devastating problem for postpartum women was the development of fistulas—abnormal connections between two parts of the body—between the vagina and bladder or rectum. In this time before cesarean sections or drugs to induce contractions and hasten labor, childbirth sometimes went on for days. The prolonged pushing and contractions could tear the vaginal wall. More commonly, the constant pressure of the baby's head crushed the delicate vasculature of the birth canal, leading to areas of ischemia and later necrosis with tissue breakdown. Because the bladder, vagina, and rectum are tightly juxtaposed at the floor of the pelvis, these injuries could lead to *vesicovaginal* fistulas, a connection between the vagina and bladder, and *rectovaginal* fistulas, between vagina and rectum.

Fistulas involving the reproductive tract were extremely debilitating, resulting in constant drainage of urine or feces out of the vaginal opening. Women had no ability to control these emissions. The ensuing malodorous incontinence, flatulence, and discomfort resulted in social isolation, vaginal and sexual dysfunction, infections, and shame. A surgeon named Phineas Miller Kollock described the condition like this at a meeting of the Georgia State Medical Society in 1857:

> The poor woman [with a vesicovaginal fistula] is now reduced to a condition of the most piteous description, compared with which, most of the other physical evils of life sink into utter insignificance. The urine passing into the vagina as soon as it is secreted, inflames and excoriates its mucous lining . . . It trickles constantly down her thighs, irritates the integument with its acrid qualities, keeps her clothing constantly soaked, and exhales without cessation its peculiar odour.

At first, Marion Sims had no interest in such female conditions. Many nineteenth-century physicians considered examination of female genitalia undignified or beneath them, believing educated gentlemen should not perform work that had always been done by illiterate women midwives. In his autobiography, Sims wrote, "If there was anything I hated, it was investigating the organs of the female pelvis." He had no interest in lifting the taboo, nor solving the mysteries, associated with women's anatomy.

However, one day a white woman named Mrs. Merrill came to Sims after traumatizing her sacrum, or tailbone, after being thrown off a pony. She complained of back pain and severe pressure on her bladder and rectum. One can picture Sims's disdain as he reluctantly prepared to examine her. He used his fingers to inspect her vagina and realized the uterus was inverted; it had turned partially inside out. In an attempt to correct this, he had the patient get on all fours, with her head low to the floor and her pelvis up high. With her vagina in this somewhat upside-down position, he inserted two fingers into the vagina, but he couldn't feel the uterus or anything to push against. Rather aimlessly, he poked about and swept his fingers in circles, not sure he was accomplishing anything.

Then, suddenly, Mrs. Merrill said, "Why, doctor, I am relieved."

Sims was surprised. He had no idea what he had done to affect this cure but concluded that pressure and air introduced by his manipulation had forced the uterus to right itself. The patient felt enormously better and credited Sims with curing her. But there was something even more remarkable about this unique encounter. Sims's examination had, for the first time, made him realize it might be possible to fully examine and view the inside of the vagina clearly. To him, this was terra incognita. If one could access the vagina, he reasoned, there was no reason it could not be the site of beneficial surgeries to help correct anatomical disorders or injuries.

Newly inspired, Sims invented one of the first vaginal speculums using a pewter spoon; and he sometimes employed a mirror to direct sunlight into the vagina to increase visibility. Using his speculum to examine a woman with a fistula, he saw the cervix for the first time. Conveying his amazement, he later wrote:

> I saw everything, as no man had ever seen before. The fistula was as plain as the nose on a man's face ... The walls of the vagina could be seen closing

in every direction; the neck of the uterus was distinct and well-defined, and even the secretions from the neck could be seen as a tear glistening in the eye, clear even and distinct, and as plain as could be. I said at once, "Why can not these things be cured?"

He determined to do so and assiduously set to work. Between 1845 and 1849, he experimented with various ways of suturing inside the vagina, working almost exclusively on at least ten enslaved women, who had no say in their treatment. Instead, Sims made an arrangement with their owners, agreeing to feed and house the slaves at no cost.

Sims performed the most surgeries on a trio of women known to history as Betsey, Anarcha, and Lucy. These patients underwent multiple operations over time, without anesthesia—in part because anesthesia use was not yet common practice, but perhaps also because of a widely held belief that Black people did not experience pain as severely as whites. Anarcha endured at least thirty suturing operations to repair her fistulas as Sims worked to perfect his technique. Early on, his silk sutures often caused infection or inflammation. To remedy this, he began to use very fine silver wire sutures instead. He also used catheters to divert the flow of urine away from his wound closures until they had fully healed.

Aware that he had made an important contribution to medicine, Sims published a report of his methods in 1852, and began growing his practice by performing fistula operations on white women. He relocated to New York City, where he founded the Woman's Hospital, the first hospital for women in the U.S., in 1855. He published often, became wealthy, and rose to the presidency of the American Medical Association. During the Civil War he decamped to Europe because his Southern sympathies would have damaged his professional standing in New York. While touring European capitals, he visited medical schools and demonstrated fistula surgery. In 1863, he treated Emperor Napoleon III's wife, Empress Eugénie, for a fistula—further increasing his fame and transforming him into a bona fide medical celebrity. When he died, he was venerated as the benefactor of women in newspapers and public speeches. Statues of him were erected in Central Park and on the grounds of the State House of South Carolina, the state of his birth. Today, Marion Sims is regarded as the "father of gynecology."

However, the fact that his advances came from the exploitation of enslaved Black women has made him a controversial figure. Since the 1970s, Sims has been vilified by multiple historians, journalists, and activists. He was criticized for seeking fame and fortune, treating slaves like animals to be experimented on, and withholding anesthesia. In 2017, his statue in New York was the site of protests; it was removed in 2018.

At the same time, other historians and writers have taken pains to judge Sims's legacy by the context in which he lived. They point out that, in Sims's time, it was considered normal to try new medical treatments on slaves. Thomas Jefferson, for example, tested the smallpox vaccine on two hundred slaves in 1801, before using it on his family and white neighbors. At a time when many doctors may have refused to care for Black patients, Sims provided a valuable service, and by many accounts truly cared for the women he treated. Vaginal fistulas were a terrible affliction that any woman would wish to be cured of, and in the late 1840s anesthesia was not widely used regardless of the patient's race.

In one speech, at a meeting of the Academy of Medicine in New York, Sims publicly thanked the enslaved women, saying: "To the indomitable courage of these long-suffering women, more than any one other single circumstance, is the world indebted for the results of these persevering efforts. Had they faltered, then would woman have continued to suffer from the dreadful injuries produced by protracted parturition . . ." In his autobiography, he wrote that when he became discouraged and paused his work, the enslaved women were "clamorous" for him to continue and even assisted in his surgeries because they, too, hoped Sims's methods would cure them.

Marion Sims ultimately succeeded, but for too long, the full history of his work and its reliance on the abhorrent institution of slavery was ignored or downplayed. Today that is no longer the case. In 2021, statues of Anarcha, Betsey, and Lucy were erected in a monument named "Mothers of Gynecology" in Montgomery, Alabama.

BABY "ANAESTHESIA"

Though mid-nineteenth-century obstetricians unwisely ignored Ignaz Semmelweis's recommendations to reduce outbreaks of puerperal fever, their adoption of anesthesia to ease labor pains was somewhat more successful. The advent of ether, discovered in America in 1842, and first demonstrated publicly by William Morton in 1846, was a watershed moment in medical history. But ether wasn't perfect. It was irritating to the mouth and nasal passages. It smelled bad and often produced nausea. It could also cause lung inflammation and scarring. And worst of all, it was flammable. In an era when surgeons used gas lamps for illumination, this made ether a perpetual hazard. Furthermore, in obstetrics, ether use was complicated by the need to provide anesthesia over many hours. It took a lot of work to haul several large bottles of ether out to family homes, or up flights of stairs in urban apartment buildings, in order to administer it to laboring mothers.

A Scottish obstetrician named James Young Simpson set out to find a better alternative. Simpson experimented by inhaling numerous chemicals himself—anything that might be reported to possibly render animals or human subjects groggy, unconscious, or insensate to pain. A pharmacist recommended he try chloroform, a colorless liquid chemical with a pleasant, fruity scent that had been invented in 1831, and tried as an asthma treatment. On the evening of November 4, 1847, Simpson and two friends inhaled chloroform in Simpson's dining room. They all experienced an initial giddiness and cheerful mood. They laughed and chatted happily for a while. And then, the next thing they knew, it was morning. They had all collapsed to the floor, unconscious, for hours.

Simpson was extremely fortunate. Chloroform could be dangerous when taken in excessive amounts. If he had inhaled too much it could have killed him, and the chemical would thereafter have been deemed too toxic for human use. If Simpson had taken too little it would have had no effect, and he would have moved on to other substances. As it happened, he realized he had found something far more suitable than ether to relieve labor pains;

chloroform was inexpensive, easy to transport, nonflammable, and simple to administer.

After a few more days of experiments with chloroform, Simpson felt ready to try it for the first time on a laboring mother. His patient was a woman preparing to deliver her second child. He poured half a teaspoon of the liquid onto a handkerchief and placed the moistened cloth over the woman's mouth and nose. He did this repeatedly until she was asleep. The birth proceeded smoothly. When the mother woke up, she was awestruck. Elated, she named her baby "Anaesthesia."

Despite his success, Simpson's advance was not quickly adopted by many of his peers. They considered anesthesia in childbirth to be dangerous and unnecessary—after all, this was a normal, physiological part of life, not a disease. A main opponent was the Church of England. Religious leaders vociferously opposed anesthesia because the Bible said that women were to experience pain in childbirth. Genesis 3:16 reads: "Unto the woman he said, I will greatly multiply thy sorrow and thy conception; in sorrow thou shalt bring forth children." Some supercilious male doctors even claimed that women who could not endure the discomfort of childbirth would never be selfless enough to be good mothers. Such attitudes had become entrenched through centuries of male chauvinism. In 1591, a woman named Eufame MacAlyane asked her midwife to ease her labor pain by giving her a potion. King James VI ordered MacAlyane burned alive for requesting something considered anti-biblical; her midwife was executed as well, for witchcraft.

In response to his religious critics, Simpson pointed out that God had anesthetized Adam when his rib was taken to create Eve. Simpson published a pamphlet titled "Answer to the Religious Objections Advanced Against the Employment of Anaesthetic Agents in Midwifery and Surgery." In it, he cited James 4:17, which said that it was sinful to fail to do something one knows is good.

Controversy swirled for years, with Simpson and his supporters unable to gain widespread obstetric adoption of anesthesia. It didn't help that patients given an excessive amount of chloroform sometimes died, which perpetuated the notion that anesthesia in childbirth was an unnecessary risk. Physicians willing to try it questioned how much or how often the drug should be given. How much less should a child receive compared to an adult? How concentrated should the chloroform mixture be? How saturated the handkerchief? Early adopters like Simpson were imprecise in their administration of the drug. One physician described his technique like this: "I have the expectant mother hold a drinking glass with the bottom filled with cotton and upon which the chloroform is poured, then have them hold the glass over their nose. When their hands become unsteady and the glass falls away from the nose, I know they are sufficiently asleep to give them relief and I continue to accelerate the delivery."

It took the meticulous work of John Snow, the doctor who would famously identify a London well as the source of a cholera outbreak in 1854, to help convince the skeptics. Snow endeavored to study chloroform's effects more precisely and standardize its delivery. He recognized that chloroform was far more potent than ether, and worked to make chloroform safer by examining its physiological impact at various doses and by making dosage tables. He found that, without caution, it was easy to tip a patient over the line from mere unconsciousness into respiratory cessation and death. Body weight certainly affected proper dosing, but also factors as fickle as anxiety level. Anxious patients could throw off predictions if they hyperventilated, or sometimes the opposite—if a fearful patient held her breath but then inhaled very deeply, she might rapidly ingest such a high dose of chloroform that her breathing and heartbeat stopped. Snow addressed these challenges and fabricated a brass vaporizer to administer the drug in a standarized, measurable way. Whereas Simpson often initiated anesthesia in the first stage of labor, keeping women unconscious for hours, Snow recognized that giving chloroform too soon could slow contractions. He used lower doses than Simpson and initiated anesthesia in the second stage of labor, when the baby's arrival was imminent.

Obstetric anesthesia would eventually prevail, if only because women demanded it—their fear of pain in childbirth ranked second only to their fear of death. The successful care of a celebrity patient would do much to convince the public that anesthesia was safe and effective. Reprising her role as a model patient (after helping Joseph Lister demonstrate the value of antisepsis), Queen Victoria helped John Snow by inviting him to administer anesthesia during her eighth delivery, the birth of Prince Leopold, who was named after Victoria's favorite uncle, King Leopold of Belgium—Princess Charlotte's widower.

On April 7, 1853, as Victoria entered the second stage of labor, Snow had her inhale a small amount of chloroform with each contraction over the course of fifty-three minutes in total. Chloroform relieved the queen's pain but never caused her to fully lose consciousness. After a smooth and painless delivery, the Queen expressed delight with "the blessed chloroform." Snow also anesthetized her for her ninth and final pregnancy, the birth of Princess Beatrice, in 1857. Soon aristocrats and commoners began to seek the drug, which became known as *anesthesia à la reine*.

A MAN'S WORLD

The end of the nineteenth century saw the beginning of two massive shifts in obstetric care—a move from midwives to male obstetricians, and a change from giving birth at home to delivering in hospitals. These shifts were intimately related. Whereas most male doctors had previously deemed obstetrics beneath them, many came to realize that the field, with its ever-present, repeat customer base, could be a lucrative addition to their practices. It was not difficult for men to push out women as providers of choice. After all, only male doctors could perform medical procedures like bloodletting, administer drugs like opium, or use instruments such as forceps. Doctors, who were finally coming to understand the physiologic mysteries of reproduction, the uterus, and childbirth, also promoted the idea that pregnancy was akin to a disease state, one that required medical attention and was not just a normal part of life. Mothers, attracted by the promise of pain relief and purportedly better care, increasingly turned to doctors. By the turn of the twentieth century, midwives had lost half their business to physicians and became increasingly relegated to delivering babies of the poor. By 1930, midwives delivered only 15 percent of babies in America.

This sea change in obstetric care did not help mothers. In fact, it proved detrimental. Doctors were not well trained in obstetrics, and hospitals remained centers of infection. Medical interventions were commonly harmful, not helpful. Bloodletting decreased fetal oxygenation, opiates slowed contractions, and misuse of forceps traumatized baby and mother. Physicians were more likely than midwives to rupture a woman's membranes to speed labor, perform episiotomies, or spread infection due to their frequent interactions with sick patients. To avoid implicating themselves or tarnishing their reputations, physicians often underreported puerperal fever deaths—it was easy to substitute diagnoses like peritonitis, "blood-poisoning," tuberculosis, or "bowel inflammation" for the disease.

In 1920, eight out of every 1,000 births in the U.S. resulted in maternal death—a rate approaching 1 percent. History rightly condemns the field of obstetrics during the fifty-year period after the advent of antisepsis in the 1880s. Tens of thousands of American mothers, and many more worldwide, died each year from avoidable infection. In the decade 1920-1929, more than 250,000 American women died in childbirth, with about half likely to have perished from puerperal sepsis. Other causes of death included hemorrhage, pulmonary embolism, botched abortions, cesarean sections, and sometimes excessive anesthesia. Because of "modern" obstetrical practices and the shift to hospital deliveries, childbirth was actually safer for U.S. mothers in 1800 than in 1930.

Still, increasing numbers of women sought the care of male obstetricians in hospitals. Doctors liked having all their patients in one place—this saved a great deal of time and allowed them to care for more patients. These circumstances resulted in a surprising fact: wealthy women were more likely than poor women to die in childbirth because the latter could not afford to pay hospital fees and usually delivered at home. Alarmed by the rising rate of maternal deaths in hospitals, Dr. Sara Josephine Baker, a respected official in the New York City Department of Health, wrote in 1927, "The United States today comes perilously near to being the most unsafe country in the world for the pregnant woman as far as her chance of living through childbirth is concerned."

Baker's claim was later supported by a startling 1933 study titled "Maternal Mortality in New York City," which unambiguously confirmed that most maternal and fetal deaths were caused by poor physician care. The report stated that, over a three-year period, 65.8 percent of the more than 2,000 maternal deaths in the city would have been preventable if "the care of the woman had been proper in all respects." The study also revealed that 24.3 percent of women in labor had undergone an "operative intervention" (including forceps deliveries) when it was estimated that only 5 percent needed one; and women who had endured such interventions were five times more likely to die. According to historian Judith Leavitt, who evaluated rates of multiple pregnancies (particularly in poor and immigrant communities), about one in thirty women could expect to die in childbirth during their reproductive years in the early twentieth century.

Honest observers could not deny that obstetrics was taking a large step backward. The history of this era should not be interpreted to imply that midwives were more knowledgeable or skilled than obstetricians, but that this was a time when the risks of intervention often outweighed the benefits, and since physicians were more likely to intervene, their results suffered. Many appalled obstetricians took to medical journals to criticize their own field and decry the poor care American mothers were receiving.

Finally, the arrival of antibiotics in the 1930s allowed doctors to defeat the enemy they could not overcome through simple hygiene and handwashing. Deadly maternal infections diminished and hospital births increased further. At last, the promise of the hospital as a safer and more comfortable place to deliver one's baby was coming true. In 1940, 55 percent of American mothers gave birth in hospitals. By 1950, this had increased to 88 percent, and was close to 100 percent in 1960. The age of delivering babies at home, surrounded by family and the local midwife, was over. Birth in more impersonal, institutional hospital rooms, with pain relief and advanced care close at hand, has been the norm ever since.

Modern medicine would turn the tragic, centuries-long saga of obstetric ignorance, ineptitude, and arrogance into a story of success. In the mid-twentieth century, the maternal death rate plummeted. The use of sterile gloves reduced transmission of infection. The drug ergometrine was used to contract the uterus and limit hemorrhaging, while blood transfusions saved the lives of mothers who did bleed profusely. Synthetic oxytocin became popular to induce labor or treat stalled labor. Epidural anesthesia made labor more comfortable. The widespread use of birth control also reduced unwanted pregnancies and back alley abortions, a significant contributor to mortality rates.

Perhaps the most important advance of all was the development of fetal ultrasound, which allowed doctors to monitor fetal growth, confirm gestational age, and verify number of fetuses. It revealed placental location and fetal position, knowledge helpful in planning safer deliveries. Ultrasound technology originated from SONAR—SOund Navigation And Ranging—first developed after the sinking of the *Titanic* as a way to detect submerged icebergs and improved during the First World War to hunt submarines. During the Second World War, future Scottish obstetrician Ian Donald served in British Coastal Command, where he learned to use SONAR. Donald went on to pioneer the use of ultrasound in pregnancy, publishing a seminal paper on the topic in 1958. He once observed, "There is not much difference between a fetus in utero and a submarine at sea. It is simply a question of refinement."

The earliest practice of fetal ultrasound involved placing mothers in bathtubs full of water, but the discovery that water-soluble gel could transmit sound waves just as well made the procedure much simpler. Focus began to turn from the mother's health to the fetus's health, with advances in prenatal monitoring, diagnosis, and treatment spawning the new discipline of maternal-fetal medicine.

And yet, despite these advances, there were still frustrating steps backward that served to remind all that childbirth is always dangerous and never something to take lightly. The use of X-rays in pregnancy was an example of unintended harm. For decades, pregnant women were subjected to X-rays to check fetal growth, until it was finally discovered in the 1950s that the radiation could cause leukemia in babies.

Far worse than this were two humbling and unforgettable episodes which illustrated that sometimes, great medical "breakthroughs" were not the discoveries of new cures but instead the recognition of great harms whose evil effects could wreak devastating, and long-lasting, consequences.

A BIOLOGICAL TIME BOMB

A drug called diethylstilbestrol (also known as DES, or stilbestrol) will forever rank among the most dreadful iatrogenic and widespread pharmaceutical disasters in history. Prescribed between 1938 and 1971, DES was a synthetic form of estrogen that reportedly reduced the risk of miscarriage and premature labor. But the science supporting this was scant; most advocates pointed to a Boston research team's study that associated DES with beneficial effects. In 1947, the FDA officially approved DES for pregnant women and its use became widespread. Pharmaceutical companies seeking to increase profits promoted DES as a healthful pregnancy supplement, one sure to increase mothers' chances of delivering fit and strong babies. DES was distributed under dozens of product names and even put into prenatal vitamins, creams, and vaginal suppositories, so that many women were not even aware they were taking it. As DES's popularity grew, some mothers were receiving a lifetime's amount of estrogen in nine months, taking what amounted to a nine-month supply of birth control pills each day.

CHILDBIRTH: *The Mysterious Killer*

But there were early signs of a problem, like reports that persistent and high doses of estrogen caused diseases in animals. Scientists who tried to replicate the positive results of the initial Boston study failed to do so. As early as 1953, a University of Chicago study of over 2,000 patients—that included a control group receiving placebos—concluded DES conferred no benefit (a weakness of the Boston study was that it had not included a control group). Three other clinical trials also showed no proof that DES was efficacious. Despite this, doctors continued to prescribe DES. Many pregnant mothers insisted on taking it, wanting to do everything possible to assure a positive outcome and believing such a widely used medication couldn't cause any harm.

Then, in the late 1960s, cases of a rare adenocarcinoma of the vagina began to appear in doctors' clinics. Vaginal cancers typically arose in elderly women, but these cancers occurred in young patients, mere teenagers, some whom had not even started menstruating. No one could understand why this was happening in young girls—until the mother of one asked a doctor if it might have anything to do with the fact that she had taken DES while pregnant with her daughter. This physician, Arthur Herbst, decided to dig deeper. His 1971 study, published in the *New England Journal of Medicine*, confirmed a link between in utero DES exposure and vaginal cancer. Today, we know that DES is an endocrine-disrupting chemical, one that alters hormone levels during embryological development, with uncertain and adverse effects. In an era when astounding medical advances were being made across disciplines, this horrible and humbling discovery underscored doctors' ignorance of many of the body's mysteries.

The FDA promptly banned the use of DES, but it was impossible to ignore the fact that greater scrutiny of DES should have arisen in the early 1950s, when studies began to contradict the rosy DES image pharmaceutical companies wanted the public to believe. The consequences of this oversight were severe. Treatment of vaginal adenocarcinoma often necessitated hysterectomy and removal of the vagina. Those who did not display evidence of disease had to undergo repetitive screening for vaginal cancer, living with anxiety and fear that the condition might one day develop. News of DES's harmful effects panicked women across the country. Daughters urgently called their mothers to ask if they had been exposed to DES in utero. Not uncommonly, these women were horrified to learn their mothers could not remember, or were not sure, if they had taken the drug. It is estimated that approximately five to ten million American mothers and fetuses were exposed to DES during pregnancy.

And then, just when it seemed the DES tragedy couldn't get any worse, it did. It later became clear that women whose mothers took DES, known as "DES daughters," had increased risk for a wide array of additional medical problems, including breast and cervical cancer. Some suffered maldevelopment of their reproductive tracts, leading to higher rates of premature birth, ectopic pregnancy, incompetent cervix, and miscarriage. This has resulted in many DES daughters becoming infertile and unable to have children of their own. Today, some studies indicate that DES sons may have increased likelihood of testicular abnormalities, and that DES grandchildren could be at risk for infertility or cancer because animal studies suggest DES can cause DNA changes that are inheritable.

For eighteen years after a 1953 study proved DES was ineffective, pharmaceutical companies continued to sell and market it to millions. Today, many call DES a "biological time bomb," and women around the world still wonder if the drug has harmed, or will harm, their health in ways we have yet to realize or understand.

A "WONDER" DRUG

On the heels of the DES debacle followed another tragedy, though one with a somewhat more fortunate ending in the United States. In 1960, forty-six-year-old Frances Kelsey, M.D., Ph.D., was a brand-new hire at the FDA, then a far smaller agency than it is today. She was one of only seven full-time physicians responsible for evaluating the safety of drugs seeking approval to be brought to market. One of Kelsey's first assignments was a drug called thalidomide, a sedative that was commonly used abroad to treat severe morning sickness in the first trimester of pregnancy.

Thalidomide's American maker, a company called Richardson-Merrell, submitted an application to the FDA in September 1960 and expected swift approval. After all, the drug had already been approved for use in Europe and Canada, and was even available over the counter in West Germany. But Kelsey did not feel the data provided was enough to prove thalidomide was safe. She, as well as the FDA pharmacologist and chemist reviewing the application, found the proffered toxicity data inadequate, the animal studies poorly documented, and the clinical trial length insufficient. Kelsey refused to rubber-stamp the approval and instructed the company to provide more information. She shared concerns about reports the drug caused peripheral neuropathy (weakness, numbness, or painful tingling in the arms and legs), and also wanted more evidence it was not harmful to the fetus when pregnant women took it.

A fourteen-month struggle, primarily between Kelsey and Richardson-Merrell's chief representative, Dr. Joseph Murray, ensued—conducted via letters, meetings, and phone calls. The company made contact with Kelsey and her division within the FDA more than fifty times to promote their interests; this record of communication displays pressure Murray tried to exert on Kelsey to expedite approval of the drug. Murray repeatedly emphasized his company's need to initiate marketing and prepare thalidomide for distribution—it was, after all, a "wonder" drug. He appeared to consider Kelsey little more than a meddlesome, nitpicky bureaucrat whose questions and requests were excessive and unreasonable. He hoped the company could simply revise warnings on the drug label to address some of Kelsey's concerns.

Time and again, Kelsey pointed out the lack of scientific data presented in the thalidomide application. The company's claims about the drug were not supported by rigorous clinical studies. Instead, their submission emphasized dozens of doctors' reports, essentially testimonials from practitioners who lauded the drug and its effects. Because thalidomide was a sedative and not a lifesaving drug, Kelsey felt that public safety demanded the highest level of proof that it did not cause dangerous side effects—any "untoward reactions would be highly inexcusable," she wrote. Company executives tried to go over Kelsey's head, but Kelsey's superiors supported her refusal to be pressured into approving a drug she was not convinced was safe. She continued to recommend further studies be conducted in order to gain more data, denying their repeated applications six times.

This unwavering resolve would make Frances Kelsey a hero.

In November 1961, reports emerged from Europe of babies born with foreshortened, flipper-like appendages instead of arms and legs—a condition termed *phocomelia*, or "seal limb." It was confirmed that thalidomide crossed the placental barrier and caused severe birth defects. By preventing the drug from being used in America, Kelsey had almost single-handedly prevented thousands of cases. It is estimated that up to 100,000 fetuses were exposed to thalidomide worldwide, with many such pregnancies resulting in miscarriage or babies that died shortly after birth. Approximately 10,000 babies survived and suffered debilitating physical disability. In the U.S., there were only seventeen phocomelia cases, all resulting from mothers who had been given drug samples by their doctors for off-label use or who had obtained thalidomide abroad.

After thalidomide, the FDA gained a far more sweeping mandate, with greater powers to demand extensive testing of new drugs to ensure their safety and efficacy. The impact of drugs on developing fetuses is now a prime focus of the agency. For her accomplishment, Frances Kelsey was presented with the President's Award for Distinguished Federal Civilian Service in 1962. She was inducted into the National Women's Hall of Fame in Seneca Falls, New York, and today, the Frances O. Kelsey Award for Excellence and Courage in Protecting Public Health is awarded by the FDA to outstanding employees or groups that demonstrate the qualities Kelsey exhibited.*

---

* Thalidomide's anti-angiogenic properties in extremities have since found a role in treating leprosy and multiple myeloma. It was FDA-approved to treat these conditions in 1998 and 2006, respectively.

Frances Kelsey receiving the President's Medal for Distinguished Federal Civilian Service from John F. Kennedy in August 1962

CROSSING THE RUBICON

The changes that have taken place in the field of obstetrics are profound. Less than a century ago, parents prayed for a healthy baby but accepted maternal and fetal death as not uncommon facts of life. Today, maternal and fetal mortality is so low that most parents go to the hospital fully expecting to return home with a perfect, healthy baby. Neonatal medicine and neonatal intensive care treatments are so advanced that it is common for premature babies as young as twenty-four weeks' gestation to survive.

The cesarean section, now an exceedingly safe and routine operation, has done much to save lives in cases of fetal distress and obstructed labor. What was once an operation that invariably resulted in maternal death, performed only as a desperate, last-gasp attempt to save the fetus's life, is now a preferred method of delivery for many patients. In 1965, 4.5 percent of births in the U.S. occurred by C-section; today the C-section rate is approximately 32 percent. It is even more popular in countries like Brazil, where the C-section rate is approximately 55 percent overall, and as high as 84 percent in private hospitals.

Surgeons' abilities have now progressed far beyond the scope of routine procedures like C-sections. Every day, intrauterine operations are performed to address fetal diseases before birth. This can be accomplished via minimally invasive techniques using endoscopes and fetoscopes, or by open fetal surgery in which the fetus is operated upon directly through a small incision in the uterus. Conditions commonly treated by fetal surgery include spina bifida, congenital diaphragmatic hernia, and congenital heart disease.

Great advances in infertility treatments represent another incredible medical breakthrough of recent decades. About 10 percent of couples hoping to become pregnant will fail to do so after having regular, unprotected sex for a year. In the 1970s, British physiologist Robert Edwards and gynecologist Patrick Steptoe studied methods to extract a woman's eggs and incubate them with sperm in a petri dish. The resulting fertilized eggs could be implanted in a woman's uterus, a process known as *in vitro* fertilization (IVF).* A media frenzy erupted in 1978, when a woman named Lesley Brown from Manchester, England, used IVF to give birth to the world's first "test-tube baby." Brown and her husband had been trying to become pregnant for nine years, without success. Their daughter, Louise, appeared completely healthy and normal. Later, intracytoplasmic sperm injection, a technique in which a single sperm is injected straight into an egg under high magnification, became a popular alternate method of IVF.

---

* Edwards won a Nobel Prize in 2010; by that time Steptoe had passed away.

CHILDBIRTH: *The Mysterious Killer* 291

At present, women undergoing IVF take medicines to stimulate their ovaries to release many more eggs than normal. Several eggs are fertilized, raising complex ethical questions about how many embryos should be implanted in the uterus and what should be done with the unused ones, which could be donated to research or another infertile couple, frozen and stored indefinitely, or destroyed. Since 1978, over eight million IVF babies have been born. The rapid growth of assisted reproductive technologies has given rise to entirely new markets involving sperm banks, egg donors, surrogacy, and long-term embryo storage.

In many ways, IVF has proved a catalyst that has shifted the public's attention past the goal of preserving maternal life, and even beyond the goal of making healthy babies, to the more futuristic aim of making perfect babies. Influencing natural reproductive outcomes is not new. Every day, prenatal screening tests compel some mothers to undergo more invasive diagnostic procedures, like amniocentesis or chorionic villus sampling, that yield fetal genetic information to diagnose severe disorders like Trisomy 13 or Trisomy 18. Infants with these conditions generally expire within days to months after birth, prompting many would-be parents to opt for termination. About two-thirds of expectant parents who discover their fetus has Trisomy 21, or Down syndrome, a less severe genetic disorder, also abort their fetuses.

IVF takes these powers of selection a step further. Since IVF yields many more fertilized eggs than a couple can possibly use, parents can now genetically test their embryos and choose which ones to implant—these are the ones that will have a chance at life. In cases where a parent carries a gene mutation for a monogenic (caused by a single abnormal gene) disease such as cystic fibrosis or Huntington's disease, doctors can select embryos that do not carry the culprit gene, which ensures a child will be spared the condition. These methods are termed *preimplantation genetic diagnosis*, or PGD.

Though few might object to the use of PGD to achieve these ends, greater controversy surrounds the prospect of using PGD when the medical benefits are less definitive, or when the aim may be to select for desirable, non-disease-related human characteristics. Most diseases are not monogenic; they are polygenic, or caused by many gene variants. As a result, scientists are learning to associate combinations of genes with the likelihood of developing various conditions, including heart disease, cancer, diabetes, and mental health disorders like Alzheimer's disease. Several for-profit companies now offer genetic testing that yields what is termed one's *polygenic risk score*, and have advertised their ability to use such scores to screen embryos for attributes like short stature, intellectual disability, and cognitive ability. We now confront the possibility that polygenic risk scoring, in concert with PGD, will allow parents to determine not just a fetus's propensity to develop a disease but also its likelihood of having various physical and intellectual attributes.

If parents become able to intentionally select their offspring to be healthier and smarter, or perhaps even stronger and more beautiful, how would that affect our society? How might that exacerbate existing health disparities between rich and poor?

Those who believe these practices will be part of a distant future are mistaken. It is already beginning to happen.

In 2021, the world learned that the first baby selected for its enhanced polygenic risk score had already been born, in 2020. The father, a North Carolina neurologist, had used IVF to produce sixteen embryos, of which six were chosen to undergo genetic testing by a company called Genomic Prediction. The embryo selected for implantation was the one whose polygenic risk score displayed the lowest probability of developing heart disease, diabetes, and cancer as an adult. In an April 2021 online Genomic Prediction panel discussion, the father explained that he believed that parents have a responsibility to provide their children with the best chance at a healthy life. "Part of that duty," he stated in a later interview, "is to make sure to prevent disease—that's why we give vaccinations." He considers PGD and polygenic risk scores to have enormous potential to reduce human diseases, with "compounding benefits" across generations.

Many scientists question the true predictive value of private companies' polygenic risk scores, but there are likely few who doubt that this technology will become more mature and popular as time goes on. At present, its promise and peril are exhilarating to some, and horrifying to others. Proponents and opponents argue the potential benefits against risks like inadvertently selecting for undesirable traits, artificially shifting natural demographics, and creating a less empathetic world in which characteristics considered less desirable are undervalued and diminished. For over a decade, these complex issues have also played a role in the larger ethical debate surrounding another technological miracle—the ability to alter the human genome through DNA modification using CRISPR/Cas9.

In 2012, scientists Jennifer Doudna and Emmanuelle Charpentier pioneered a powerful gene-editing tool using CRISPR technology.* Prior researchers had identified repetitive and clustered portions of bacterial DNA, known as CRISPRs, that were found to constitute a bacterial defense mechanism against viruses—which have attacked bacteria since time immemorial.† These CRISPR sequences enabled bacteria to identify, excise, and incorporate the DNA of attacking viruses, an adaptation that allowed them to remember, and counterattack, viruses that had assailed them in the past. It was as if bacteria had developed a form of immunity against viruses. Studying this process, Doudna and Charpentier harnessed the same molecular components to localize and target specific DNA sequences in an organism's genome. Cas9 is an enzyme protein that cleaves the DNA strand so that unwanted gene mutations can be excised, and, if desired, replaced with normal copies of a gene. This gene-editing technology transformed the nascent field of gene therapy, and hastened society's need to confront ethical questions regarding the creation of "designer babies."

How far should humanity go in its quest to eliminate disease and disability?

Instead of simply *selecting* the best embryos via PGD, we must now ponder something far more consequential: the ability to *create* the best embryos using CRISPR/Cas9 gene modification. Here, methods to eradicate illness might easily open the door to future gene modification for nontherapeutic

---

* CRISPR stands for "clustered regularly interspaced short palindromic repeats." In 2020, Doudna and Charpentier were awarded a Nobel Prize for their revolutionary discovery.
† A fact we learned in chapter three that proponents of "phage" therapy hope to exploit.

aims such as human enhancement related to strength, intelligence, or beauty. Articles on these topics abound with metaphors referencing slippery slopes or Pandora's box. The dystopian settings of novels like Aldous Huxley's *Brave New World*, or movies like *GATTACA*, in which society is divided between those who are genetically enhanced and those who are not, no longer sound so far-fetched.

At the outset, few opposed the use of gene modification technology to cure diseases caused by a single mutation, through efforts to replace an abnormal gene with a normal copy of that gene. This strategy has been used to treat conditions like retinitis pigmentosa and sickle cell disease, and could be used in the future to cure afflictions like Huntington's disease, Duchenne muscular dystrophy, or cystic fibrosis. But because most diseases aren't caused by only one gene variant, the best way to approach them quickly grows fraught with ethical dilemmas. Take, for example, a gene that appears to be associated with an increased chance of developing anxiety, depression, and alcoholism. Not everyone who carries this gene will be adversely affected. Should parents be able to correct or excise this gene from their embryo's genome? Who gets to define what condition or potential condition is worthy of "treatment" or modification?

Meanwhile, an even more controversial advance has captured the public's attention—one that has the potential to change the human genome forever.

The most urgent biological debate of our time concerns our ability to make gene modifications, not just in mere *somatic* cells but also in *germ line* cells that will pass genetic changes on to one's progeny in perpetuity. Most of the body's cells are termed somatic cells—including skin, muscle, and nerve cells, for example. Using CRISPR/Cas9 to correct a gene mutation in a somatic cell to cure a patient's disease will only affect that person, and the genomic change will disappear when that patient dies one day.

In contrast, there are only two kinds of germ line cells: egg and sperm cells. These are the reproductive cells by which our DNA is passed to our offspring. If one were to use gene modification to change the DNA of egg or sperm cells, or single-celled embryos known as zygotes, these cells could later become human beings whose altered genome will be passed on to their progeny indefinitely. Therefore, using CRISPR/Cas9 to change the genome of a germ line cell could forever change the human genome and the genetic diversity of our species.

This is why the development of CRISPR/Cas9 is no less momentous than the splitting of the atom. The latter led to the nuclear age and gave humanity the capacity to destroy itself with nuclear weapons. Gene editing also has the potential to change humanity as we know it. CRISPR/Cas9 could even be used to combine human DNA with animal DNA to create chimeric organisms—advocates could propose this as a way to mass-produce transplantable human organs. The myriad possibilities are daunting to contemplate, and our understanding of gene modification technology is in its infancy. For all its precision, CRISPR methods sometimes cut DNA in places investigators do not expect, or may cause unintended genetic changes that do not manifest in the near term. We might believe a certain genetic modification is beneficial, only to learn later that it changed our genome in some unforeseen, detrimental way.

And if we ever did inhabit a world in which gene modification was made safe, routine, and infallible, who would get to decide what human characteristics are considered normal or abnormal? Once defined, there would be increasing pressure for parents to ensure their children benefit from an accumulation of favorable traits, and a dearth of unfavorable ones. A 2017 position statement by the American Society of Human Genetics warned, "In identifying some individuals and their traits as 'unfit,' we experience a collective loss of our humanity." Francis Collins, past director of the NIH, said in 2015, "Evolution has been working toward optimizing the human genome for 3.85 billion years. Do we really think that some small group of human genome tinkerers could do better without all sorts of unintended consequences?"

What we are contemplating is nothing less than altering evolution itself.

Not surprisingly, about seventy-five nations, including the United States, ban experiments to alter germ line cells in order to produce genetically modified humans. But just as the world's superpowers fear the development of nuclear weapons by rogue nations, responsible scientists fear those who may flout international norms or limits on germ line experimentation. If one country's scientists forge ahead, it may become difficult for international scientists to restrain themselves for fear of falling behind in the equivalent of a genetic therapy arms race.

Already, there is concern that that rogue nation will be China.

In 2015, researchers at Sun Yat-sen University in Guangzhou used CRISPR/Cas9 to attempt to modify a gene that causes beta thalassemia, a hereditary blood disorder, in eighty-five embryos obtained from an IVF clinic. This disturbing act alarmed the international community. Even worse, it confirmed scientists' fears, because the experiment was a failure and resulted in many genetic errors. Only four of the embryos were found to have the desired gene modification. About a third of those tested contained unintended, off-target gene changes—some of which could have impacted normal human development. The laboratory's experiments were stopped, but it was easy for many scientists to see where the techniques used could be improved and made more effective. Western scientists became increasingly fearful that additional Chinese investigators might ignore international norms and someday even create a genetically modified human.

In 2018, those fears were validated. A Chinese scientist named He Jiankui crossed the Rubicon. He had earned his Ph.D. from Rice University and studied CRISPR at Stanford before becoming an associate professor at a university in Shenzhen, China. In November 2018, he announced to the world that he had created the first genetically modified humans, twin girls named Lulu and Nana, who had received a new, mutated *CCR5* gene that would purportedly make them resistant to HIV infection. The protein product of the normal *CCR5* gene enables HIV to enter cells. By altering *CCR5*, He hoped to confer lifetime protection from the virus. To conduct this experiment, he had recruited couples in which one partner was HIV positive. Employing IVF, genetically modified embryos were implanted in a mother's uterus, and the twins, born in 2018, appeared healthy and normal.

He's announcement sparked a firestorm of criticism across the world.* Scientists, journalists, and ethicists condemned He's actions and renewed calls for an international moratorium on gene editing in germ line cells. In article after article, authors excoriated He for violating fundamental ethical guidelines; some feared there was no predicting what unintended genetic modifications might someday affect these twin girls' lives, or the lives of their children and children's children.

To the great relief of many observers, a group of over one hundred Chinese scientists co-signed a letter rebuking He, as did the Chinese Academy of Medical Sciences. Two months later, He's university fired him, and in December 2019, he was convicted in a Chinese court and sentenced to three years in prison for "illegal medical practice." Denounced for seeking "fame and profit," He was fined three million yuan ($430,000 USD), forever barred from conducting research in the field of human reproduction, and banned from applying for future research funding. These events lend some hope that China will be a responsible international actor in the field of biomedical research.

But the genie is out of the bottle, and now humanity will face one of its greatest tests. It must be acknowledged that there are some who do not fear advances in germ line gene modification. Huntington's disease, which causes progressive neurological degeneration with severe physical and mental deficits before middle age, is a prime target for future treatment via CRISPR modification in somatic cells. But this will only help patients one at a time. If it is possible to completely eradicate Huntington's disease from the face of the earth using germ line modification, so that no child would ever have it again, many might say—why wouldn't we do that? Today we celebrate the eradication of smallpox and the near-eradication of polio—would this be any different? In

---

* Criticisms of He included the charge that his procedure was medically unnecessary. There are other ways for an HIV positive parent to ensure his or her child is born HIV negative, including sperm washing and preimplantation screening. Moreover, the CCR5 gene is likely to have several other unknown functions. One of its known roles is as a protective factor against West Nile virus infection.

fact, some argue that it is unethical *not* to pursue germ line experimentation, or to fail to do everything possible to alleviate human suffering and disease.

There are probably few who believe that genetic engineering of humans will never be allowed to proceed. Most likely it will, first to combat severe diseases, and perhaps later for nontherapeutic purposes in ways that may make us cringe today. Many reputable scientists urge extreme caution and the widespread adoption of internationally agreed-upon guidelines but do not feel germ line experimentation should be banned outright. As science evolves and capabilities expand, scientific and regulatory bodies must remain vigilant, ever wary of the risks of unintended, off-target genetic changes, and adverse impact on the diversity of the human genome.

How long before we live in a society that permits parents to modify embryos to become resistant to disease or less likely to develop cancer, diabetes, or heart disease?

How long before we are able to pay to make our children taller or more intelligent?

Time will tell. Nothing else in physicians' armamentarium brings them closer to the role of playing God. This is new territory, and the ethical and societal ramifications are profound. Of all the advances detailed in this book, gene modification is likely to be the one to have the most enduring impact on humanity. It has the potential to be the greatest medical breakthrough of all time. It also threatens to change the human species forever. It will not be a question of if this is possible. It is. The question is how we will allow it to happen.

CONCLUSION

The Masters of Medicine

> It helps a man immensely to be a bit of a hero-worshipper, and the stories of the lives of the masters of medicine do much to stimulate our ambition and rouse our sympathies.
>
> —William Osler, M.D.

The history of medicine is a fluid, living discipline that is, by definition, changing as rapidly as contemporary science. A distillation of the long arc of medical progress down to our most crucial discoveries offers an opportunity to examine those factors most likely to result in future medical advances in this century and beyond. What sense can be made of these scores of triumphs? A bird's-eye view reveals that such manifold breakthroughs, and the visionaries who made them, bear much in common.

Not surprisingly, many innovations resulted from years of hard work by dedicated scientists who displayed incredible grit and determination. John Gibbon's two-decade quest to invent the heart-lung machine, Marie Curie's years of toil to isolate a modicum of radium, and George Papanicolaou's mastery of the ignoble art of biopsying guinea pig cervixes each exemplifies humankind's admirable tenacity and diligence. But at the same time—at least in the realm of medical discoveries—this does not convey the full picture. Far from it, because as we have seen, many of the most significant breakthroughs did not happen this way.

Some discoveries were literally epiphanies. Like unexpected bolts of lightning out of the clear blue sky, heroes like Frederick Banting and Wilhelm Röntgen were struck by sudden, unanticipated thoughts or ideas that changed the world. The randomness of these moments can only humble the vast majority of scientists who dedicate entire careers to solving problems in their fields, but who will never come close to emulating the accomplishments of pioneers who often were not even experts in the disciplines they transformed. Frederick Banting was not an endocrinologist. He was not even a doctor of internal medicine. Is it any wonder that a world's expert like John Macleod might have laughed Banting out of his office at their first meeting? Meanwhile, Röntgen's moment of epiphany would spark the invention of a totally new specialty, radiology; yet, he was a physicist, not a physician. This is part of the mystery of medical discovery. Many crucial innovations are not planned; they are not anticipated, nor made by individuals assiduously working on research germane to the problem. Sometimes an investigator looking for something entirely different simply stumbles upon a world-changing observation. This scenario typifies another category of discovery, one that never fails to delight and enthrall. One word describes it best.

Serendipity.

Serendipitous discoveries make the best bedtime stories, because sometimes earth-shattering breakthroughs really do come from pure luck. Several seemingly random events had to occur for Alexander Fleming to notice an odd pattern of clearing around specks of mold in his petri dish. If Louis Pasteur's assistant had not forgotten to inoculate chickens with fresh, chicken cholera-causing microbes, he and Pasteur would not have inadvertently produced an attenuated organism that they later discovered could confer immunity without disease. If Sidney Farber had not tragically hastened the deaths of leukemic children by unwisely treating them with

CONCLUSION: *The Masters of Medicine*

folates, he would not have realized that anti-folate treatment might help them. And if the mustard gas disaster at Bari had not occurred, doctors might not have zeroed in on nitrogen mustard as the first effective cancer chemotherapeutic agent.

Tales such as these are great anecdotes that fill us with wonder and awe. But if we, like scientists, are data-driven, we must admit that serendipity and lucky mistakes are often given greater credit than they deserve in the history of discoveries. They certainly can't be counted on for future advances, for by their very nature, there is no predicting when, or if, good fortune will strike again for the benefit of humanity.

Instead, there is a greater common denominator that has played the major role in more breakthroughs than any other. That is the role of the medical maverick. Mavericks are people with the imagination and creativity to think outside the box. Their minds are receptive to the possibility that an inquiry might yield answers to questions neither they, nor anyone else, has previously thought to ask. They view problems from a new perspective, and have the courage to follow their convictions even when it goes against convention. They are, in short, brilliant.

This is a beautiful characteristic of the human species—that there are those among us with the potential to make breathtaking intellectual leaps, the courage to take astounding risks, and the fortitude to withstand ridicule and ostracism for their beliefs. These unique qualities confer hope to those who fear the ways in which humans may one day be supplanted by advanced robots or artificial intelligence. For all our flaws and failings, it is our ability to conceive new ideas, and sometimes take irrational risks, that makes us irreplaceable. Werner Forssmann defied orders, risked his life, and literally fought off a friend in a preposterous scheme to catheterize his own heart. Edward Jenner vaccinated a boy with cowpox and then, in a supremely audacious, arrogant, and dangerous act, intentionally inoculated the boy with lethal smallpox as a test. William Coley purposely infected patients with deadly bacteria. Ignaz Semmelweis's pursuit of truth and science culminated in his death.

Mavericks like Semmelweis are not insane. But they believe so strongly in a concept, practice, or idea that they devote themselves entirely, and with obsessive tenacity, to furthering it and sharing it with others. The outside observer may see this and call it grit or determination. The maverick is poised to benefit from serendipity because, as Louis Pasteur once famously said, "Chance favors the prepared mind." The most important question we can ask ourselves today is, where will the next medical mavericks come from and how can we support their development?

Many have pondered this question. One exceptionally qualified to do so was Ernst Chain, the co-discoverer of penicillin. Chain emphasized that future innovators will likely be eccentrics, radicals, and reformers. They may not be top students, or be popular, and he urged our vigilance for the emergence of such unconventional minds. "When they do appear," he said in a 1966 speech, "it is our job to recognize them and give them the opportunities to develop their talents. This is not an easy task, for they are bound to be lone wolves, awkward individualists, nonconformists, and they will not very well fit into any established organization."

Chain also emphasized that breakthroughs rarely result from bureaucratic initiatives and are seldom achieved by throwing money at a problem. This was one of the criticisms of the war on cancer and the years preceding it, when substantial government funds were deployed in a scattershot, all-of-the-above strategy to conduct laborious studies of myriad soil, plant, and animal samples. Many of these were low-yield efforts made before our understanding of cancer's molecular underpinnings and the immune system enabled research to become more targeted. Chain's point was that mankind's return on investment would be higher if we learned to identify the mavericks, and give them the freedom to inquire, investigate, and create. This point should never be wielded to belittle the importance of funding for science. That will always be necessary. Even the war on cancer produced some winning therapies. But the smartest money will be spent on finding and nurturing brilliant minds.

It is unquestionably difficult to predetermine which young scientists will become future Nobel Prize winners, but no one can doubt that it serves the

CONCLUSION: *The Masters of Medicine*

interest of humanity to maximize the number of physicians and scientists engaged in research and the fight against disease. On this measure it appears we have significant room for improvement. Although it is fashionable for politicians and policymakers to advocate for improved STEM education—our acronym for science, technology, engineering, and math—a look at the highest levels of scientific research, where the greatest discoveries are likely to be made, reveals conditions that deter many young scientists from staying in their fields.

For one thing, the pendulum has swung from U.S. government largesse in the days of the war on cancer to a relative paucity of funding for scientific research. It is now harder than ever for investigators to receive research funding from the National Institutes of Health. In the year 2000, slightly more than 30 percent of NIH grant applications were approved. This funding rate has steadily declined and in 2020 had dropped to approximately 20 percent. This means only one in five projects proposed by scientists moves ahead. As a result, investigators are spending more and more of their time writing research grants instead of performing actual research, and the intense competition for funding means scientists are less likely to take risks and propose novel research ideas. Projects closely adhering to traditional lines of inquiry are more likely to receive funding than unconventional proposals. The system also appears to favor well-known or experienced scientists over up-and-coming young ones, and may be biased against women and minority researchers.

Another, potentially greater, hindrance to future discoveries is the attrition of scientists in academia as countless Ph.D. graduates become discouraged and disillusioned by a traditional system of advancement that seems to disfavor them at every turn. In every college in America, promising science students are encouraged to pursue their doctorates. They spend years working toward their degrees, with many in their early thirties by the time they attain them. Unfortunately, these new graduates quickly learn that there are far too many of them vying for far too few tenure-track university positions. In the field of biomedicine, it has been estimated that fewer than one in six

Ph.D. graduates will succeed in attaining such a post. This means that over 80 percent will never get the job they trained for: to pursue their own original research and lead a team of younger investigators. Instead, many scientists find themselves spending years in limbo, working as "postdocs," a term denoting junior scientists who work for tenured professors as they wait to secure a permanent position that will likely never materialize.

Postdoctoral researchers essentially serve as the supremely well-educated, cheap labor that works at the direction of those vaunted few professors atop the research pyramid—professors who sometimes compound the problem if they become reluctant to retire. A 2018 study that surveyed data on almost 14,000 postdoc employees revealed a median annual salary of $47,484, indicating that half earned less than this, some potentially as low as $23,660. These income levels prove untenable for many in their thirties and forties, at a time when they are likely to be having children and building their families. Add low pay to poor job security and no guarantee of advancement, and it's no wonder that many of our most brilliant young scientists leave academia to seek jobs in industry or on Wall Street that pay salaries five or ten times higher.

A 2020 survey by the journal *Nature* revealed that 55 percent of biomedical researchers held a negative view of their job prospects. These are dedicated, well-trained, accomplished professionals—top students who devoted over a decade of their lives to the pursuit of science. Even luminaries like Emmanuelle Charpentier, who won a Nobel Prize in 2020 for her CRISPR/Cas9 research, spent twenty-five years of her career bouncing around among various positions at nine different institutions before landing as a director at the Max Planck Institute for Infection Biology in Berlin in 2015. Katalin Karikó, whose mRNA research led to the revolutionary vaccines that protect so many of us today, labored in obscurity for years with little support from the NIH or her university. The fact that she didn't abandon her calling along the way is a testament to her determination—many others would have.

Facing long odds, numerous young researchers burn out, quit, or move on to other jobs. As a result, approximately half of America's postdoctoral workers are foreigners from countries like China. These immigrants appear more willing to accept the modest wages and dim prospects that many Americans will not. But now, even this is beginning to change. As countries like China catch up to, and in some arenas even surpass, the U.S. in science and technology, fewer and fewer foreigners are willing to come. Instead, they stay in their home countries, which benefit from their talents.

America has a well-worn tradition of importing great scientists. European professors fleeing the Nazis were integral to the Manhattan Project. The Cold War prompted a relaxation of immigration restrictions to allow a wave of physicians, engineers, and scientists from Asia in the 1960s and 1970s. But in this century, it would be unwise to count on importing our geniuses. It is more prudent to develop and nurture homegrown talent. It is not enough to merely encourage students to become scientists—we must also show our commitment to their success by paying them appropriately, funding more of their proposals, and giving them secure paths to careers in which they continue to apply their talents toward making discoveries.

The news is not all bad. Our capitalist society generates profit-driven enterprises that can be powerful forces for innovation. Companies like Moderna, Pfizer and BioNTech, and Johnson & Johnson, not government labs, developed effective Covid-19 vaccines faster than anyone previously thought possible. The system works best when public and private entities work in concert; Covid-19 vaccine development, for example, was greatly hastened by large federal grants to the aforementioned private companies. Government will always play an important role because private companies have little interest in researching rare diseases whose cures will never be profitable. We should also depend on government-funded research to investigate lines of basic science inquiry that do not have obvious applications today but may in the future.

There is also another very bright spot to consider. Until the recent past, roughly half of humanity was excluded from the ranks of scientific discoverers. Practically speaking, only men were allowed to become doctors and scientists for centuries. Women like Marie Curie and Frances Kelsey were pioneers twice over because they succeeded in a man's world. The year 2019 marked the first time that women comprised more than half the medical students in the United States. This century will see scores of important discoveries by generations of women scientists eager to make their mark. For inspiration, they will look to innovators like Karikó, Charpentier, and Doudna. It is well past time humanity doubles its assets in the war against disease. Until now, we have been fighting with one hand tied behind our backs.

A delay in granting women entry into the sciences has not been humanity's only misstep. We have seen how medical discoveries are often subject to human frailty and failings. Envy, self-interest, and paranoia ruined relations between insulin discoverers Banting, Best, Macleod, and Collip. Crawford Long, William Morton, Horace Wells, and Charles Jackson battled for years in vain attempts to secure sole credit as the inventor of anesthesia. Howard Florey faced an overwhelming tide of misinformation and press adoration for Alexander Fleming as he futilely hoped to receive proper credit for the discovery of penicillin. How much more productive would these talented minds have been had they devoted themselves entirely to science, instead of upstaging one another or seeking credit?

At the same time, it should be acknowledged that not every human vice causes science to suffer. Rivalry, for one, can spur competitors to excel. Jonas Salk and Albert Sabin resented one another, but there is little doubt that their rivalry prompted each to dedicate themselves to develop polio vaccines as quickly as possible. Louis Pasteur's response to Robert Koch's belligerence was to out-discover his nemesis. Even after his discovery of microbes and the anthrax vaccine solidified his position atop the scientific world, Pasteur pressed on and rose to even greater heights with discoveries like the rabies vaccine.

That said, in the twenty-first century there is far more to gain from scientists working cooperatively than in competition with one another. As globalization makes the world smaller in every way, international collaboration becomes increasingly important because the stakes are now higher than ever. Pandemics can leap across oceans in a single day—it only takes one plane

CONCLUSION: *The Masters of Medicine*

flight. The cooperation of the Chinese scientists who quickly shared Covid-19's genetic sequence with the world enabled the rapid development of an effective mRNA vaccine in only forty-two days. Conversely, the actions of rogue Chinese scientists have threatened to upend international norms pertaining to the production of genetically modified humans. There has never been a more important time for the world's scientists to work together and adhere to a fundamental code of ethics that may even supersede the policies of their national governments.

We are now at the cusp of attaining healthcare advances that will dwarf even the amazing accomplishments of the past two centuries. We aim for a world in which regenerative medicine will enable victims of spinal cord injuries to walk again; artificial intelligence will be harnessed to provide rapid, accurate diagnosis of myriad conditions; gene therapy will provide cures for inheritable diseases and reduce risk of heart disease; vaccines will treat and prevent cancer; and donor organs will be grown using stem cells. A hundred years from now, we will be better at slowing aging and postponing death. A "Methuselah gene" (a variant of the cholesteryl ester transfer protein gene) has been identified that is more prevalent in centenarians and has been found to raise HDL cholesterol (the good kind). Studying this gene may ultimately result in therapeutic medicines that mimic its actions. Anti-aging researchers are also interested in telomeres—caps of repetitive DNA located at the ends of chromosomes that shield chromosomes from damage. Telomeres become shorter with every cell division, so their length is like a cell's life-timer, one that will eventually run out, signaling a cell's death. By lengthening or preserving telomeres, possibly by enhancing the activity of a natural repair enzyme called *telomerase*, scientists hope to discover ways to extend the life of cells, and ultimately human longevity.

The possibilities will defy our imagination. Just as it would have been impossible for Louis Pasteur to contemplate face transplants, in utero surgery, or CRISPR, inventions of the future may be beyond our present ability to comprehend. Now, more than ever, we must encourage students to devote themselves to science. At the same time, humanism will become increasingly important; as scientific progress necessitates urgent consensus on ethical questions like germ line gene modification, we must preserve and defend those unique characteristics that make us human. We cannot plunge impetuously toward scientific advancement if doing so makes us lose even an ounce of our humanity.

The success of the human race is not guaranteed. World history, as we know it today, comprises only the briefest of moments in the lifespan of our planet. Humankind has already taken numerous steps backward, from the fall of empires to the cataclysmic destruction of world wars. It is now within our power to destroy the Earth and ourselves. Problems such as climate change and obesity make many believe we are well on our way to doing just that.

It will fall to members of the next generation to take up the mantle of scientific innovation, knowing that medical progress is far from assured. It takes the dedication, persistence, and imagination of new mavericks to push forward into novel arenas of discovery. Our bequest to them must be the gift of education. To educate them about health, disease, and ethics. To share the stories of past medical heroes who improved all our lives. Our children and grandchildren must meet the challenges of this century and the next to expand upon the astounding progress we've made. In the blink of an eye, they will be our physicians and surgeons. And there is no predicting which one of them will make the next crucial breakthrough.

Acknowledgments

In writing this narrative of humanity's greatest medical breakthroughs, I am deeply indebted to the scores of authors and historians whose prior books elucidated such discoveries in far greater detail than I. Scanning the notes section will reveal those works that have earned their place as the definitive accounts of their subjects, and I gratefully acknowledge the gifts these authors have bequeathed to me and all the scholars that follow them.

Like all researchers, I have benefited from the generosity of numerous professionals who devote themselves to the libraries, museums, and universities they serve. I am enormously grateful to the staff of the Baystate Medical Center Health Sciences Library in Springfield, Massachusetts, for their kind assistance over many years. I am particularly indebted to Sandra Savenko, who, at my request, acquired hundreds of articles (several over a century old) from medical journals, and scores of rare books, with unfailing graciousness and alacrity. I could not have completed this project without her help.

I am also indebted to the staff of the Richard Salter Storrs Library in Longmeadow, Massachusetts, for their ever-present readiness to track down all manner of books to aid my research, and to the following institutions that generously provided needed information, documents, and photos: Thomas Fisher Rare Book Library at the University of Toronto, Alexander Fleming

Laboratory Museum, The Royal Society, University of British Columbia Library, University of California San Diego Library, Dana-Farber Cancer Institute, The Royal College of Surgeons of England, Queen Victoria Hospital, and East Grinstead Museum.

My career in the highly subspecialized world of retina surgery has carried me far away from the everyday treatment of the afflictions covered in this book, and I have greatly benefited from the help and expertise of many doctors and friends who kindly gave of their time to review or critique portions of the manuscript along the way. Thank you to: Mary-Alice Abbott, Vivian Lam Braciale, Bernard Chang, Ray Chen, Richard Engelman, Jonathan Green, Gerald Hausman, Kati Karikó, Esther Lam, Janice Lam, Wilfred Lam, Evan Lau, Jonathan Lee, Alexander Lin, Calvin Lu, Peter Lu, Stephen Lu, Tom Mennella, Christopher Ollari, and Vivian Pao.

Many thanks to my agent, Steve Harris, for believing in this book and confidently shepherding its path to BenBella Books. I'll remain ever grateful to the entire BenBella team, including Glenn Yeffeth, Adrienne Lang, Alicia Kania, Sarah Avinger, Madeline Grigg, Jessika Rieck, and Jennifer Canzoneri, for their passion for this book and all they have done to help produce it and bring it to readers everywhere. I owe a special thank-you to my erudite copy editor Leah Baxter, for doing such a meticulous job scrutinizing and fact-checking every page of this book, and for improving it in numerous ways.

Most of all, I have been blessed to benefit from the skill and patience of my incomparable editor Vy Tran, who guided this project from beginning to end. She spent countless hours reviewing this narrative and skillfully helped me streamline and improve every story in it. The hardest thing about writing a book like this is deciding what to leave out, because many more volumes could easily be filled with additional fascinating details about the lives of each visionary depicted in these pages. I could not have made the tough decisions alone. Thank you, Vy.

Finally, my gratitude to my wife, Christina, grows and compounds with each passing season. She readily takes on the unenviable task of being the first reviewer of all my writing and supports me in all the ways that make it possible for me to pursue this project and many others. Her counsel and opinion mean more to me than even she might suspect. Everything I do is more meaningful and joyful because of her.

Notes

INTRODUCTION

1 *"To wrest from nature"*: William Osler, "Chauvinism in Medicine," *Montreal Medical Journal* 1902; 31(9): 684–699. Italics added by author. Osler was one of the most influential physicians in modern history. He was a founding professor of the Johns Hopkins Hospital, helped establish internal medicine as an academic discipline, and launched the first medical residency training program.

2 *average life expectancy*: Charles Kenny, *The Plague Cycle: The Unending War Between Humanity and Infectious Disease* (New York: Scribner, 2021), 145–146.

2 *today average worldwide longevity . . . 73.4 years*: World Health Organization, "Global Health Estimates: Life Expectancy and Healthy Life Expectancy," https://www.who.int/data/gho/data/themes/mortality-and-global-health-estimates/ghe-life-expectancy-and-healthy-life-expectancy, accessed June 25, 2022.

2 *77.8 in the United States*: Elizabeth Arias, Betzaida Tejada-Vera, and Farida Ahmad, "Provisional Life Expectancy Estimates for January Through June, 2020," *National Center for Health Statistics* 2021, Report No. 010, February 2021. This figure averages the life expectancies of men, at 75.1 years, and women, at 80.5 years.

2 *conquer death due to disease*: Yuval Harari, *Homo Deus: A Brief History of Tomorrow* (New York: Harper Collins, 2017), 21–29.

CHAPTER ONE: HEART DISEASE

6 *tingling sensation*: Dick Cheney, *In My Time: A Personal and Political Memoir* (New York: Threshold Editions, 2011), 119.

6 *His physician prescribed*: Dick Cheney and Jonathan Reiner, *Heart: An American Medical Odyssey* (New York: Scribner, 2013), 27, 36. This book written by Cheney and his cardiologist was the main source for this brief retelling of Cheney's medical journey. It is an excellent account of how the vice president benefited from numerous advances in cardiology.

6 *stopped smoking*: Richard Cheney, "Reflections of a Former Vice President on Long-time Cardiac Experiences," *Baylor Medical Center Proceedings* 2009; 22(3): 276–278.

6 *50 percent narrowing*: Cheney, *Heart*, 58.
7 *second, mild heart attack*: Ibid., 69.
7 *clot in his RCA*: Ibid., 76-77.
7 *"no functional limitations"*: Ibid., 103.
8 *performed an angioplasty*: Cheney, *In My Time*, 292.
8 *implantable cardioverter-defibrillator*: Ibid., 524.
8 *blacked out*: Cheney, *Heart*, 225.
9 *222 beats per minute*: Ibid., 241.
10 *called his daughter*: Ibid., 280.
10 *twenty months*: Ibid., 309.
11 *25 percent*: Centers for Disease Control and Prevention, "Heart Disease Facts," September 27, 2021, https://www.cdc.gov/heartdisease/facts.htm, accessed June 26, 2022.
13 *He postulated*: James Herrick, "Clinical Features of Sudden Obstruction of Coronary Arteries," *Journal of the American Medical Association* 1912; 59: 2015-2020; Richard Ross, "A Parlous State of Storm and Stress: The Life and Times of James B. Herrick," *Circulation* 1983; 67(5): 955-959.
13 *"It fell like a dud"*: James Herrick, "An Intimate Account of My Early Experience with Coronary Thrombosis," *American Heart Journal* 1944; 27(1): 1-18.
14 *"I cannot possibly"*: Werner Forssmann, *Experiments on Myself: Memoirs of a Surgeon in Germany*, trans. Hilary Davies (New York: St. Martin's, 1974), 83-85. The quotations presented in this recounting of Forssmann's story come from this portion of his autobiography, which was first published in Germany in 1972.
16 *The X-ray showed*: Werner Forssmann, "The Catheterization of the Right Side of the Heart," *Wiener Klinische Wochenschrift* 1929; 8: 2085-2087.
16 *eight more times*: Lawrence Altman, "Daring Experiment Aided Heart Care," *New York Times*, July 10, 1979, C3.
17 *On October 30, 1958*: David Monagan and David Williams, *Journey into the Heart: A Tale of Pioneering Doctors and Their Race to Transform Cardiovascular Medicine* (New York: Gotham Books, 2007), 36-37.
17 *under fire from*: Forssmann, *Experiments on*, 250.
17 *like a village pastor*: Felix Belair, "3 Win Nobel Prize for Heart Study," *New York Times*, October 19, 1956, 1, 8. Subsection of article contributed by United Press under heading "Forssmann Hails Americans."
18 *"We've killed him!"*: Thomas Ryan, "The Coronary Angiogram and Its Seminal Contribution," *Circulation* 2002; 106: 752-756.
18 *"When the injection began"*: William Sheldon, "F. Mason Sones, Jr.-Stormy Petrel of Cardiology," *Clinical Cardiology* 1994; 17: 405-407.
18 *"After three or four explosive"*: Ibid., 406.
19 *"Again?"*: Monagan, *Journey into*, 37.
19 *President Dwight D. Eisenhower*: Thomas Lee, "Seizing the Teachable Moment—Lessons from Eisenhower's Heart Attack," *New England Journal of Medicine* 2020; 383: e100(1)-e100(3); Robert Gilbert, "Eisenhower's 1955 Heart Attack: Medical Treatment, Political Effects, and the 'Behind the Scenes' Leadership Style," *Politics and the Life Sciences* 2008; 27(1): 2-21.

20 *30 to 40 percent*: Thomas Lee and Lee Goldman, "The Coronary Care Unit Turns 25: Historical Trends and Future Directions," *Annals of Internal Medicine* 1988; 108: 887–894.
20 *inadvertently opened*: Richard Mueller and Timothy Sanborn, "The History of Interventional Cardiology: Cardiac Catheterization, Angioplasty, and Related Interventions," *American Heart Journal* 1995; 129(1): 146–172.
20 *first insert the smallest*: Charles Dotter and Melvin Judkins, "Transluminal Treatment of Arteriosclerotic Obstruction: Description of a New Technique and a Preliminary Report of Its Application," *Circulation* 1964; 30: 654–670.
20 *eighty-two-year-old diabetic*: Ibid., 657–658.
21 *"I've been standing here"*: Misty Payne, "Charles Theodore Dotter: The Father of Intervention," *Texas Heart Institute Journal* 2001; 28(1): 28–38.
21 *"Visualize but do not"*: Charles Dotter, "Transluminal Angioplasty: A Long View," *Radiology* 1980; 135(3): 561–564.
21 *climbed Mount Hood's*: Payne, "Charles Theodore Dotter," 31.
22 *experimenting with ways*: Monagan, *Journey into*, 90–92.
22 *"a sausage-shaped distensible"*: Spencer King, "Angioplasty from Bench to Bedside to Bench," *Circulation* 1996; 93(9): 1621–1629.
22 *polyvinyl chloride*: Matthias Barton, Johannes Grüntzig, Marc Husmann, et al., "Balloon Angioplasty: The Legacy of Andreas Grüntzig, M.D. (1939–1985)," *Frontiers in Cardiovascular Medicine* 2014; 1: 1–25.
22 *In 1974, he performed*: Ibid., 8; Alfred Bollinger and Maria Schlumpf, "The Beginning of Balloon Conception and Application in Peripheral Arterial Disease," *Journal of Invasive Cardiology* 2008; 20(3): E85–E87.
23 *poster displaying*: Monagan, *Journey into*, 111–112; King, "Angioplasty from," 1623.
23 *Dr. Richard Myler*: James Forrester, *The Heart Healers: The Misfits, Mavericks, and Rebels Who Created the Greatest Medical Breakthrough of Our Lives* (New York: St. Martin's, 2015), 225.
23 *Adolf Bachmann*: Monagan, *Journey into*, 123–129.
24 *"To the surprise of us all"*: Spencer King, "The Development of Interventional Cardiology," *Journal of the American College of Cardiology* 1998; 31(4 Suppl B): 64B–88B.
24 *spontaneous round of applause*: Ibid., 67B.
27 *"A surgeon who"*: G. Wayne Miller, *King of Hearts: The True Story of the Maverick Who Pioneered Open Heart Surgery* (New York: Times Books, 2000), 51.
27 *"Surgery of the heart"*: Stephen Paget, *The Surgery of the Chest* (Bristol, UK: John Wright & Co., 1896), 121.
27 *In September 1896*: Forrester, *The Heart Healers*, 28–30.
27 *"He was deathly pale"*: James Blatchford, "Ludwig Rehn: The First Successful Cardiorrhaphy," *Annals of Thoracic Surgery* 1985; 39(5): 492–495.
28 *"The sight of the heart"*: Ibid., 494.
28 *even introduced him*: Orla Werner, Christian Sohns, Aron Popov, et al., "Ludwig Rehn (1849–1930): The German Surgeon Who Performed the Worldwide First Successful Cardiac Operation," *Journal of Medical Biography* 2012; 20: 32–34.
29 *Dwight Harken*: Forrester, *The Heart Healers*, 25–28.

Notes

30 *"The only moment of panic"*: Ibid., 33.

30 *removed metal fragments*: Dwight Harken and Paul Zoll, "Foreign Bodies in and in Relation to the Thoracic Blood Vessels and Heart," *American Heart Journal* 1946; 32(1): 1-19.

32 *Walter Stockton*: Forrester, *The Heart Healers*, 41-42.

32 *"The purse-string suture"*: Charles Bailey, "The Surgical Treatment of Mitral Stenosis (Mitral Commissurotomy)," *Diseases of the Chest* 1949; 15(4): 377-393.

32 *"backward cutting punch"*: Ibid., 386.

32 *"It is my Christian duty"*: David Cooper, *Open Heart: The Radical Surgeons Who Revolutionized Medicine* (New York: Kaplan, 2010), 74.

33 *"the Butcher"*: Kevin Fong, *Extreme Medicine: How Exploration Transformed Medicine in the Twentieth Century* (New York: Penguin, 2012), 88.

33 *to schedule two operations*: Bailey, *The Surgical Treatment*, 388-390; Forrester, *The Heart Healers*, 44-46.

34 *successfully performed mitral valve*: Dwight Harken, Laurence Ellis, Paul Ware, et al., "The Surgical Treatment of Mitral Stenosis: I. Valvuloplasty," *New England Journal of Medicine* 1948; 239(22): 801-809.

34 *"Dear Dr. Harken"*: Forrester, *The Heart Healers*, 53.

35 *He tried using monkey lungs*: Ibid., 56-57.

35 *tried this on dogs*: Ibid., 63-65. Lillehei's full name was Clarence Walton Lillehei.

36 *Gregory Glidden*: Miller, *King of Hearts*, 128-139.

36 *Pamela Schmidt*: Ibid., 141-143.

36 *"Queen of Hearts"*: Ibid., 158.

36 *series of seven cases*: Ibid., 151-152.

36 *the child's mother suffered*: Ibid., 157-158. In total, forty-five operations were completed employing cross-circulation; twenty-eight of the patients survived. The longest time a patient's circulation was maintained by the parent-donor was forty minutes. See also Cooper, *Open Heart*, 190.

37 *as a research fellow*: John Gibbon, "Development of the Artificial Heart and Lung Extracorporeal Blood Circuit," *Journal of the American Medical Association* 1968; 206(9): 1983-1986.

37 *"During that long night"*: Ibid., 1983.

38 *built a prototype machine*: John Gibbon, "The Maintenance of Life During Experimental Occlusion of the Pulmonary Artery Followed by Survival," *Surgery, Gynecology and Obstetrics* 1939; 69: 602-614.

38 *Watson learned about*: John Gibbon, "The Development of the Heart-Lung Apparatus," *Review of Surgery* 1970; 27(4): 231-244.

38 *cascading down a series*: Mark Kurusz, "May 6, 1953: The Untold Story," *ASAIO Journal* 2012; 58(1): 2-5.

38 *May 6, 1953*: John Gibbon, "Application of a Mechanical Heart and Lung Apparatus to Cardiac Surgery," *Minnesota Medicine* 1954; 37(3): 171-177; John Gibbon, "The Development of the Heart-Lung Apparatus," *American Journal of Surgery* 1978; 135: 608-619; Cooper, *Open Heart*, 152.

39 *Dr. René Favaloro*: Faisal Bakaeen, Eugene Blackstone, Gosta Pettersson, et al., "The Father of Coronary Artery Bypass Grafting: René Favaloro and the 50th Anniversary of Coronary Artery Bypass Grafting," *Journal of Thoracic and Cardiovascular Surgery* 2018; 155(6): 2324-2327.

40 *Shumway and his colleague*: Forrester, *The Heart Healers*, 171-172; Cheney, *Heart*, 287-288.

41 *debate in America*: Forrester, *The Heart Healers*, 176.
41 *Louis Washkansky*: Christiaan Barnard and Curtis Pepper, *One Life* (Toronto: Macmillan, 1969), 304-305.
41 *a patient was deemed*: Forrester, *The Heart Healers*, 174-175; Raymond Hoffenberg, "Christiaan Barnard: His First Transplants and Their Impact on Concepts of Death," *British Medical Journal* 2001; 323: 1478-1480. In his autobiography (and thereafter) Barnard stated that he waited for the donor heart to stop beating before opening the chest to remove the heart, but in another account given by his brother Marius, also a surgeon who was present at the history-making operation, Barnard removed the heart while it was still beating. If this is true, it can be assumed that Barnard wished to avoid criticism, both ethically and perhaps legally, for removing a beating heart. See also Cooper, *Open Heart*, 334.
41 *"Inserting my hand"*: Barnard, *One Life*, 371-372.
42 *Barnard restarted it*: Christiaan Barnard, "The Operation. A Human Cardiac Transplant: An Interim Report of a Successful Operation Performed at Groote Schuur Hospital, Cape Town," *South African Medical Journal* 1967: 41(48): 1271-1274.
42 *"Dit lyk of dit gaan werk!"*: Barnard, *One Life*, 378-379.
42 *"Eyes over masks"*: Ibid., 379.
42 *second transplant patient*: David Cooper, "Christian Barnard—the Surgeon Who Dared: The Story of the First Human to Human Heart Transplant," *Global Cardiology Science and Practice* 2018: 1-16.
43 *first combined heart-lung*: Robert Robbins, "Norman E. Shumway," *Clinical Cardiology* 2000; 23: 462-466.
43 *about 3,800*: United Network for Organ Sharing (UNOS), 2021 Data, https://unos.org/data/transplant-trends/, accessed June 29, 2022.
43 *about 805,000 Americans*: Centers for Disease Control and Prevention, "Heart Disease Facts."
43 *improving cardiac function*: Konstantinos Malliaras, Raj Makkar, Rachel Smith, et al., "Intracoronary Cardiosphere-Derived Cells After Myocardial Infarction: Evidence of Therapeutic Regeneration in the Final 1-Year Results of the CADUCEUS Trial," *Journal of the American College of Cardiology* 2014; 63(2): 110-122.
44 *In 2008, scientists decellularized*: Harald Ott, Thomas Matthiesen, Saik-Kia Goh, et al., "Perfusion-decellularized Matrix: Using Nature's Platform to Engineer a Bioartificial Heart," *Nature Medicine* 2008; 14(2): 213-221.
44 *to produce functional cardiac tissue*: Jacques Guyette, Jonathan Charest, Robert Mills, et al., "Bioengineering Human Myocardium on Native Extracellular Matrix," *Circulation Research* 2016; 118(1): 56-72.

CHAPTER TWO: DIABETES

46 *Elizabeth Hughes*: Caroline Cox, *The Fight to Survive: A Young Girl, Diabetes, and the Discovery of Insulin* (New York: Kaplan, 2009), ix-xvii.
46 *seventy-five to sixty-five*: Ibid.,1.
47 *750 calories*: Ibid., 40.
47 *one egg for breakfast*: Thea Cooper and Arthur Ainsberg. *Breakthrough: Elizabeth Hughes, the Discovery of Insulin, and the Making of a Medical Miracle* (New York: St. Martin's, 2010), 80. Cox and Cooper's

books are engaging reads that relate far more detail about Elizabeth Hughes's remarkable medical journey.

48 *dropped to fifty-two pounds*: Ibid., 108.
50 *Around the fifth century*: Marianna Karamanou, Athanase Protogerou, Gregory Tsoucalas, et al., "Milestones in the History of Diabetes Mellitus: The Main Contributors," *World Journal of Diabetes* 2016; 7(1): 1-7.
50 *"pissing evil"*: Ibid., 3.
51 *because deceased patients*: Michael Bliss, *The Discovery of Insulin* (Chicago: University of Chicago Press, 1982), 25. I and anyone interested in the history of insulin's discovery are deeply indebted to Canadian historian and author Michael Bliss, whose 1982 classic remains the definitive account of the saga.
52 *Some did try this*: Ibid., 28-33.
52 *Dr. Frederick Banting*: Ibid., 45-48; Frederick Banting, *The Story of the Discovery of Insulin* (Unpublished manuscript, 1940), F. G. Banting Papers, Thomas Fisher Rare Book Library, University of Toronto; Box 1, Folders 9-13: 67-77, 95-99. Henceforth documents from the Banting Collection at the University of Toronto will be referred to as "Banting Papers."
52 *Banting served as*: Seale Harris, *Banting's Miracle: The Story of the Discoverer of Insulin* (Philadelphia: Lippincott, 1946), 28-32.
53 *October 30, 1920*: Banting, *The Story of*, 89; Bliss, *The Discovery*, 48-49. The article Banting read was authored by Dr. Moses Barron and published in the journal *Surgery, Gynecology and Obstetrics*. Barron had published in this journal because lithiasis—the formation of a stone in the body such as the one he had found in the pancreatic duct—was germane to the practice of surgery.
53 *For Banting, this article*: Banting, *The Story of*, 91-95.
53 *"Diabetus"*: Frederick Banting, "Note dated Oct 31/20 from loose leaf notebook 1920/21," Banting Papers; Folder 1.
54 *did not impress Macleod*: Bliss, *The Discovery*, 52.
54 *how to best prepare*: Frederick Banting and Charles Best, "The Internal Secretion of the Pancreas," *Journal of Laboratory and Clinical Medicine* 1922; 7(5): 251-266.
54 *to 0.12 percent*: Ibid., 255.
55 *"stand & walk"*: Bliss, *The Discovery*, 68-70.
55 *one to three dollars each*: Ibid., 61.
56 *"I scarcely know"*: Frederick Banting, "F. G. Banting's Draft of Letter to J. R. R. Macleod, August 9, 1921," Banting Papers; Box 62, Folder 2a, 1-5.
56 *"no possibility of mistake"*: J. R. R. Macleod, "Letter to F. G. Banting, August 23, 1921," Banting Papers; Box 62, Folder 4, 1-4.
56 *"excellent condition"*: Bliss, *The Discovery*, 75.
56 *"I have seen patients die"*: Banting, *The Story of*, 189.
56 *asked Macleod for*: Bliss, *The Discovery*, 82.
57 *"I told him that"*: Frederick Banting, "F. G. Banting: Account of the Discovery of Insulin," Banting Papers; Box 37, Folder 2, 4.
57 *"began to froth"*: Bliss, *The Discovery*, 83.
57 *In his view*: J. R. R. Macleod, "Letter to Col. Gooderham: History of the Researches Leading to

the Discovery of Insulin, Sept. 20, 1922," J. B. Collip Papers, Thomas Fisher Rare Book Library, University of Toronto; Box 37, Folder 3: 3-4, 9-10.
58 *Physiological Journal Club*: Bliss, *The Discovery*, 90-91.
59 *"students were talking"*: Banting, "F. G. Banting: Account of," 5.
59 *"Had I been told"*: J. R. R. Macleod, "Letter to Col. Gooderham," 11.
59 *newborn or fetal animals*: Frederick Banting and Charles Best, "Pancreatic Extracts," *Journal of Laboratory and Clinical Medicine* 1922; 7: 464-472.
60 *"Banting asked me"*: Macleod, "Letter to Col. Gooderham," 12.
60 *initially suggested by Macleod*: Ibid., 9.
60 *The dog's blood sugar*: Banting, "F. G. Banting: Account of," 5.
61 *performed an important experiment*: Bliss, *The Discovery*, 102-103.
62 *December 30, 1921*: Ibid., 104-108.
62 *"When I was called upon"*: Banting, *The Story of*, 200.
63 *"I did not sleep a wink"*: Ibid., 200-201.
63 *Banting now wanted*: Bliss, *The Discovery*, 111.
63 *Leonard Thompson*: Frederick Banting, Charles Best, James Collip, et al., "Pancreatic Extracts in the Treatment of Diabetes Mellitus," *Canadian Medical Association Journal* 1922; 2: 141-146; Bliss, *The Discovery*, 112-113.
64 *110 mg/dL to 320*: "Patient Records for Leonard Thompson," Banting Papers, Box 8B, Folder 17B.
64 *"I experienced then"*: Bliss, *The Discovery*, 117.
64 *"The worst blow fell"*: Banting, *The Story of*, 210-211.
65 *"announced to me"*: Charles Best, "Letter to Sir Henry Dale, February 22, 1954," Feasby Papers, Thomas Fisher Rare Book Library, University of Toronto; Box 3, Folder 5, 4-5.
65 *far more effective*: Banting, "Pancreatic Extracts," 144-145.
65 *Clowes eventually convinced*: Cooper, *Breakthrough*, 166-173.
66 *"My daughter"*: Antoinette Hughes, "Letter to Dr. Frederick Banting, July 3, 1922," Banting Papers; Box 8A, Folder 26A, 2-3, 6.
66 *"wt 45 lbs."*: Frederick Banting, "Notes on First Examination of Elizabeth Hughes," Banting Papers; Box 8A, Folder 25A, 3.
67 *first slice of white bread*: Bliss, *The Discovery*, 154.
67 *"I can't express my"*: Elizabeth Hughes, "Letter to Mother and Father, September 24, 1922," Hughes (Elizabeth) Papers, Thomas Fisher Rare Book Library, University of Toronto; Box 1, Folder 36, 5-6.
67 *"Oh, it is simply too wonderful"*: Elizabeth Hughes, "Letter to Mumsey, October 1, 1922," Hughes (Elizabeth) Papers; Box 1, Folder 39, 4.
67 *"Dr. Allen said with his mouth"*: Elizabeth Hughes, "Letter to Mother, November 28, 1922," Hughes (Elizabeth) Papers; Box 1, Folder 53, 4.
67 *"By Christmas of 1922"*: Bliss, *The Discovery*, 164. The first ten verses of Ezekiel, chapter thirty-seven, describe the prophet in a valley filled with dry bones that God resurrects into a great host of men.
68 *"A man carried his wife"*: Banting, *The Story of*, 304-308.
70 *$10,000*: George Ross, "Letter to Prime Minister Mackenzie King, May 8, 1923," Banting Papers, Box 1, Folder 29, 1.

70 *$7,500*: W. L. Mackenzie King, "Letter to F. G. Banting, Esq., July 23, 1923," Banting Papers; Box 62, Folder 25, 2.
70 *enormous sums*: Banting, *The Story of*, 263-268.
71 *Canadian National Research Council*: James Collip, "Recollections of Sir Frederick Banting," *Canadian Medical Association Journal* 1942; 47(5): 401-403.
71 *February 20, 1941*: Harris, *Banting's Miracle*, 221-230.
71 *fiancé did not learn*: Cooper, *Breakthrough*, 239.
71 *42,000 lifesaving insulin*: Ibid., 244.
72 *34 percent concordance rate*: Jaakko Kaprio, Jaakko Tuomilehto, Markku Koskenvuo, et al., "Concordance for Type 1 (Insulin-Dependent) and Type 2 (Non-Insulin-Dependent) Diabetes in a Population-Based Cohort of Twins in Finland," *Diabetologia* 1992; 35(11): 1060-1067. Probandwise concordance rate reported.
72 *80 to 90 percent*: Roch Nianogo and Onyebuchi Arah, "Forecasting Obesity and Type 2 Diabetes Incidence and Burden: The ViLA-Obesity Simulation Model," *Frontiers in Public Health* 2022; 10(818816): 1-13.
72 *422 million people*: World Health Organization, "Diabetes," https://www.who.int/health-topics/diabetes, accessed July 2, 2022.
72 *37.3 million Americans*: Centers for Disease Control and Prevention, "The Facts, Stats, and Impacts of Diabetes," https://www.cdc.gov/diabetes/library/features/diabetes-stat-report.html, accessed July 2, 2022.
72 *eighth leading cause*: Centers for Disease Control and Prevention, "Leading Causes of Death," https://www.cdc.gov/nchs/fastats/leading-causes-of-death.htm, accessed July 2, 2022.
72 *about ninety-six million*: Centers for Disease Control and Prevention, "The Facts, Stats."
72 *up to 70 percent*: Adam Tabak, Christian Herder, Wolfgang Rathmann, et al., "Prediabetes: A High-Risk State for Developing Diabetes," *Lancet* 2012; 379(9833): 2279-2290.
73 *eight thousand pounds*: Siddhartha Mukherjee. *The Gene: An Intimate History*. (New York: Scribner, 2016), 239.
74 *In November 2021*: Gina Kolkata, "A Cure for Severe Diabetes? For an Ohio Patient, It Worked," *New York Times*, November 28, 2021, 1.

CHAPTER THREE: BACTERIAL INFECTION

75 *seventy-seven deaths*: Jez Gale, "Southampton Blitz—City Remembers on 75th Anniversary," *The Southern Daily Echo* (Southampton, UK), November 30, 2015.
75 *Albert Alexander*: Bill Sullivan, "Guns, Not Roses—Here's the True Story of Penicillin's First Patient," *The Conversation*, March 11, 2022, https://theconversation.com/guns-not-roses-heres-the-true-story-of-penicillins-first-patient-178463, accessed July 3, 2022; Penny Schwartz, "Local Artists Share Childhood Bond," *The Press Enterprise* (Riverside, CA), November 2, 2012. Although many prior accounts of Albert Alexander's injury perpetuated a myth that he cut his face on a rose thorn working in his garden, interviews with Alexander's daughter, Sheila LeBlanc, in the 2010s confirmed that the injury occurred during the bombing raid in Southampton.
76 *The doctors cultured*: Edward Abraham, Ernst Chain, Charles Fletcher, et al., "Further Observations on Penicillin," *Lancet* 1941; 238(6155): 177-189.

77 *"a thousand times smaller"*: Eric Lax, *The Mold in Dr. Florey's Coat* (New York: Henry Holt & Co., 2004), 4. Lax's comprehensive retelling of the entire penicillin story is an excellent resource for anyone interested in delving deeper into the saga.

77 *"All the people"*: Elmer Bendiner, "The Man Who Did Not Invent the Microscope," *Hospital Practice*, August 1984: 168.

78 5×10^{30}, *gram of soil*: William Rosen, *Miracle Cure: The Creation of Antibiotics and the Birth of Modern Medicine* (New York: Viking, 2017), 24.

78 *Contagionists believed*: Lindsey Fitzharris, *The Butchering Art: Joseph Lister's Quest to Transform the Grisly World of Victorian Medicine* (New York: Scientific American/Farrar, Straus and Giroux, 2017), 53-54.

79 *Even cases that*: Ibid., 153.

80 *Florence Nightingale*: Jeannette Farrell, *Invisible Enemies: Stories of Infectious Diseases* (New York: Farrar, Straus and Giroux, 1998), 173-174.

80 *Max von Pettenkofer*: Ibid., 189.

80 *a brewer*: Patrice Debré, *Louis Pasteur*, trans. Elborg Forster (Baltimore: Johns Hopkins University Press, 1998), 87.

81 *These microorganisms*: Thomas Goetz, *The Remedy: Robert Koch, Arthur Conan Doyle, and the Quest to Cure Tuberculosis* (New York: Gotham Books 2014), 58.

81 *devised an experiment*: René Dubois, *Louis Pasteur: Free Lance of Science* (Boston: Little, Brown, 1950), 169-170.

81 *France's silkworm trade*: Goetz, *The Remedy*, 59.

82 *This highly contagious disease*: Debré, *Louis Pasteur*, 302-303.

82 *Using his microscope*: Steve Blevins and Michael Bronze, "Robert Koch and the 'Golden Age' of Bacteriology," *International Journal of Infectious Diseases* 2010; 14: 744-751; Goetz, *The Remedy*, 23-29.

83 *Koch had isolated*: K. Codell Carter, trans., *Essays of Robert Koch* (New York: Greenwood Press, 1987), 1-17. Carter's book is an invaluable resource—an English translation of Koch's most important academic articles.

83 *he was filled with self-doubt*: Thomas Brock, *Robert Koch: A Life in Medicine and Bacteriology* (Washington, D.C.: ASM Press, 1999), 36-38. Microbiologist Dr. Thomas Brock translated much of Robert Koch's correspondence and academic papers. His book was the first English-language biography of Koch.

83 *"I had been receiving"*: Hubert Lechevalier and Morris Solotorovsky, *Three Centuries of Microbiology* (New York: McGraw Hill, 1965), 69.

83 *at 1 a.m.*: Brock, *Robert Koch*, 44-45.

84 *"Within the very first hour"*: Lechevalier, *Three Centuries of*, 69.

84 *"I consider this"*: Carter, *Essays of Robert Koch*, xiv.

84 *"My experiments were"*: Brock, *Robert Koch*, 45.

85 *"Every one of my works"*: Louise Robbins, *Louis Pasteur and the Hidden World of Microbes* (New York: Oxford University Press, 2001), 63.

85 *Pasteur isolated*: Debré, *Louis Pasteur*, 343-344.

85 *commonly told version*: Bernard Dixon, "The Hundred Years of Louis Pasteur," *New Scientist*, October 2, 1980, 30-32.

86 *old, stale culture fluid*: Debré, *Louis Pasteur*, 379; Lechavalier, *Three Centuries of*, 52–54; Louis Pasteur, "Sur les Maladies Virulentes et en Particulier Su la Maladie Appelée Vulgairement Choléra des Poules," *Comptes Rendus de l'Académie des Sciences* 1880; 90: 239–248. ("On Virulent Diseases and Particularly on the Disease Commonly Called Fowl Cholera")

86 *On May 5, 1881*: Louis Pasteur, "Summary Report of the Experiments Conducted at Pouilly-le-Fort, Near Melun, on the Anthrax Vaccination," trans. Tina Dasgupta, *Yale Journal of Biology and Medicine* 2002; 75: 59–62.

87 *9.01 percent to 0.65 percent*: André Eyquem, "One Century After Louis Pasteur's Victory Against Rabies," *American Journal of Reproductive Immunology* 1986; 10: 132–134. Pasteur has generally received the credit for developing the anthrax vaccine; but, in truth, a veterinarian named Jean Joseph Henri Toussaint actually did it first a year before, creating a heat-attenuated vaccine. Still, it was Pasteur's improvement of the vaccine and public demonstration that led to its widespread adoption.

87 *numerous methodological improvements*: Alex Sakula, "Robert Koch: Centenary of the Discovery of the Tubercle Bacillus, 1882," *Thorax* 1982; 37: 246–251.

87 *one-quarter of deaths*: Goetz, *The Remedy*, x.

89 *"the most important"*: Clifford Pickover, *The Medical Book* (New York: Sterling Publishing, 2012), 228.

89 *For the crowd's perusal*: Brock, *Robert Koch*, 128.

89 *"If the number"*: Carter, *Essays of Robert Koch*, 83.

89 *"The color contrast"*: Ibid., 84.

89 *"All who were present"*: Ibid., xvi.

90 *Pasteur complimented him*: Brock, *Robert Koch*, 116.

90 *Two of Koch's assistants*: Blevins, "Robert Koch and the 'Golden Age,'" 746; Wolfgang Hesse, "Walther and Angelina Hesse—Early Contributors to Bacteriology," trans. Dieter Gröschel, *American Society for Microbiology News* 1992; 58(8): 425–428.

90 *"The assumptions from which"*: Carter, *Essays of Robert Koch*, 64.

90 *"Only a few of Pasteur's"*: Ibid., 65–67.

90 *When a translation*: Blevins, "Robert Koch and the 'Golden Age,'" 748.

91 *The animal demonstration*: Debré, *Louis Pasteur*, 406–407.

91 *"Yet however blazingly clear"*: Ibid., 407–408.

91 *a fuming Koch*: H. H. Mollaret, "Contribution to the Knowledge of Relations Between Koch and Pasteur," trans. E. T. Cohn, B. H. Fasciotto-Dunn, U. Kuhn, et al. *NTM-Schriftenr. Gesch. Naturwiss, Technik, Med, Leipzig* 1983; 20(1), S57–65.

91 *"When I saw in the program"*: Brock, *Robert Koch*, 174.

91 *"recueil allemande"*: Mollaret, "Contribution to the Knowledge," S57–65; Goetz, *The Remedy*, 77–78.

92 *"Koch acted ridiculous"*, *"It was a triumph"*: Brock, *Robert Koch*, 174–175.

92 *"I was anxious to hear"*: Carter, *Essays of Robert Koch*, 97–115.

92 *"You do not acknowledge"*: Debré, *Louis Pasteur*, 408.

92 *"You ascribe to me errors"*: Goetz, *The Remedy*, 77.

93 *In August 1883, a cholera outbreak*: Lechavalier, *Three Centuries of*, 144–146.

93 *There, from a twenty-two-year-old man*: Brock, *Robert Koch*, 159–160.

94 *he sought to weaken it*: Leonard Hoenig, "Triumph and Controversy: Pasteur's Preventative Treatment of Rabies as Reported in JAMA." *Archives of Neurology* 1986; 43: 397-399.

94 *He then devised*: Dubos, *Louis Pasteur*, 334.

95 *Pasteur's vaccination method*: Louis Pasteur, "Prevention of Rabies," in *The Founders of Modern Medicine*, edited by Elie Metchnikoff (Freeport, NY: Books for Libraries Press, 1939), 379-387; Dubos, *Louis Pasteur*, 335-336.

95 *nineteen Russians*: Debré, *Louis Pasteur*, 445-446.

95 *By October 1886*: Eyquem, "One Century After Louis Pasteur's," 132. Pasteur was not without his critics. Various contemporaries and historians accused him of deceit (purposely keeping his methods secret to prevent others from using them), taking credit for the ideas of others (such as Toussaint), and unethical behavior (e.g., using the rabies vaccine on humans before adequate testing). However, the wider view of Pasteur's contributions reveals him to be an undeniably brilliant man of science who played a crucial role in the adoption of the germ theory.

96 *"lymph"*: Thomas Daniel, "Robert Koch, Tuberculosis, and the Subsequent History of Medicine," *American Review of Respiratory Disease* 1982; 125(3): 1.

96 *the disease in guinea pigs*: Christoph Gradmann, "Robert Koch and the White Death: From Tuberculosis to Tuberculin," *Microbes and Infection* 2006; 8: 297-299.

96 *numerous physicians*: B. Lee Ligon, "Robert Koch: Nobel Laureate and Controversial Figure in Tuberculin Research," *Seminars in Pediatric Infectious Diseases* 2002; 13(4): 289-299.

98 *The first children saved*: John Gravenstein, "Toxoid Vaccines," in *Vaccines: A Biography* (New York: Springer 2010), 107.

99 *"Number 606"*: Rosen, *Miracle Cure*, 55. It has commonly been stated that the compound was named "606" because it was the 606th compound devised and tested, but this is inaccurate. In Ehrlich's organizational system, the first digit specified a unique compound under investigation, and the following digits denoted variations of that compound. So salvarsan was simply the sixth version of what had been classified as the "sixth" compound.

99 *"salvarsan"*: Robert Schwartz, "Paul Ehrlich's Magic Bullets," *New England Journal of Medicine* 2004; 350(11): 1079-1080. There were, however, limits to salvarsan's effectiveness. It contained arsenic, and this conferred a degree of toxicity that limited how often it could be given. It was also difficult to administer; a single dose of powdered salvarsan had to be highly diluted—dissolved in 600 ml of fluid. Receiving this massive bolus was a difficult experience for patients to endure. See also Rosen, *Miracle Cure*, 58.

100 *"He was not a conversationalist"*: Lax, *The Mold*, 8.

100 *growth had been inhibited*: Gwyn MacFarlane, *Alexander Fleming: The Man and the Myth* (Cambridge, MA: Harvard University Press, 1984), 99-101.

101 *in the range of forty to fifty*: V. D. Allison, "Personal Recollections of Sir Almroth Wright and Sir Alexander Fleming," *Ulster Medical Journal* 1974; 43(2): 89-98.

101 *"That's funny"*: Frank Diggins, "The True History of the Discovery of Penicillin by Alexander Fleming, with Refutation of the Misinformation in the Literature," *British Journal of Biomedical Science* 1999; 56: 83-93.

102 *The more likely source*: Ronald Hare, *The Birth of Penicillin, and the Disarming of Microbes* (London; George Allen & Unwin, 1970), 84.

102 *the London weather*: Ibid., 76-79.

103 *In a petri dish*: Alexander Fleming, "On the Antibacterial Action of Cultures of a Penicillium, with Special Reference to Their Use in the Isolation of B. influenzae." *British Journal of Experimental Pathology* 1929; 10, 226-236.

103 *He gave a presentation*: B. Lee Ligon, "Penicillin: Its Discovery and Early Development," *Seminars in Pediatric Infectious Diseases* 2004; 15(1): 52-57.

103 *He thought it might be useful*: Gwyn MacFarlane, *Howard Florey: The Making of a Great Scientist* (London: Oxford University Press, 1979), 189.

103 *Second, he wrote a paper*: Fleming, "On the Antibacterial Action," 226-236.

105 *"a temperamental Continental"*: Ronald Clark, *Ernst Chain: Penicillin and Beyond* (New York: St. Martin's, 1985), 1.

105 *Florey and Chain developed*: Lax, *The Mold*, 66-67.

105 *Chain came across*: Clark, *Ernst Chain*, 33.

105 *Margaret Campbell-Renton*: Ronald Bentley, "Leslie A. (Epstein) Falk (1915-2004) and Penicillin Production at Oxford," *Journal of Medical Biography* 2007; 15: 93.

105 *Unfortunately, Heatley bristled*: Lax, *The Mold*, 102.

106 *invited Heatley to stay on*: MacFarlane, *Howard Florey*, 302-303.

106 *He experimented by adding*: Lax, *The Mold*, 101.

106 *"I obtained . . ."*: Clark, *Ernst Chain*, 43.

107 *drops of urine*: Lax, *The Mold*, 113.

107 *Even when penicillin*: Ibid., 115.

107 *"It looks quite promising"*: MacFarlane, *Alexander Fleming*, 175-176.

107 *These experiments were reported*: Ernst Chain, Howard Florey, Arthur Gardner, et al., "Penicillin as a Chemotherapeutic Agent," *Lancet* 1940; Aug: 226-231.

107 *To Florey's chagrin*: Lax, *The Mold*, 139.

108 *Yet the prospect of losing*: Norman Heatley, "In Memoriam, H. W. Florey: An Episode," *Journal of General Microbiology* 1970; 61: 297.

109 *sending their two children*: MacFarlane, *Howard Florey*, 320-321.

109 *"striking improvement"*: Abraham, "Further Observations," 185. Alexander was not actually the first human patient to receive penicillin. Unbeknownst to Florey, a New York physician named Martin Henry Dawson had read the Oxford team's *Lancet* article and prepared his own penicillin, also from one of Fleming's original mold samples that had been sent to another American doctor in the mid-1930s. Dawson injected penicillin in a patient with bacterial endocarditis four months before constable Alexander received it. Dawson's patient died, but he was heartened by the lack of toxicity in this first, human test. See also Lennard Bickel, *Rise Up to Life: A Biography of Howard Walter Florey Who Made Penicillin and Gave It to the World* (New York: Charles Scribner's Sons, 1972), 124-126.

110 *"This is the sort of"*: Lax, *The Mold*, 155.

110 *"P-patrol"*: Bickel, *Rise Up to Life*, 122. There were also other members of the Oxford team, including: Edward Abraham, Arthur Gardner, Arthur Gordon Sanders, Jean Orr-Ewing, Mary Ethel Florey (Howard Florey's wife), and Margaret Jennings (whom Florey married after Mary Ethel's death). See also Robert Bud, *Penicillin: Triumph and Tragedy* (New York: Oxford University Press 2007), 30.

110 *The case of another patient*: MacFarlane, *Alexander Fleming*, 185; Abraham, "Further Observations," 185–186.
110 *Three and a half weeks*: Abraham, "Further Observations," 177–189.
110 *in order to convince*: MacFarlane, *Alexander Fleming*, 177.
111 *"This project . . ."*: Lax, *The Mold*, 159.
111 *not to share samples*: Clark, *Ernst Chain*, 66.
111 *to Japan via submarine*: Ibid., 68.
112 *"I left the room silently"*: Ibid.
113 *"I saw a whole tremendous"*: Ibid., 57.
113 *Chain also argued*: MacFarlane, *Alexander Fleming*, 206.
113 *"It is quite clear"*: Clark, *Ernst Chain*, 115.
114 *"The Professor spun"*: Lax, *The Mold*, 174.
114 *One of the first innovations*: Ligon, "Penicillin," 55.
114 *lab member Mary Hunt*: MacFarlane, *Alexander Fleming*, 211.
114 *six times more*: Ligon, "Penicillin," 55–56. The Peoria team ultimately tested about 1,000 mold samples over a five-year period, hoping to find the best penicillin-producing mold. Of these, only Fleming's *penicillium notatum*, the Peoria team's cantaloupe mold, and one other mold were found to emit significant amounts of penicillin. The impossibly low odds that one of these rare species would find its way in a culture plate in Alexander Fleming's lab truly elevates the story of penicillin to one of the most serendipitous occurrences in the history of humankind.
115 *"a carpet bag salesman"*: MacFarlane, *Howard Florey*, 341.
115 *the support of Dr. Alfred Newton Richards*: Lax, *The Mold*, 186–187; Kevin Brown, *Penicillin Man: Alexander Fleming and the Antibiotic Revolution* (Gloucestershire, UK: Sutton Publishing, 2004), 173–174.
115 *Florey published*: Mary Ethel Florey and Howard Florey, "General and Local Administration of Penicillin," *Lancet* 1943; 1: 387–397.
115 *By 1944, twenty-two*: Clark, *Ernst Chain*, 74.
115 *paying royalties*: Lax, *The Mold*, 251; Bernard Dichek, "The Chain Reaction," *Jerusalem Post*, January 22, 2013.
116 *"with my old penicillin"*: Bickel, *Rise Up to Life*, 110.
116 *They even gave Fleming*: Lax, *The Mold*, 144.
116 *Fleming called Florey*: Bickel, *Rise Up to Life*, 166–168.
117 *"Sir, In the leading article"*: MacFarlane, *Alexander Fleming*, 198.
117 *Fleming appeared to enjoy*: Ligon, "Penicillin," 56.
117 *It was not uncommon for*: MacFarlane, *Howard Florey*, 350–351.
117 *title of a New York Times*: Associated Press, "Fleming and Two Co-workers Get Nobel Award for Penicillin Boom," *New York Times*, October 26, 1945, 21.
118 *Time magazine*: Time, May 15, 1944, cover.
118 *started a fund*: Lax, *The Mold*, 232.
119 *"I have now quite good evidence"*: Howard Florey, "Letter to Sir Henry Dale, December 11, 1942," Royal Society, HF/1/3/4/3/1.
119 *"It has long been a source"*: Howard Florey, "Letter to E. Mellanby, June 19, 1944," Royal Society, HF/1/3/2/18/107.

120 *Author Eric Lax*: Lax, *The Mold*, 251.
121 *As early as 1940*: Edward Abraham and Ernst Chain, "An Enzyme from Bacteria Able to Destroy Penicillin. 1940," *Review of Infectious Diseases* 1988; 10(4): 677–678.
121 *"educate them to"*: Alexander Fleming, "Penicillin: Nobel Lecture, December 11, 1945," Nobelprize .org, 93, www.nobelprize.org/prizes/medicine/1945/fleming/lecture, accessed July 6, 2022.
121 *more than 80 percent*: Mariya Lobanovska and Giulia Pilla, "Penicillin's Discovery and Antibiotic Resistance: Lessons for the Future?" *Yale Journal of Biology and Medicine* 2017; 90: 135–145.
121 *2.8 million Americans*: Centers for Disease Control and Prevention, "Biggest Threats and Data," *2019 AR Threats Report*, https://www.cdc.gov/drugresistance/index.html, accessed July 6, 2022.
122 *One study of antibiotic prescriptions*: Katherine Fleming-Dutra, Adam Hersh, Daniel Shapiro, et al. "Prevalence of Inappropriate Antibiotic Prescriptions Among US Ambulatory Care Visits 2010–2011," *Journal of the American Medical Association* 2016; 315(17): 1864–1873.
122 *70 percent of antibiotics*: Food and Drug Administration, "2015 Summary Report on Antimicrobials Sold or Distributed for Use in Food-Producing Animals," December 2016, https://www.fda.gov /media/102160/download, accessed July 6, 2022.
124 *the GAIN Act did not*: Jonathan Darrow and Aaron Kesselheim, "Incentivizing Antibiotic Development: Why Isn't the Generating Antibiotic Incentives Now (GAIN) Act Working?" *Open Forum Infectious Diseases* 2020; 7(1): 1–3.
124 *genetically engineered bacteria*: Lobanovska, "Penicillin's Discovery," 142.
124 *some metals have*: Elena Sanchez-Lopez, Daniela Gomes, Gerard Esteruelas Bonilla, et al., "Metal-Based Nanoparticles as Antimicrobial Agents: An Overview," *Nanomaterials* 2020; 10(2): 292.
124 *To describe phage therapy*: Dmitriy Myelnikov, "An Alternative Cure: The Adoption and Survival of Bacteriophage Therapy in the USSR: 1922–1955," *Journal of the History of Medicine and Allied Sciences* 2018; 73(4): 385–411.

CHAPTER FOUR: VIRAL INFECTION

126 *In October 1952*: Nina Seavey, Jane Smith, and Paul Wagner, *A Paralyzing Fear: The Triumph Over Polio in America* (New York: TV Books, 1998), 253–265. Arvid Schwartz and his medical history are profiled in Seavey's book, and in a 1998 film documentary of the same name.
127 *Completely helpless*: Daniel Wilson, *Living with Polio: The Epidemic and Its Survivors* (Chicago: University of Chicago Press, 2005), 46–47.
129 *A child who was unable*: Jeffrey Kluger, *Splendid Solution: Jonas Salk and the Conquest of Polio* (New York; G. P. Putnam's Sons, 2004), 2.
129 *72,000 stray cats*: Seavey, *A Paralyzing Fear*, 21.
130 *The death rate rose as high as*: Wilson, *Living with Polio*, 46.
131 *It had to be infectious*: Charlotte Jacobs, *Jonas Salk: A Life* (New York: Oxford University Press, 2015), 67.
131 *"We failed utterly"*: Simon Flexner and Paul Lewis, "The Nature of the Virus of Epidemic Poliomyelitis," *Journal of the American Medical Association* 1909; 53(25): 2095.
132 *27,000 cases*: Joseph Melnick, "Current Status of Poliovirus Infections," *Clinical Microbiology Reviews* 1996; 9(3): 293–300.
132 *In New York City alone*: David Oshinsky, *Polio: An American Story* (New York: Oxford University

Press, 2005), 22. Oshinsky's book provides an excellent and comprehensive account of the quest to defeat polio—recommended for anyone who wishes to delve deeper into the topic.
132 *A 1955 survey*: Joe Coffey, "History Happenings: Before COVID-19 Came Polio and, Finally, a Vaccine," *The Gazette* (Cedar Rapids, IA), April 20, 2021.
132 *the steam yacht Pocantico*: James Tobin, *The Man He Became: How FDR Defied Polio to Win the Presidency* (New York: Simon & Schuster, 2013), 15-17.
133 *"regular, old-fashioned"*: Ibid., 29.
133 *On August 10, 1921*: Ibid., 47-51.
133 *uniquely poised to contract*: Oshinsky, *Polio*, 27.
134 *Each summer approximately 75,000*: Tobin, *The Man He Became*, 28-29.
134 *so much buoyancy*: Oshinsky, *Polio*, 37.
134 *$200,000, approximately two-thirds of his fortune*: Jacobs, *Jonas Salk*, 70.
135 *Far more children died*: Oshinsky, *Polio*, 5.
135 *found in a Chinese book*: Arthur Boylston, "The Origins of Inoculation," *Journal of the Royal Society of Medicine* 2012; 105(7): 309-313.
135 *China since the tenth century*: Simon Winchester, *The Man Who Loved China: The Fantastic Story of the Eccentric Scientist Who Unlocked the Mysteries of the Middle Kingdom* (New York: Harper, 2008), 276.
136 *killed 400,000 people worldwide*: Paul Offit, *The Cutter Incident: How America's First Polio Vaccine Led to the Growing Vaccine Crisis* (New Haven, CT: Yale University Press, 2005), 12.
136 *The first Westerner*: Andrew Artenstein, "Smallpox," in *Vaccines: A Biography* (New York: Springer 2010), 11-13.
137 *"I shall never have smallpox"*: Stefan Riedel, "Edward Jenner and the History of Smallpox and Vaccination," *Baylor University Medical Center Proceedings* 2005; 18: 21-25.
137 *Jenner gradually collected*: Alfredo Morabia, "Edward Jenner's 1798 Report of Challenge Experiments Demonstrating the Protective Effects of Cowpox Against Smallpox," *Journal of the Royal Society of Medicine* 2018; 111(7): 255-257.
137 *In May 1796*: Kendall Smith, "Edward Jenner and the Small Pox Virus," *Frontiers in Immunology* 2011; 2(21): 1-6. Though Jenner deserves credit for investigating and promoting vaccination, he was not the first to vaccinate for smallpox using cowpox. Unbeknownst to him, an English farmer named Benjamin Jesty had successfully vaccinated his wife and two sons with cowpox twenty-two years before—though Jesty did not seek to disseminate or popularize his method.
138 *In France*: Debré, *Louis Pasteur*, 384.
139 *In the 1880s, Mayer*: Adolf Mayer, "Concerning the Mosaic Disease of Tobacco," in *Phytopathological Classics Number 7*, trans. James Johnson (St. Paul, MN: American Phytopathological Society Press, 1942), 9-24.
139 *In 1892, a Russian*: Dmitri Ivanowski, "Concerning the Mosaic Disease of the Tobacco Plant," in *Phytopathological Classics Number 7*, trans. James Johnson (St. Paul, MN: American Phytopathological Society Press, 1942), 25-30.
139 *in 1898, a Dutch*: Martinus Beijerinck, "Concerning a Contagium Vivum Fluidum as a Cause of the Spot Disease of Tobacco Leaves," in *Phytopathological Classics Number 7*, trans. James Johnson (St. Paul, MN: American Phytopathological Society Press, 1942), 33-52.
140 *In the 1930s, Max Theiler*: Offit, *The Cutter Incident*, 14; Max Theiler and Hugh Smith, "The Use

of Yellow Fever Virus Modified by In Vitro Cultivation for Human Immunization," *Journal of Experimental Medicine* 1937; 65: 787–800.
142 *"Polio Panic" and "Polio's Deadly Path"*: Oshinsky, *Polio*, 85.
142 *25,000 polio cases*: Ibid., 81.
142 *42,000 cases*: Ibid., 128.
142 *58,000 cases*: Ibid., 81.
143 *virus was uncommonly found*: Ibid., 125.
144 *Many virus families*: Jane Smith, *Patenting the Sun: Polio and the Salk Vaccine* (New York: William Morrow, 1990), 109.
144 *If that monkey was injected*: Richard Carter, *Breakthrough: The Saga of Jonas Salk* (New York: Trident Press, 1966), 79. Considered an indispensable biography of Jonas Salk, Carter's book benefited from his extensive, weeks-long interviews with Salk, and interviews with other major figures in the effort to defeat polio.
144 *In 1949, a third type*: Oshinsky, *Polio*, 117. Of the 196 poliovirus strains tested, 161 were type 1, twenty were type 2, and fifteen were type 3. See also Jacobs, *Jonas Salk*, 89.
144 *$1.19 million*: Carter, *Breakthrough*, 73.
145 *In 1948, Enders*: John Enders, Thomas Weller, and Frederick Robbins, "Cultivation of the Lansing Strain of Poliomyelitis Virus in Cultures of Various Human Embryonic Tissues," *Science* 1949; 109: 85–87.
146 *"Now, Dr. Salk"*: Carter, *Breakthrough*, 81.
146 *"like being kicked", "I could feel"*: Ibid.
146 *"It became obvious"*: Oshinsky, *Polio*, 151–152.
146 *"There is no valid reason"*: Carter, *Breakthrough*, 92.
147 *He could be arrogant*: Ibid., 137.
147 *"There were sixteen or seventeen"*: Ibid., 107–108.
148 *A single kidney*: Oshinsky, *Polio*, 154.
148 *It was a delicate balance*: Smith, *Patenting the Sun*, 132.
148 *If a monkey became ill*: Oshinsky, *Polio*, 156.
148 *he and his lab team first*: Smith, *Patenting the Sun*, 136.
149 *"When you inoculate children"*: Carter, *Breakthrough*, 139.
149 *the vaccine appeared to work*: Jonas Salk, "Studies in Human Subjects on Active Immunization Against Poliomyelitis," *Journal of the American Medical Association* 1953; 151(13): 1081–1098.
149 *"It was a tense meeting", "It was almost as if"*: Carter, *Breakthrough*, 144.
149 *Impatient for a vaccine*: Jacobs, *Jonas Salk*, 116.
149 *"So far as anyone knows"*: John Troan, *Passport to Adventure* (Pittsburgh: Neworks Press, 2000), 198; *Pittsburgh Press*, January 27, 1953, 2. John Troan was a reporter for the *Pittsburgh Press* newspaper whom Salk befriended and allowed to cover his vaccine development closely.
150 *"Researcher Salk"*: "Vaccine for Polio," *Time*, February 9, 1953, 43.
150 *"Although it was nice"*: Albert Sabin, "Letter to Jonas Salk, February 9, 1953," Jonas Salk Papers, Mandeville Special Collections, University of California, San Diego, Box 93, Folder 5.
150 *"Polio Conquest Nearer" and "Hint Polio Vaccine Ready"*: Oshinsky, *Polio*, 171.
151 *"In the studies that are"*: Carter, *Breakthrough*, 162.
151 *"Told me I was"*: Ibid., 156.

151 *"Jonas E. Christ"*: Ibid., 214.
151 *He testified before*: Jacobs, *Jonas Salk*, 128.
151 *14,000 schools*: Troan, *Passport to Adventure*, 219.
151 *623,972 first, second, and*: Marcia Meldrum, "'A Calculated Risk': The Salk Polio Vaccine Field Trials of 1954," *British Medical Journal* 1998; 317: 1233-1236.
152 *"Attention everyone!"*: Carter, *Breakthrough*, 231.
152 *Approximately 95 percent*: Oshinsky, *Polio*, 199.
152 *$9 million*: Carter, *Breakthrough*, 242.
152 *The first three words*: Kluger, *Splendid Solution*, 296.
152 *"They brought the report"*: Seavey, *A Paralyzing Fear*, 189.
152 *The vaccine was safe*: Thomas Francis, "Evaluation of the 1954 Poliomyelitis Vaccine Field Trial," *Journal of the American Medical Association* 1955; 158(14): 1266-1270. In placebo-controlled areas, the vaccine was 68% effective against type 1, 100% effective against type 2, and 92% effective against type 3. This was far better than many expected. See also Kluger, *Splendid Solution*, 296.
153 *The total effort*: Troan, *Passport to Adventure*, 223.
153 *"POLIO IS CONQUERED"*: "Polio Is Conquered," *Pittsburgh Press*, April 12, 1955, 1.
153 *"POLIO ROUTED!"*: "Polio Routed!" *New York Post*, April 13, 1955, 1; Kluger, *Splendid Solution*, 301.
153 *"TRIUMPH OVER POLIO"*: "Triumph Over Polio," *South China Morning Post*, April 13, 1955, 1; Jacobs, *Jonas Salk*, 167.
153 *"Hi Billy, I'm back from"*: Seavey, *A Paralyzing Fear*, 208.
153 *"The worst tragedy"*: Carter, *Breakthrough*, 3.
154 *He tried to get the press to stop*: Jacobs, *Jonas Salk*, 135.
154 *"Who holds the patent"*: Carter, *Breakthrough*, 283-284.
154 *"Young man, a great"*: Ibid., 285.
155 *On April 24, 1955*: Offit, *The Cutter Incident*, 83.
155 *On April 25*: Neal Nathanson and Alexander Langmuir, "The Cutter Incident: Poliomyelitis Following Formaldehyde-Inactivated Poliovirus Vaccination in the United States During the Spring of 1955," *American Journal of Hygiene* 1963; 78: 16-27.
155 *An extensive investigation*: David Bodian, Thomas Francis, Carl Larson, et al. "Interim Report, Public Health Service Technical Committee on Poliomyelitis Vaccine," *Journal of the American Medical Association* 1955; 159(15): 1445; Offit, *The Cutter Incident*, 67, 110. The failure to completely kill the virus was presumably due to sediment consisting of monkey kidney cell debris that had formed over months while the vaccine mixtures were in storage. Viruses caught in the sediment were shielded from exposure to the formaldehyde intended to kill them. New protocols including filtration to remove sediment were instituted at all manufacturing centers.
155 *The botched Cutter vaccine*: Oshinsky, *Polio*, 237.
155 *"This was the first and only time"*: Carter, *Breakthrough*, 323.
155 *15,000 cases*: Oshinsky, *Polio*, 255.
156 *Sabin weakened the poliovirus*: Ibid., 245.
156 *Sabin tested his vaccine*: Ibid., 245-246.
157 *difficulty with administering the Salk vaccine*: Seavey, *A Paralyzing Fear*, 229.
157 *77 million people*: Oshinsky, *Polio*, 253.

157 *"hundreds of children"*: Ibid., 265.
157 *proved longer-lasting*: Jonas Salk, "Persistence of Immunity After Administration of Formalin-Treated Poliovirus Vaccine," *Lancet* 1960; 2(7153): 715–723.
157 *92 percent in five years*: Carter, *Breakthrough*, 370.
157 *97 percent*: Jacobs, *Jonas Salk*, 227.
157 *"Not only is scientific justification"*: Carter, *Breakthrough*, 376.
157 *He believed his vaccine*: Oshinsky, *Polio*, 266–267.
158 *"Normally, my father tried"*: Ibid., 268.
158 *From 1952 to 1981*: Melinda Moore, Peter Katona, Jonathan Kaplan, et al., "Poliomyelitis in the United States, 1969–1981," *Journal of Infectious Diseases* 1982; 146(4): 558.
159 *The risk was about*: Neal Nathanson, "Eradication of Poliomyelitis in the United States," *Reviews of Infectious Diseases* 1982; 4(5): 943.
159 *In 1996, the CDC*: Oshinsky, *Polio*, 278–279.
159 *In late December 2019*: Andrew Green, "Li Wenliang," *Lancet* 2020; 395: 682.
160 *"7 confirmed cases"*: Li Wenliang, WeChat posts, *Wuhan University Clinical Medicine 2004 WeChat Group*, December 30, 2019, https://web.archive.org/web/20200206144253/http://www.bjnews.com.cn/feature/2020/01/31/682076.html, accessed July 11, 2022.
160 *December 31, 2019*: Derrick Bryson Taylor, "A Timeline of the Coronavirus Pandemic," *New York Times*, March 17, 2021.
160 *"I will join medical workers"*: Editorial, "He Warned of Coronavirus. Here's What He Told Us Before He Died," *New York Times*, February 7, 2020.
161 *By April, almost 10 million*: Taylor, "A Timeline."
162 *On June 28, 1802*: Carlos Franco-Paredes, Lorena Lammaoglia, and Jose Santos-Preciado, "The Spanish Royal Philanthropic Expedition to Bring Smallpox Vaccination to the New World and Asia in the 19th Century," *Clinical Infectious Diseases* 2005; 41: 1285–1289. Those who research this mission will often encounter an alternate spelling of Balmis's name: Francisco Xavier de Balmis.
163 *vaccination of 1.5 million*: Kenny, *The Plague Cycle*, 130.
163 *career of Katalin Karikó*: Gina Kolata, "Kati Kariko Helped Shield the World from the Coronavirus" *New York Times*, April 9, 2021, 6; Carolyn Johnson, "A One-Way Ticket. A Cash-Stuffed Teddy Bear. A Dream Decades in the Making," *Washington Post*, October 1, 2021; Damian Garde and Jonathan Saltzman, "The Story of mRNA," *Stat*, November 10, 2020, https://www.statnews.com/2020/11/10/the-story-of-mrna-how-a-once-dismissed-idea-became-a-leading-technology-in-the-covid-vaccine-race/, accessed September 17, 2022.
164 *"I felt like a god"*: Kolata, "Kati Kariko."
164 *repeated grant rejections*: Author interview with Katalin Karikó, September 30, 2022.
165 *"I am an RNA scientist"*: Kolata, "Kati Kariko."
165 *red blood cell counts in mice*: Author interview with Katalin Karikó, September 30, 2022.
167 *two-thirds of child deaths*: Kenny, *The Plague Cycle*, 191.
167 *A 1997 Hong Kong outbreak*: Paul Chan, "Outbreak of Avian Influenza A(H5N1) Virus Infection in Hong Kong in 1997," *Clinical Infectious Diseases* 2002; 34: S58–S64.
167 *A 2003 outbreak*: Arjan Stegeman, Annemarie Bouma, Armin Elberts, et al., "Avian Influenza A Virus (H7N7) Epidemic in the Netherlands in 2003: Course of the Epidemic and Effectiveness

of Control Measures," *Journal of Infectious Diseases* 2004; 190: 2088-2095; John Barry, *The Great Influenza: The Story of the Deadliest Plague in History* (New York: Viking, 2004), 114.

CHAPTER FIVE: CANCER

169 *Einar Gustafson*: This retelling of Einar Gustafson's childhood history and lymphoma diagnosis is derived from the following sources: Siddhartha Mukherjee, *The Emperor of All Maladies: A Biography of Cancer* (New York: Scribner, 2010), 96; Douglas Martin, "Einar Gustafson, 65, 'Jimmy' of Child Cancer Fund, Dies," *New York Times*, January 24, 2001, 17; Pamela Ferdin, "'This Is Jimmy. Heard You Were Lookin' for Me,'" *Washington Post*, May 22, 1998. Any student of cancer's history would do well to study Mukherjee's Pulitzer Prize-winning book. It is an excellent and comprehensive history of oncology from antiquity to 2010 that highlights the stories of Einar Gustafson and Sidney Farber.

170 *Ninety percent*: David Nathan, *The Cancer Treatment Revolution: How Smart Drugs and Other Therapies Are Renewing Our Hope and Changing the Face of Medicine* (Hoboken, NJ: John Wiley & Sons, 2007), 45.

170 *Each year*: Rebecca Siegel, Kimberly Miller, Hannah Fuchs, et al., "Cancer Statistics, 2021," *CA: A Cancer Journal for Clinicians* 2021; 71: 7-33. The figure, 21 percent, is data from 2018.

170 *one out of every three*: Robin Hesketh, *Betrayed by Nature: The War on Cancer* (New York: Palgrave Macmillan, 2012), 20.

171 *second most prolific killer*: Clifton Leaf, *The Truth In Small Doses. Why We're Losing the War on Cancer— and How to Win It*. (New York: Simon & Schuster, 2013), 35-36.

172 *inconvenient addiction to cocaine*: Gerald Imber, *Genius on the Edge: The Bizarre Double Life of Dr. William Stewart Halsted* (New York: Kaplan, 2010), 55-57.

173 *cancer would spring up again*: Mukherjee, *The Emperor*, 59.

173 *spread directly outward*: Michael Osborne, "William Stewart Halsted: His Life and Contributions to Surgery," *Lancet Oncology* 2007; 8: 256-265.

173 *an extensive operation*: Imber, *Genius on the Edge*, 120-121.

173 *"a mistaken kindness"*: Osborne, "William Stewart Halsted," 259-260.

173 *On the evening of November 8, 1895*: K. T. Claxton, *Wilhelm Röntgen* (London: Heron Books, 1970), 40-44.

174 *X-rays were not blocked*: Wilhelm Röntgen, "On a New Kind of Rays," *Science* 1896; 3(59): 227-231.

174 *swollen and painfully*: Paul Hodges, *The Life and Times of Emil H. Grubbe* (Chicago: University of Chicago Press, 1964), 23-24.

174 *Grubbé treated the tumor*: Mukherjee, *The Emperor*, 75-76.

175 *endured four years of arduous*: Eve Curie, *Madame Curie*, trans. Vincent Sheean (New York: Da Capo Press, 2001), 169, 175; Robert Reid, *Marie Curie* (New York: E. P. Dutton & Co., 1974), 95.

175 *a new way to treat cancer*: Reid, *Marie Curie*, 126.

175 *so tightly packed that*: Glenn Infield, *Disaster at Bari* (New York; Macmillan, 1971), 2.

176 *only one battery*: Jennet Conant, *The Great Secret: The Classified World War II Disaster That Launched the War on Cancer* (New York; W. W. Norton & Company, 2020), xiii.

176 *"I would regard it as"*: Ibid., x.

176 *105 Junkers Ju 88 bombers*: Guy Faguet, *The War on Cancer: An Anatomy of a Failure. A Blueprint for the Future* (New York: Springer, 2005), 70.

Notes

176 *seventeen ships were sunk*: Infield, *Disaster at Bari*, 141.
176 *An oil pipeline*: Conant, *The Great Secret*, xv.
176 *The total number of dead*: Infield, *Disaster at Bari*, xi.
176 *"Second Pearl Harbor"*: Ibid., 141.
176 *the distinctive smell of garlic*: Ibid., 62.
176 *sixty to seventy pounds*: Ibid., 17. The ship was named for John Harvey, a member of the Continental Congress in 1777, and signer of the Articles of Confederation. Every crewman on the ship was lost in the attack.
177 *628 military personnel*: Faguet, *The War on Cancer*, 71.
177 *An intact American bomb casing*: Conant, *The Great Secret*, 89-90.
177 *The 1925 Geneva Protocol*: Ibid., 16.
177 *their white blood cell counts*: Ibid., 101. Mustard gas was used in the First World War, and low white blood cell counts had been noted in some victims, but few in the medical community had grasped the importance of this effect.
178 *"If mustard could do this"*: Ibid.
178 *noticed by Colonel Cornelius Rhoads*: Ibid., 164. Rhoads's interest in Alexander's Bari data had been primed by his knowledge of a secret government research project conducted by two Yale University pharmacologists named Louis Goodman and Alfred Gilman. In August 1942, Goodman and Gilman gave nitrogen mustard to a series of seven lymphoma patients. The results were mixed. By mid-1943, this study was discontinued and deemed too small to determine if the benefits of treatment outweighed the chemical's adverse effects. See also Louis Goodman, Maxwell Wintrobe, William Dameshek, et al., "Nitrogen Mustard Therapy," *Journal of the American Medical Association* 1946; 132: 126-132; Conant, *The Great Secret*, 207-208.
178 *clinical trial of 160 cancer patients*: Cornelius Rhoads, "Nitrogen Mustards in the Treatment of Neoplastic Disease," *Journal of the American Medical Association* 1946; 131(8): 656-658; Cornelius Rhoads, "Report on a Cooperative Study of Nitrogen Mustard (HN2) Therapy of Neoplastic Disease," *Transactions of the Association of American Physicians* 1947; 60(1): 110-117.
178 *Sidney Farber*: Mukherjee, *The Emperor*, 11-12, 18-19.
179 *deterred many young*: John Laszlo, *The Cure of Childhood Leukemia: Into the Age of Miracles* (New Brunswick, NJ: Rutgers University Press, 1995), 182.
179 *Wills found a nutrient*: Mukherjee, *The Emperor*, 28-29. Because folic acid was first extracted from leafy vegetables, its name was derived from the Latin word for "leaf," *folium*. See also Laszlo, *The Cure*, 27.
179 *In 1945, scientists*: R. Leuchtenberger, C. Leuchtenberger, D. Laszlo, et al., "The Influence of 'Folic Acid' on Spontaneous Breast Cancers in Mice," *Science* 1945; 101(2611): 46.
179 *it actually worsened the disease*: Mukherjee, *The Emperor*, 29.
180 *Yellapragada Subbarao*: Ibid., 30-31. Subbarao's name is sometimes spelled "Subbarow."
180 *On December 28, 1947*: Sidney Farber, Louis Diamond, Robert Mercer, et al., "Temporary Remissions in Acute Leukemia in Children Produced by Folic Acid Antagonist, 4-Aminopteroyl-glutamic Acid (Aminopterin)," *New England Journal of Medicine* 1948; 238(23): 787-793.
180 *Sandler's leukemia relapsed*: Mukherjee, *The Emperor*, 35.
180 *severe disapproval*: Denis Miller, "A Tribute to Sidney Farber—the Father of Modern Chemotherapy,"

British Journal of Haematology, 2006; 134: 20–26; Robert Cooke, *Dr. Folkman's War: Angiogenesis and the Struggle to Defeat Cancer* (New York: Random House, 2001), 114; Mukherjee, *The Emperor*, 34.

180 *In a 1948 article*: Farber, "Temporary Remissions," 787–793.
182 *the Variety Club, $45,456*: Mukherjee, *The Emperor*, 95–96.
182 *The National Foundation inspired*: Gretchen Krueger, "'For Jimmy and the Boys and Girls of America': Publicizing Childhood Cancers in Twentieth-Century America," *Bulletin of the History of Medicine* 2007; 81: 70–93.
183 *he found the perfect one*: Mukherjee, *The Emperor*, 96.
183 *"Tonight we take you"*: The original broadcast of Edwards's interview with Einar Gustafson and the Boston Braves players can be found on the Jimmy Fund website: http://www.jimmyfund.org/about-us/about-the-jimmy-fund/einar-gustafson-jimmy-was-inspiration-for-the-jimmy-fund/, accessed July 14, 2022. Partial transcripts and descriptions of the interview can also be found in: Saul Wisnia, *The Jimmy Fund of Dana-Farber Cancer Institute* (Charleston, SC: Arcadia Publishing, 2002), 18–19; Mukherjee, *The Emperor*, 97–99.
185 *That very evening*: Mukherjee, *The Emperor*, 99. In establishing the Jimmy Fund, Farber was greatly aided by the work of philanthropist and socialite Mary Lasker, as well as celebrities like Ted Williams.
186 *cure choriocarcinoma*: Min Chiu Li, Roy Hertz, and Donald Spencer, "Effect of Methotrexate Therapy upon Choriocarcinoma and Chorioadenoma," *Proceedings of the Society for Experimental Biology and Medicine* 1956; 93(2): 361–366; Mukherjee, *The Emperor*, 135–138.
187 *In 1951, two biochemists*: Leaf, *The Truth*, 234–235. Hitchings and Elion shared a Nobel Prize in 1988.
187 *Even the dirt*: Vincent DeVita and Elizabeth DeVita-Raeburn, *The Death of Cancer* (New York: Farrar, Straus and Giroux, 2015), 66.
187 *periwinkle plant*: James Wright, "Almost Famous: E. Clark Noble, the Common Thread in the Discovery of Insulin and Vinblastine," *Canadian Medical Association Journal* 2002; 167(12): 1391–1396. As a medical student, Clark Noble lost a coin flip with Charles Best to determine who would first work with Frederick Banting (and discover insulin) during the fateful summer of 1921. Clark's brother, Robert, worked in the lab of James Collip, also of insulin fame.
187 *Robert Noble, injected*: Robert Noble, Charles Beer, and Harry Cutts, "Role of Chance Observations in Chemotherapy: *Vinca Rosea*," *Annals of New York Academy of Sciences* 1958; 76(3): 882–894. Robert Noble later found that Eli Lilly & Company had also investigated the periwinkle plant and identified the same phenomenon.
187 *Noble later cooperated*: Robert Noble, "The Discovery of the Vinca Alkaloids—Chemotherapeutic Agents Against Cancer," *Biochemistry and Cell Biology* 1990; 68: 1344–1351.
188 *tested actinomycin D in humans*: Mukherjee, *The Emperor*, 122–123.
188 *"maximum tolerated dose"*: Cooke, *Dr. Folkman's War*, 53.
189 *Since each drug caused different*: Nathan, *The Cancer Treatment*, 48–49. The idea of combination chemotherapy was also promoted by Abraham Goldin and Lloyd Law at the National Cancer Institute, and Howard Skipper and Frank Schabel at the Southern Research Institute in Alabama.
189 *Many of Frei and Freireich's*: DeVita, *The Death of Cancer*, 47; Laszlo, *The Cure*, 182–183.
189 *"It's fine for rats and mice"*: Nathan, *The Cancer Treatment*, 50.
190 *"This is a meat market!", "It was embarrassing"*: DeVita, *The Death of Cancer*, 49.

190 *advised to avoid*: Ibid., 50.
190 *toxic VAMP cocktail*: Mukherjee, *The Emperor*, 144-145.
190 *"For me, this was a nightmare"*: Nathan, *The Cancer Treatment*, 57. David Nathan later became the president of the Dana-Farber Cancer Institute.
190 *60 percent*: Domenico Ribatti, "Sidney Farber and the Treatment of Childhood Acute Lymphoblastic Leukemia with a Chemotherapeutic Agent," *Pediatric Hematology and Oncology* 2012; 29: 299-302.
191 *By 1975, the five-year*: Leaf, *The Truth*, 238.
191 *all-cancer five-year survival rate*: DeVita, *The Death of Cancer*, 36.
192 *8.7 percent*: John Bailar and Elaine Smith, "Progress Against Cancer?" *New England Journal of Medicine* 1986; 314(19): 1226-1232.
192 *"a black box that we're trying"*: Cooke, *Dr. Folkman's War*, 157.
192 *a macabre joke*: Robert Bazell, *Her-2: The Making of Herceptin, a Revolutionary Treatment for Breast Cancer* (New York: Random House, 1998), xvi.
193 *studied the prostate glands*: Charles Huggins, Lillian Eichelberger, and James Wharton, "Quantitative Studies of Prostatic Secretion: I. Characteristics of the Normal Secretion; The Influence of the Thyroid, Suprarenal, and Testis Extirpation and Androgen Substitution on the Prostatic Output," *Journal of Experimental Medicine* 1939; 70(6): 543-556.
193 *when he removed their testicles*: Hesketh, *Betrayed by Nature*, 15.
193 *a Scottish surgeon named George Beatson*: Bazell, *Her-2*, 25.
193 *shown in 1971*: Mary Cole, C. Jones, and I. Todd, "A New Anti-Oestrogenic Agent in Late Breast Cancer: An Early Clinical Appraisal of ICI46474," *British Journal of Cancer* 1971; 25(2): 270-275.
194 *how heavily many tumors were invested*: Leaf, *The Truth*, 208-209.
194 *"purifying dirt"*: Cooke, *Dr. Folkman's War*, 117.
194 *"You're making a mockery"*: Ibid., 183.
194 *"Judah made presentations almost"*: DeVita, *The Death of Cancer*, 277.
194 *published in Science*: Yuen Shing, Judah Folkman, R. Sullivan, et al., "Heparin Affinity: Purification of a Tumor-Derived Capillary Endothelial Cell Growth Factor," *Science* 1984; 223: 1296-1298.
195 *surgeon named Percivall Pott*: Mukherjee, *The Emperor*, 237-239.
195 *cytologist named George Papanicolaou*: Barron Lerner, *The Breast Cancer Wars: Hope, Fear, and the Pursuit of a Cure in Twentieth Century America* (New York: Oxford University Press, 2001), 48-49; Mukherjee, *The Emperor*, 286-289.
195 *"New Cancer Diagnosis"*: George Papanicolaou, "George Nicholas Papanicolaou's New Cancer Diagnosis Presented at the Third Race Betterment Conference, Battle Creek, Michigan, January 2-6, 1928, and Published in the Proceedings of the Conference," *CA: A Cancer Journal for Clinicians* 1973; 23(3): 174-179. Part of the reason Papanicolaou's work was ignored for so long may have been because he had not been able to publish his research in a major medical journal. His first paper was presented at a conference on eugenics, a field that became discredited after the Second World War.
196 *large-scale trial*: Mukherjee, *The Emperor*, 289-290.
196 *six million women*: Michael Kinch, *The End of the Beginning: Cancer, Immunity, and the Future of a Cure* (New York: Pegasus Books, 2019), 114.
196 *virologist named Peyton Rous*: James Patterson, *The Dread Disease: Cancer and Modern American Culture* (Cambridge, MA: Harvard University Press, 1987), 59; Mukherjee, *The Emperor*, 173.

197 *he excised a tumor*: Hesketh, *Betrayed by Nature*, 111.
197 *The search was fruitless until*: Leaf, *The Truth*, 261–262.
197 *a band across sub-Saharan*: Ibid., 269.
197 *a used station wagon*: Ibid., 274.
197 *a region's temperature*: Ibid., 276–277; Kinch, *The End*, 109–110.
198 *In 1963, Burkitt sent*: Kinch, *The End*, 110–111.
198 *"New Evidence That"*: *Life*, June 22, 1962, cover.
198 *To prove bacteria*: Mukherjee, *The Emperor*, 283.
199 *In 1970, two virologists*: Geoffrey Cooper, Rayla Greenberg-Temin, and Bill Sugden, eds. *The DNA Provirus: Howard Temin's Scientific Legacy* (Washington, D.C.: ASM Press, 1995), xiii, xx, 47; Howard Temin and Satoshi Mizutani, "RNA-Dependent DNA Polymerase in Virions of Rous Sarcoma Virus," *Nature* 1970; 226(5252): 1211–1213; David Baltimore, "RNA-Dependent DNA Polymerase in Virions of RNA Tumor Viruses," *Nature* 1970; 226(5252): 1209–1211.
199 *four different genes*: J. Michael Bishop, *How to Win a Nobel Prize* (Cambridge, MA: Harvard University Press, 2003), 161.
199 *An even greater discovery*: Ibid., 164.
200 *termed* proto-oncogenes: Mukherjee, *The Emperor*, 362.
200 *If a suppressor gene*: Mukherjee, *The Emperor*, 368. Each person has two copies of each tumor suppressor gene (one from each parent), and children only develop retinoblastoma if both copies become mutated, which can occur spontaneously or by inheritance.
200 *at least a hundred oncogenes*: Ibid., 386.
201 *An oncologist named*: Bazell, *Her-2*, 42.
202 *Pancreatic cancer has twelve*: DeVita, *The Death of Cancer*, 265.
202 *A 2006 study of*: Tobias Sjoblom, Sian Jones, Laura Wood, et al., "The Consensus Coding Sequences of Human Breast and Colorectal Cancers," *Science* 2006; 314: 268–274.
202 *"The actual course of research"*: William Hahn and Robert Weinberg, "Rules for Making Human Tumor Cells," *New England Journal of Medicine* 2002; 347(20): 1593–1603.
203 *Dr. William Coley*: Stephen Hall, *A Commotion in the Blood* (New York; Henry Holt & Co., 1997), 22–24; Charles Graeber, *The Breakthrough: Immunotherapy and the Race to Cure Cancer* (New York; Twelve, 2018), 36–39; Kinch, *The End*, 124–125.
203 *"one of the most malignant"*: William Coley, "The Diagnosis and Treatment of Bone Sarcoma," *Glasgow Medical Journal* 1936; 8(2): 82.
203 *German immigrant*: William Coley, "The Treatment of Malignant Tumors by Repeated Inoculations of Erysipelas: With a Report of Ten Original Cases," *Clinical Orthopedics and Related Research* 1991; 262: 3–11 (reprinted from the *American Journal of the Medical Sciences* 1893; 105: 487).
203 *"absolutely hopeless"*: Hall, *A Commotion*, 40.
204 *roamed the Lower East Side*: Ibid., 29.
204 *prominent neck scar*: Stephen Hoption Cann, Johannes van Netten, and Chris van Netten, "Dr. William Coley and Tumour Regression: A Place in History or in the Future," *Postgraduate Medical Journal* 2003; 79(938): 672–680.
204 *"If erysipelas"*, *"determined to try"*: Hall, *A Commotion*, 42. Coley later learned a European physician named Friedrich Fehleisen had previously tested a similar idea, injecting bacteria into seven

Notes

patients in 1882. Also, a German scientist named W. Busch had tried the method in one patient, in 1866. See also Hall, *A Commotion*, 47–48.

205 *It did not take long*: Coley, "The Treatment," 3–11; Hall, *A Commotion*, 51–57; Graeber, *The Breakthrough*, 46–50.

205 *a colleague who was traveling*: William Coley, "The Treatment of Inoperable Sarcoma by Bacterial Toxins (The Mixed Toxins of the *Streptococcus Erysipelas* and the *Bacillus Prodigiosus*)," *Proceedings of the Royal Society of Medicine* 1910; 3: 1–48. Coley monitored Zola, who remained cancer-free for eight years before dying from a recurrence while living in Italy.

205 *tried the bacterial inoculations*: Graeber, *The Breakthrough*, 51. At a medical conference in 1896, Coley reported a series of 160 patients and stated "nearly one-half" of ninety-three sarcoma cases showed some improvement. See also Hall, *A Commotion*, 72.

205 *could not replicate*: Editorial, "The Failure of the Erysipelas Toxins," *Journal of the American Medical Association* 1894; 23(24): 919.

205 *concocted a heat-killed*: Edward McCarthy, "The Toxins of William B. Coley and the Treatment of Bone and Soft-Tissue Sarcomas," *Iowa Orthopedic Journal* 2006; 26: 154–158.

206 *Though he would endure*: Hall, *A Commotion*, 53.

206 *for sale until 1952*: Graeber, *The Breakthrough*, 56.

206 *Coley treated more than a thousand*: McCarthy, "The Toxins," 157.

207 *D'Angelo, was slated for*: Steven Rosenberg and John Barry, *The Transformed Cell: Unlocking the Mysteries of Cancer* (New York: G. P. Putnam's Sons, 1992), 11–23. Note: in his book, Rosenberg states that he changed the names of patients, such as "James D'Angelo," to protect patient privacy.

208 *extract their T cells*: Ibid., 87.

208 *He treated sixty-six patients*: Andrew Pollack, "Setting the Body's 'Serial Killers' Loose on Cancer," *New York Times*, August 1, 2016, 1; Rosenberg, *The Transformed Cell*, 193–194.

208 *grew billions of Taylor's*: Rosenberg, *The Transformed Cell*, 203–208, 213. In his book, Rosenberg calls this patient "Linda Granger," but her identity as Linda Taylor became known after the publication of numerous media stories about her as the first cancer patient to have been cured by immunotherapy.

208 *cover of* Newsweek: Hall, *A Commotion*, 294, 296.

208 *If T cells could recognize*: Graeber, *The Breakthrough*, 90.

209 *tumor cells in mice*: Matt Richtel, *An Elegant Defense: The Extraordinary New Science of the Immune System* (New York: William Morrow, 2019), 308; Graeber, *The Breakthrough*, 105.

209 *Allison's CTLA-4 inhibitor*: Dana Leach, Matthew Krummel, and James Allison, "Enhancement of Antitumor Immunity by CTLA-4 Blockade," *Science* 1996; 271: 1734–1736.

210 *former president Jimmy Carter*: Kinch, *The End*, 242–243.

210 *Scientists have developed*: Gideon Gross, Tova Waks, and Zelig Eshhar, "Expression of Immunoglobulin-T-Cell Receptor Chimeric Molecules as Functional Receptors with Antibody-Type Specificity," *Proceedings of the National Academy of Sciences of the United States of America* 1989; 86: 10024–10028; Kinch, *The End*, 192–193.

211 *Emily Whitehead*: Denise Grady, "In Girl's Last Hope, Altered Immune Cells Beat Leukemia," *New York Times*, December 9, 2012.

211 *In 1997, Karen Cummings*: Wisnia, *The Jimmy Fund*, 118; Mukherjee, *The Emperor*, 395.

212 *six feet, five inches tall*: Martin, "Einar Gustafson."

212 *"I've had that before"*: Ibid.
212 *Since the war on cancer*: DeVita, *Death of Cancer*, 244.
212 *From the 1990s to*: Ibid., 245.
212 *The five-year survival rate*: Siegel, "Cancer Statistics," 15. The five-year survival rate from this 2021 report pertains to cancers diagnosed between 2010 and 2016.

CHAPTER SIX: TRAUMA

214 *had not even sought*: Ira Rutkow, *James A. Garfield* (New York: Henry Holt and Co., 2006), 54-55.
214 *civil service reform*: Justus Doenecke, *The Presidencies of James A. Garfield & Chester A. Arthur* (Lawrence, KS: University Press of Kansas, 1981), 38-39.
215 *Consul to Paris*: Rutkow, *James A. Garfield*, 71.
215 *six feet away*: Ibid., 83.
215 *"My God!"*: Kenneth Ackerman, *Dark Horse: The Surprise Election and Political Murder of James A. Garfield* (New York: Carroll & Graf Publishers, 2003), 378.
215 *four inches to the right*: D. Willard Bliss, "Report of the Case of President Garfield, Accompanied with a Detailed Account of the Autopsy," *The Medical Record* 1881; 20(15): 393-402. This report was Dr. Bliss's official record of the president's entire course of care, from the shooting to the autopsy.
216 *"The President was deathly pale"*: Robert Reyburn, "Clinical History of the Case of President James Abram Garfield," *Journal of the American Medical Association* 1894; 22: 412. Reyburn published an extensive, day-by-day account of the president's care in 1894. This detailed record has served as a valuable primary source for medical historians ever since.
216 *Bliss had extensive experience*: Rutkow, *James A. Garfield*, 85.
216 *"Nelaton" probe*: Gustavo Colon, "President James Garfield's Death: A Criticism," *Journal of the Louisiana State Medical Society* 2001; 153: 454-456.
216 *"gently passed it"*: Bliss, "Report of the Case," 393.
216 *Bliss's parents*: Candice Millard, *Destiny of the Republic: A Tale of Madness, Medicine, and the Murder of a President* (New York: Doubleday, 2011), 141.
217 *Surgery was considered*: Fitzharris, *The Butchering Art*, 22.
217 *many candles*: Richard Hollingham, *Blood and Guts: A History of Surgery* (New York: Thomas Dunne Books, 2008), 98.
217 *Surgeons did not routinely*: J. Wesley Alexander, "The Contributions of Infection Control to a Century of Surgical Progress," *Annals of Surgery* 1985; 201: 423-428.
218 *"the gleam of [Liston's] knife"*: D. J. Coltart, "Surgery Between Hunter and Lister: As Exemplified by the Life and Works of Robert Liston (1794-1847)," *Proceedings of the Royal Society of Medicine* 1972; 65: 556-560.
218 *"Now, gentlemen, time me!"*: Reginald Magee, "Robert Liston: Surgeon Extraordinary," *ANZ Journal of Surgery* 1999; 69: 878-881; Fitzharris, *The Butchering Art*, 14.
218 *cut off a man's testicle*: Fitzharris, *The Butchering Art*, 10.
218 *One frightened man*: Ibid., 12.
218 *Long participated in*: Crawford Williamson Long, "An Account of the First Use of Sulphuric Ether by Inhalation as an Anaesthetic in Surgical Operations," *Southern Medical and Surgical Journal*

Notes

1949; 5: 705-713; Editorial, "Crawford W. Long (1815-1878): Discoverer of Ether for Anesthesia," *Journal of the American Medical Association* 1965; 194(9): 160-161.

218 *"seemed incredulous"*: Long, "An Account of," 708.

219 *"Gentlemen, this is no"*: Daniel Robinson and Alexander Toledo, "Historical Development of Modern Anesthesia," *Journal of Investigative Surgery* 2012; 25: 141-149.

219 *"We are going to try"*, *"When are you going to"*: Coltart, "Surgery Between Hunter and Lister," 559.

219 *"This Yankee dodge"*: Ibid.; Fitzharris, *The Butchering Art*, 15.

219 *Sitting in the audience*: Alex Sakula, "Lord Lister, OM PRS (1827-1912)," *Journal of Medical Biography* 2005; 13: 70.

220 *80 percent of operations*: Alexander, "The Contributions," 423.

220 *Predictably, this increased*: Fitzharris, *The Butchering Art*, 17.

220 *"Houses of Death"*: Ibid., 46.

220 *"A man laid on the operating table"*, *40 percent*: Richard Fisher, *Joseph Lister* (New York: Stein and Day, 1977), 123-124.

220 *a Philadelphia hospital*: Ibid., 52.

220 *blamed such infections*: Rhoda Truax, *Joseph Lister: Father of Modern Surgery* (New York: Bobbs-Merrill Company, 1944), 37.

220 *intentionally destroyed*: Fisher, *Joseph Lister*, 124.

221 *"11 P.M. Query"*: Ibid., 122.

221 *"It is a common observation"*: Hector Charles Cameron, *Joseph Lister: The Friend of Man* (London: Whitefriars Press, 1949), 54-55.

221 *He read about Louis Pasteur's*: Fisher, *Joseph Lister*, 121.

221 *minute organisms*: Joseph Lister, "On a New Method of Treating Compound Fracture, Abscess, Etc.," *Lancet*, March 16, 1867: 326-329.

221 *"When I read"*: Joseph Lister, speaking at the "Meeting of the International Medical Congress," *Boston Medical and Surgical Journal* 1876; 95: 328.

222 *as a preventative measure*: Rickman John Godlee, *Lord Lister* (London: Macmillan, 1917), 180. As a biographer, Godlee had the benefit of being a surgeon, and Lister's nephew and mentee.

222 *Lister tested numerous types*: Fisher, *Joseph Lister*, 155.

222 *engineers in Carlisle*: Godlee, *Lord Lister*, 182; Fitzharris, *The Butchering Art*, 161. Lister also noted that carbolic acid had been used by some physicians in continental Europe—though not, apparently, in a way that had significantly reduced infection rates on a large scale, nor how Lister wanted to try it, as a preventative. See also Fitzharris, *The Butchering Art*, 179-181.

222 *eleven-year-old boy*: Lister, "On a New Method," 327.

222 *The safest move would be*: Fisher, *Joseph Lister*, 136. In the early 1860s, about one-fourth of compound fracture patients who did not have the limb amputated died. See also Fitzharris, *The Butchering Art*, 191.

223 *set the bones*: Guy Theodore Wrench, *Lord Lister: His Life & Work* (New York: Frederick Stokes Company, 1913), 106-107.

223 *ten compound fracture patients*: Lister, "On a New Method," 327-329, 507-509; Fitzharris, *The Butchering Art*, 166-167.

223 *not a single case of sepsis*: Joseph Lister, "On the Antiseptic Principle in the Practice of Surgery," *British Medical Journal* 1867; 2(351): 246–248.

223 *"I now perform an operation"*: Godlee, *Lord Lister*, 198.

223 *not what he hoped*: Ibid., 199–208, 311–312.

224 *They could not abide*: Fitzharris, *The Butchering Art*, 185.

224 *carbolic acid poisoning*: Fisher, *Joseph Lister*, 165.

224 *"carbolic acid mania"*: D. Campbell Black, "Mr. Nunneley and the Antiseptic Treatment (Carbolic Acid)," *British Medical Journal* 1869; 2(453): 281.

224 *"Shut the door quickly"*: Laurence Farmer, *Master Surgeon: A Biography of Joseph Lister* (New York: Harper & Brothers, 1962), 111.

224 *spray carbolic acid*: Wrench, *Lord Lister*, 228.

224 *far wider acceptance*: Fitzharris, *The Butchering Art*, 210–211.

225 *"Gentlemen, I am the only"*: Fisher, *Joseph Lister*, 194.

225 *about 50 percent*: Alexander, "The Contributions," 424.

225 *hailed as a hero*: Truax, *Joseph Lister*, 180–181; Fitzharris, *The Butchering Art*, 214–215.

225 *450 physicians*: Reported in "Meeting of the International Medical Congress," *Boston Medical and Surgical Journal* 1876; 95: 323.

225 *express purpose of discrediting*: Fitzharris, *The Butchering Art*, 219–221.

225 *banned Lister's method*: Ibid., 215.

225 *"A large proportion"*: Rutkow, *James A. Garfield*, 107–108.

225 *"Little, if any faith"*: Ibid., 108. It should be noted that Lister did manage to gain one important convert at the end of his American tour—the influential Massachusetts General Hospital surgeon Henry Bigelow. Bigelow had previously banned antiseptic technique at the hospital; but he became so impressed by Lister that he subsequently embraced its use. Massachusetts General was the first American hospital to require carbolic acid antisepsis. See also Fitzharris, *The Butchering Art*, 223.

226 *Bliss remained strongly opposed*: James Herndon, "Ignorance Is Bliss," *The Harvard Orthopaedic Journal* 2013; 15: 74–77; Millard, *Destiny of*, 141.

227 *the next day*: Bradley Weiner, "The Case of James A. Garfield," *Spine* 2003; 28(10): E183–E186.

227 *wine and daily injections*: Millard, *Destiny of*, 175.

227 *"I think that we have"*: Ibid., 159.

227 *"If I can't save him"*: *Chicago Tribune*, July 4, 1881, 2.

227 *on July 23*: Reyburn, "Clinical History," 463–464; Bliss, "Report of the Case," 396.

227 *Two days later*: Reyburn, "Clinical History," 498–499.

228 *fired into the backs*: Howard Wilcox, "The President Ails: American Medicine in Retrospect," *Delaware Medical Journal* 1981; 53(4): 201–210.

228 *"induction balance"*: Weiner, "The Case of," E184.

228 *Multiple, non-sterile drainage tubes*: Reyburn, "Clinical History," 500; Ibrahim Eltorai, "Fatal Spinal Cord Injury of the 20th President of the United States: Day-by-Day Review of His Clinical Course, with Comments," *Journal of Spinal Cord Medicine*; 27(4): 330–341.

228 *on August 8*: Reyburn, "Clinical History," 545; Millard, *Destiny of*, 216.

228 *His weight dropped*: Reyburn, "Clinical History," 549.

228 *via enema*: Ibid., 547.

228 *Always retaining*: Allan Peskin, *Garfield* (Kent, OH: Kent State University Press, 1978), 600-601.
228 *Infection became so widespread that*: Reyburn, "Clinical History," 578-580.
228 *began to have hallucinations*: Eltorai, "Fatal Spinal Cord Injury," 337.
228 *"easily passed downwards"*: Reyburn, "Clinical History," 580.
229 *2,000 people*: Millard, *Destiny of*, 226.
229 *it had ricocheted*: Bliss, "Report of the Case," 401; Reyburn, "Clinical History," 665.
229 *The consensus of most*: George Paulson, "Death of a President and His Assassin—Errors in Their Diagnosis and Autopsies," *Journal of the History of the Neurosciences* 2006; 15: 77-91; Millard, *Destiny of*, 253.
229 *many Civil War veterans*: Weiner, "The Case of," E185.
230 *Bliss was roundly criticized*: Eltorai, "Fatal Spinal Cord Injury," 340; Ackerman, *Dark Horse*, 439.
230 *overly optimistic public bulletins*: Paulson, "Death of a President," 81. In Bliss's defense, using one's fingers to explore a gunshot wound was considered the standard of care in the U.S. at the time. And his optimistic bulletins were reportedly influenced by the fact that the president himself would be reading the reports in the newspapers. As a result, his physicians were reluctant to report negative information that might cause Garfield dismay. See also Reyburn, "Clinical History," 415.
230 *"more to cast distrust"*: Rutkow, *James A. Garfield*, 131.
230 *"ignorance is Bliss"*: Herndon, "Ignorance Is Bliss," 74-75.
230 *"General Garfield"*: James Clark, *The Murder of James A. Garfield: The President's Last Days and the Trial and Execution of His Assassin* (Jefferson, NC: McFarland & Company, 1993), 122-123.
230 *Pendleton Civil Service*: Ackerman, *Dark Horse*, 437.
230 *helped American physicians*: Rutkow, *James A. Garfield*, 132.
231 *"The only winner in war"*: Jeffery Howard, Russ Kotwal, Caryn Stern, et al., "Use of Combat Casualty Data to Assess the US Military Trauma System During the Afghanistan and Iraq Conflicts, 2001-2017," *Journal of the American Medical Association Surgery* 2019; 154(7): 600-608.
231 *16 percent*: Andrew Bamji, *Faces from the Front: Harold Gillies, The Queens Hospital, Sidcup, and the Origins of Modern Plastic Surgery* (Solihull, West Midlands, UK: Helion & Company, 2017), 17.
232 *Valadier was not licensed*: Donald Simpson and David David, "World War I: The Genesis of Craniomaxillofacial Surgery?" *ANZ Journal of Surgery* 2004; 74: 71-77.
232 *"I stood spellbound"*: Harold Gillies and D. Ralph Millard, *The Principles and Art of Plastic Surgery* (Boston; Little, Brown, 1957), 7.
233 *to recruit surgeons*: Harold Ellis, "Two Pioneers of Plastic Surgery: Sir Harold Delf Gillies and Sir Archibald McIndoe," *British Journal of Hospital Medicine* 2010; 71(12): 698.
233 *luggage tags*: Murray Meikle, *Reconstructing Faces: The Art and Wartime Surgery of Gillies, Pickerill, McIndoe, & Mowlem* (Dunedin, NZ; Otago University Press, 2013), 56.
233 *2,000 dreadfully disfigured*: Murray Meikle, "The Evolution of Plastic and Maxillofacial Surgery in the Twentieth Century: The Dunedin Connection," *Surgeon* 2006; 4(5): 325-334.
233 *"There were wounds"*: Reginald Pound, *Gillies: Surgeon Extraordinary* (London: Michael Joseph, Ltd., 1964), 33.
233 *He learned it was unwise*: Bamji, *Faces from*, 70.
234 *Gillies's simplest skin flap*: Harold Gillies, *Plastic Surgery of the Face* (London: Hodder and Stoughton, 1920), 19-21.
234 *so that the hair*: Bamji, *Faces from*, 88.

235 *11,752 surgeries*: D. N. Matthews, "Gillies, Mastermind of Modern Plastic Surgery," *British Journal of Plastic Surgery* 1979; 32: 68-77.
236 *"This poor sailor"*: Gillies, *Plastic Surgery*, 356-359.
236 *"father of plastic surgery"*: Gillies, *The Principles and Art*, 633. Gillies later learned that the tubed pedicle method had also been independently developed by a Russian surgeon named Vladimir Filatov, and a German named Hugo Ganzer, during the war.
236 *"plastic surgery is a constant"*: Bamji, *Faces from*, 131.
237 *August 31, 1940*: Tom Gleave, *I Had a Row with a German* (London: Macmillan, 1941), 65-66. This was Gleave's memoir, published a year after the Battle of Britain under the pseudonym "R.A.F. Casualty" because the British Air Ministry forbade pilots from publishing military accounts in their own names. See also Meikle, *Reconstructing Faces*, 147.
237 *"I went back to"*: Ibid., 66.
237 *the sun shone brightly*: James Rothwell, "The Weather During the Battle of Britain in 1940," *Weather* 2012; 67(4): 109-110.
238 *"heard a metallic click"*, *"A long spout"*: Gleave, *I Had a Row*, 68.
238 *"like the centre"*: Ibid., 69.
238 *blinding flash*: Ibid., 70.
238 *"The skin on my right leg"*: Ibid., 71-72.
239 *His nose was practically*: Fong, *Extreme Medicine*, 41.
239 *put Gleave on his back*: Tom Gleave, interviewed on *This Is Your Life* television program, Season 31, Episode 12, aired January 9, 1991, on Thames Television.
239 *a wheelbarrow*: Peter Williams and Ted Harrison, *McIndoe's Army: The Injured Airmen Who Faced the World* (London: Pelham Books, 1979), 42.
239 *tannic acid*: E. R. Mayhew, *The Reconstruction of Warriors: Archibald McIndoe, the Royal Air Force and the Guinea Pig Club* (London: Greenhill Books, 2004), 58-59.
239 *"What on earth"*, *"I had a row"*: Gleave, *I Had a Row*, 80.
239 *McIndoe was passionate, headstrong*: Leonard Mosley, *Faces from the Fire: The Biography of Sir Archibald McIndoe* (Englewood, NJ: Prentice Hall, 1962), 47, 144; Mayhew, *The Reconstruction*, 75.
240 *4,500 Allied airmen*: Mosley, *Faces from*, 9.
240 *gone down in the salt water*: Ibid., 83-84; Meikle, *The Evolution*, 332.
240 *In the salt baths*: Mayhew, *The Reconstruction*, 62-63.
240 *large heat lamps*: Fong, *Extreme Medicine*, 43-44.
241 *"waltzing"*: Ibid., 45-46.
241 *Sir Stafford Cripps*: Mosley, *Faces from*, 146.
241 *"One day"*: Ibid., 102.
242 *"You need a new nose"*: Gleave, *I Had a Row*, 97-98.
242 *at least ten operations*: Meikle, *Reconstructing Faces*, 133-134.
242 *"Imagine how they feel"*: Mosley, *Faces from*, 95.
242 *Flowers and live music*: Alexandra Macnamara and Neil Metcalfe, "Sir Archibald Hector McIndoe (1900-1960) and the Guinea Pig Club: The Development of Reconstructive Surgery and Rehabilitation in the Second World War (1939-1945)," *Journal of Medical Biography* 2014; 22(4): 224-228.

Notes

242 *Rank among his patients*: Meikle, *The Evolution*, 332.
243 *"the most exclusive"*: Mayhew, *The Reconstruction*, 78.
243 *"the town that never stared"*: Williams, *McIndoe's Army*, 36.
243 *game of soccer*: Macnamara, "Sir Archibald," 226.
243 *"The first time you see"*: Mayhew, *The Reconstruction*, 165-166.
244 *"We have now arrived"*: Ibid., 76.
244 *an Australian surgeon*: Fong, *Extreme Medicine*, 50-51.
244 *"The body is a three-dimensional"*: G. Ian Taylor and John Palmer, "The Vascular Territories (Angiosomes) of the Body: Experimental Study and Clinical Applications," *British Journal of Plastic Surgery* 1987; 40: 113-141.
245 *France's president*: Christopher Dente and David Feliciano, "Alexis Carrel (1873-1944)," *Archives of Surgery* 2005; 140: 609-610.
245 *owned a lace factory*: Sheldon Levin, "Alexis Carrel's Historic Leap of Faith," *Journal of Vascular Surgery* 2013; 61(3): 832-833.
245 *cigarette paper*: Hollingham, *Blood and Guts*, 172-173.
246 *surgeon Bohdan Pomahač*: Fong, *Extreme Medicine*, 54-65. Fong's book contains excellent sections on Archibald McIndoe and Bohdan Pomahac's operation on Dallas Wiens.
247 *"I could not bear"*: Katie Moisse and Angela Hill, "Dallas Wiens Reunites with Daughter After Full Face Transplant," May 8, 2011, https://abcnews.go.com/Health/Wellness/full-face-transplant-recipient-dallas-weins-reunites-daughter/story?id=13558167, accessed July 17, 2022.
247 *look nothing like the donor*: Fong, *Extreme Medicine*, 54.
247 *Another major ethical consideration*: Branislav Kollar and Bohdan Pomahač, "Facial Restoration by Transplantation," *Surgeon* 2018; 16: 245-249.
247 *seventeen hours*: Bohdan Pomahač, Julian Pribaz, Elof Eriksson, et al., "Three Patients with Full Facial Transplantation," *New England Journal of Medicine* 2012; 366(8): 715-722.
247 *postponed until the donor's heart*: Fong, *Extreme Medicine*, 61-62.
248 *"I wouldn't even know"*: Moisse, "Dallas Wiens Reunites."
248 *March 30, 1981*: Oliver Beahrs, "The Medical History of President Ronald Reagan," *Journal of the American College of Surgeons* 1994; 178: 86-96; David Rockoff and Benjamin Aaron, "The Shooting of President Reagan: A Radiologic Chronology of His Medical Care," *Radiographics* 1995; 15(2): 407-418.
249 *"You not only broke a rib"*: Ronald Reagan, *An American Life* (New York: Simon & Schuster, 1990), 260.
249 *The president's appearance*: Benjamin Aaron and David Rockoff, "The Attempted Assassination of President Reagan," *Journal of the American Medical Association* 1994; 272(21): 1689-1693.
249 *"Honey, I forgot to duck"*: Reagan, *An American Life*, 260.
249 *"I hope you're a Republican"*: Ibid., 261.
249 *a professed Democrat*: Aaron, "The Attempted Assassination," 1690.
250 *"Am I dead?"*: Hedrick Smith, *The Power Game: How Washington Works* (New York: Random House, 1988), 299.
250 *"the greatest love offering"*: Stuart Taylor, "Hinkley Hails 'Historical' Shooting to Win Love," *New York Times*, July 9, 1982, 10.
251 *Dominique Jean Larrey*: Panagiotis Skandalakis, Panagiotis Lainas, Odyseas Zoras, et al., "'To Afford

the Wounded Speedy Assistance'*: Dominique Jean Larrey and Napoleon," *World Journal of Surgery* 2006; 30: 1392-1399; Fong, *Extreme Medicine*, 95-96.

251 *separated into three groups*: Bamji, *Faces from*, 26.

251 *200 amputations*: David Welling and Norman Rich, "Dominique Jean Larrey and the Russian Campaign of 1812," *Journal of the American College of Surgeons* 2013; 216(3): 493-500.

252 *0.25 percent*: Arthur Kellermann, Eric Elster, and Todd Rasmussen, "How the US Military Reinvented Trauma Care and What This Means for US Medicine," *Health Affairs*, July 3, 2018, https://www.healthaffairs.org/do/10.1377/hblog20180628.431867/full/, accessed November 6, 2022.

252 *a portion of trachea grown from*: Paolo Macchiarini, Philipp Jungebluth, Tetsuhiko Go, et al., "Clinical Transplantation of a Tissue-Engineered Airway," *Lancet* 2008; 372: 2023-2030.

CHAPTER SEVEN: CHILDBIRTH

254 *Because Charlotte's parents*: Anne Stott, *The Lost Queen: The Life and Tragedy of the Prince Regent's Daughter* (Yorkshire, UK: Pen & Sword Books, 2020), 21.

254 *the people loved her*: Steven Parissien, *George IV: Inspiration of the Regency* (New York: St. Martin's, 2001), 232, 239.

255 *could not abide the idea*: Ibid., 235.

255 *intermittently bled*: Christopher Hibbert, *George IV: Regent and King* (New York: Harper & Row, 1973), 97.

255 *Charlotte's contractions began*: Details of Princess Charlotte's delivery are drawn from the following sources: Eardley Holland, "The Princess Charlotte of Wales: A Triple Obstetric Tragedy," *Journal of Obstetrics & Gynecology of the British Empire* 1951; 58(6): 905-919; Andrew Friedman, Ernest Kohorn, and Sherwin Nuland, "Did Princess Charlotte Die of a Pulmonary Embolism?" *British Journal of Obstetrics and Gynaecology* 1988; 95: 683-688; William Ober, "Obstetrical Events That Shaped Western European History," *Yale Journal of Biology and Medicine* 1992; 65: 201-210.

257 *hot water bottles*: Hibbert, *George IV*, 98.

258 *about 300,000*: World Health Organization, "Maternal Mortality," September 19, 2019, https://www.who.int/news-room/fact-sheets/detail/maternal-mortality, accessed July 20, 2022.

258 *about one hundred times*: J. Drife, "The Start of Life: A History of Obstetrics," *Postgraduate Medical Journal* 2002; 78: 311-315.

258 *1 to 1.5 percent*: Laura Helmuth, "The Disturbing, Shameful History of Childbirth Deaths," *Slate*, September 10, 2013, https://slate.com/technology/2013/09/death-in-childbirth-doctors-increased-maternal-mortality-in-the-20th-century-are-midwives-better.html, accessed July 20, 2022.

258 *American advice book*: Tina Cassidy, *Birth: The Surprising History of How We Are Born* (New York: Atlantic Monthly Press, 2006), 245.

258 *filled diaries and letters*: Judith Leavitt, *Brought to Bed: Childbearing in America, 1750-1950* (New York: Oxford University Press, 1986), 33.

259 *It seems paradoxical*: Mihaela Pavlicev, Roberto Romero, and Philipp Mitteroecker, "Evolution of the Human Pelvis and Obstructed Labor: New Explanations of an Old Obstetrical Dilemma," *American Journal of Obstetrics & Gynecology* 2020; 222(1): 3-16. This paper theorizes that the narrow human pelvis evolved, not only to permit bipedalism, but also to better support the weight of human viscera and relatively large fetuses over a long gestational period.

259 *walk upright with ease*: Cassidy, *Birth*, 10.

Notes

259 *Anatomical head size*: Helmuth, "The Disturbing."
259 *A study of births*: Cassidy, *Birth*, 23.
259 *At birth, we are helpless*: Ibid., 17.
260 *who actively struggled*: James Nicopoullos, "'Midwifery Is Not a Fit Occupation for a Gentleman,'" *Journal of Obstetrics and Gynaecology* 2003; 23(6): 589-593.
260 *caused the baby to be formed*: Randi Epstein, *Get Me Out: A History of Childbirth from the Garden of Eden to the Sperm Bank* (New York: W. W. Norton & Company, 2010), 10.
260 *eggs from the right*: Jacqueline Wolf, *Cesarean Section: An American History of Risk, Technology, and Consequence* (Baltimore: Johns Hopkins University Press, 2018), 70.
260 *far beyond obstetrics*: Cassidy, *Birth*, 27.
260 *burned at the stake*: Harold Ellis, "Dame Hilda Lloyd: First President of the Royal College of Obstetricians and Gynaecologists," *Journal of Perioperative Practice* 2009; 19(6): 192-193.
260 *a giant sheet*: Richard Wertz and Dorothy Wertz, *Lying-In: A History of Childbirth in America* (New York: Free Press, 1977), 43, 81.
262 *transported by carriage*: Peter Dunn, "The Chamberlen Family (1560-1728) and Obstetric Forceps," *Archives of Disease in Childhood–Fetal and Neonatal Edition* 1999; 81: F232-F234.
262 *banged bells*: Epstein, *Get Me Out*, 23.
263 *passed down from*: Dunn, "The Chamberlen Family," F233.
263 *named François Mauriceau*: Epstein, *Get Me Out*, 26.
263 *Roger van Roonhuysen*: Drife, "The Start of Life," 312.
263 *"English lock"*: Ibid.
263 *wearing a dress*: Wertz, *Lying-In*, 81.
263 *have never witnessed*: Cassidy, *Birth*, 133, 138.
263 *applied without looking*: Leavitt, *Brought to Bed*, 41.
264 *an autopsy*: Holland, "The Princess Charlotte," 915.
264 *Pulmonary embolism*: Friedman, "Did Princess Charlotte," 687.
264 *"as if by an earthquake"*: Hibbert, *George IV*, 102.
264 *Public events were canceled*: Stott, *The Lost Queen*, 239.
265 *Croft was condemned*: Holland, "The Princess Charlotte," 915.
265 *"triple obstetric tragedy"*: Ibid., 918.
265 *feeble-minded boy*: Humphrey Arthure, "Princess Charlotte of Wales—a Royal Tragedy," *Midwife, Health Visitor & Community Nurse* 1977; 13: 147-149; Ober, "Obstetrical Events," 203.
266 *"A woman could be"*: Irvine Loudon, *Death in Childbirth: An International Study of Maternal Care and Maternal Mortality 1800-1950* (New York: Oxford University Press, 1992), 54.
266 *rotting breast milk*: Epstein, *Get Me Out*, 53.
266 *painting hospital walls*: Sherwin Nuland, *The Doctor's Plague: Germs, Childbed Fever, and the Strange Story of Ignác Semmelweis* (New York: W. W. Norton & Company, 2003), 62.
266 *burn their clothes*: Ibid., 58.
266 *on the roofs of hospitals*: Epstein, *Get Me Out*, 54-55.
267 *8,000 patients*: Loudon, *Death in Childbirth*, 65.
267 *about 11 percent*: Ignaz Semmelweis, *The Etiology, Concept and Prophylaxis of Childbed Fever*, trans. K. Codell Carter (Madison, WI: University of Wisconsin Press, 1983), 64.

267 *over 18 percent*: Ibid., 72.
267 *in corridors*: Wertz, *Lying-In*, 121.
267 *"That they were afraid"*: Nuland, *The Doctor's Plague*, 85.
268 *"To me . . . it appeared"*: Semmelweis, *The Etiology*, 81.
268 *He thought overcrowding*: Ibid., 69.
268 *asked the priest*: Ibid., 71, 73.
268 *Jakob Kolletschka*: Ibid., 87-89.
269 *government rules dictated*: Nuland, *The Doctor's Plague*, 81.
269 *"Only God knows"*: Semmelweis, *The Etiology*, 98.
269 *less than 2 percent*: Ibid., 90.
270 *"milk metastasis theory"*: Nuland, *The Doctor's Plague*, 35-36.
270 *tight women's petticoats*: K. Codell Carter and Barbara Carter, *Childbed Fever: A Scientific Biography of Ignaz Semmelweis* (Westport, CT: Greenwood Press, 1994), 34.
270 *his immediate superior*: Irvine Loudon, *The Tragedy of Childbed Fever* (New York: Oxford University Press, 2000), 101.
270 *less than 1 percent*: Hollingham, *Blood and Guts*, 88.
270 *poorly written*: Irvine Loudon, "Semmelweis and His Thesis," *Journal of the Royal Society of Medicine* 2005; 98: 555.
270 *Few read it*: Semmelweis, *The Etiology*, 25.
270 *"I denounce you before God"*: Loudon, *The Tragedy*, 104.
271 *increasingly erratic*: Nuland, *The Doctors' Plague*, 162-163; Semmelweis, *The Etiology*, 57.
271 *a result of his beating*: Nuland, *The Doctors' Plague*, 168.
272 *from 1 to 6 percent*: Carter, *Childbed Fever*, 79.
273 *went on for days*: Epstein, *Get Me Out*, 35.
273 *"The poor woman"*: L. Lewis Wall, "The Medical Ethics of Dr. J. Marion Sims: A Fresh Look at the Historical Record," *Journal of Medical Ethics* 2006; 32: 346-350.
274 *"If there was anything"*: J. Marion Sims, *The Story of My Life* (New York: D. Appleton and Company, 1884), 231.
274 *"Why, doctor"*: Ibid., 233.
274 *employed a mirror*: Seale Harris, *Woman's Surgeon: The Life Story of J. Marion Sims* (New York: Macmillan, 1950), 87.
274 *"I saw everything"*: Sims, *The Story*, 234-235. Sims was actually not the first to repair a vaginal fistula. Two doctors, John Peter Mettauer and George Hayward, performed repairs in 1838 and 1839, respectively. But Sims was the one who did the most to perfect and disseminate the technique through practice and publications.
275 *at least thirty*: Epstein, *Get Me Out*, 43.
275 *silver wire sutures*: Sims, *The Story*, 245.
276 *vilified by multiple historians*: Barron Lerner, "Scholars Argue Over Legacy of Surgeon Who Was Lionized, Then Vilified," *New York Times*, October 28, 2003.
276 *Thomas Jefferson*: Epstein, *Get Me Out*, 41-42.
276 *a terrible affliction*: Wall, "The Medical Ethics," 346-350.

276 *"To the indomitable":* J. Harry Thompson, *Report: Columbia Hospital For Women and Lying-In Asylum* (Washington, D.C.: U.S. Government Printing Office, 1873), 49.
276 *"clamorous":* Ibid.; Sims, *The Story,* 243.
277 *large bottles of ether:* Harold Ellis, "Sir James Young Simpson: Pioneer of Anaesthesia in Childbirth," *British Journal of Hospital Medicine* 2020; 81(4): 1-2.
277 *They laughed and chatted:* Cassidy, *Birth,* 84.
278 *chloroform was inexpensive:* S. W. McGowan, "Sir James Young Simpson Bart, 150 Years On," *Scottish Medical Journal* 1997; 42: 185-187.
278 *her baby "Anaesthesia":* Drife, "The Start of Life," 313.
278 *never be selfless enough:* Epstein, *Get Me Out,* 84.
278 *burned alive:* Cassidy, *Birth,* 85.
278 *pointed out that God:* Ellis, "Sir James Young Simpson," 1.
278 *cited James 4:17:* Cassidy, *Birth,* 85.
279 *"I have the expectant mother":* Leavitt, *Brought to Bed,* 121.
279 *make chloroform safer:* Donald Caton, "John Snow's Practice of Obstetric Anesthesia," *Anesthesiology* 2000; 92: 247-252.
279 *held her breath:* Hollingham, *Blood and Guts,* 77.
280 *fifty-three minutes:* Epstein, *Get Me Out,* 4.
280 *"the blessed chloroform":* Caton, "John Snow's," 250.
280 *anesthesia à la reine:* Ober, "Obstetrical Events," 207.
281 *lost half their business:* Cassidy, *Birth,* 39.
281 *only 15 percent:* Ibid., 31.
281 *underreported puerperal fever:* Loudon, *Death in Childbirth,* 35.
281 *eight out of every 1,000:* Ibid., 153.
281 *more than 250,000:* Ibid., 50.
281 *safer for U.S. mothers in 1800:* Wolf, *Cesarean Section,* 20, 52.
281 *wealthy women were more likely:* Helmuth, "The Disturbing."
282 *"The United States":* S. Josephine Baker, "Maternal Mortality in the United States," *Journal of the American Medical Association* 1927; 89(24): 2016-2017.
282 *a startling 1933 study:* The Committee on Maternal Mortality of the New York Academy of Medicine, "Maternal Mortality in New York City," *Journal of the American Medical Association* 1933; 101(23): 1826-1828.
282 *one in thirty:* Leavitt, *Brought to Bed,* 25.
282 *55 percent:* Ibid., 171.
283 *sinking of the Titanic:* Olivia Gordon, *The First Breath: How Modern Medicine Saves the Most Fragile Lives* (London: Bluebird, 2019), 64-65.
283 *"There is not much difference":* Epstein, *Get Me Out,* 192.
284 *Boston research team's:* O. Watkins Smith, George Van Smith, and David Hurwitz, "Increased Excretion of Pregnanediol in Pregnancy from Diethylstilbestrol with Special Reference to the Prevention of Late Pregnancy Accidents," *American Journal of Obstetrics & Gynecology* 1946; 51: 411-415; O. Watkins Smith, "Diethylstilbestrol in the Prevention and Treatment of Complications of Pregnancy," *American Journal of Obstetrics & Gynecology* 1948; 56(5): 821-834.

284 *lifetime's amount of estrogen*: Epstein, *Get Me Out*, 136.
285 *diseases in animals*: Ibid., 140.
285 *University of Chicago study*: W. J. Dieckmann, M. E. Davis, L. M. Rynkiewicz, et al., "Does the Administration of Diethylstilbestrol During Pregnancy Have Therapeutic Value?" *American Journal of Obstetrics & Gynecology* 1953; 66(5): 1062-1081.
285 *Three other clinical trials*: Robert Hoover, Marianne Hyer, Ruth Pfeiffer, et al., "Adverse Health Outcomes in Women Exposed In Utero to Diethylstilbestrol," *New England Journal of Medicine* 2011; 365: 1304-1314.
285 *until the mother of one*: Epstein, *Get Me Out*, 144.
285 *His 1971 study*: Arthur Herbst, Howard Ulfelder, and David Poskanzer, "Adenocarcinoma of the Vagina: Association with Maternal Stilbestrol Therapy with Tumor Appearance in Young Women," *New England Journal of Medicine* 1971; 284: 878-881.
286 *five to ten million*: Casey Reed and Suzanne Fenton, "Exposure to Diethylstilbestrol During Sensitive Life Stages: A Legacy of Heritable Health Effects," *Birth Defects Research Part C: Embryo Today* 2013; 99(2): 134-146.
286 *increased risk for*: Hoover, "Adverse Health Outcomes," 1304.
286 *some studies indicate*: Taher Al Jishi and Consolato Sergi, "Current Perspective of Diethylstilbestrol (DES) Exposure in Mothers and Offspring," *Reproductive Toxicology* 2017; 71: 71-77; Retha Newbold, "Lessons Learned from Perinatal Exposure to Diethylstilbestrol," *Toxicology and Applied Pharmacology* 2004; 199: 142-150.
286 *For eighteen years*: Epstein, *Get Me Out*, 148.
286 *"biological time bomb"*: Al Jishi, "Current Perspective," 71.
286 *seven full-time*: Bridget Kuehn, "Frances Kelsey Honored for FDA Legacy," *Journal of the American Medical Association* 2010; 304(19): 2109-2112.
287 *toxicity data inadequate*: Trent Stephens and Rock Brynner, *Dark Remedy: The Impact of Thalidomide and Its Revival as a Vital Medicine* (Cambridge, MA: Perseus Publishing, 2001), 48-49. The FDA's chemist reviewing the application was Lee Geismar, and the pharmacologist was Jiro Oyama.
287 *not harmful to the fetus*: James Essinger and Sandra Koutzenko, *Frankie: How One Woman Prevented a Pharmaceutical Disaster* (North Palm Beach, FL: Blue Sparrow Books, 2018), 154.
287 *more than fifty times*: Stephens, *Dark Remedy*, 50.
287 *displays pressure*: Ibid., 50-51; Essinger, *Frankie*, 113, 149-150.
287 *nitpicky bureaucrat*: Linda Bren, "Frances Oldham Kelsey: FDA Medical Reviewer Leaves Her Mark on History," *FDA Consumer magazine* 2001; 35(2): 24-29.
287 *essentially testimonials*: Richard McFadyen, "Thalidomide in America: A Brush with Tragedy," *Clio Medica* 1976; 11(2): 79-93.
288 *"untoward reactions"*: Stephens, *Dark Remedy*, 52.
288 *tried to go over*: Essinger, *Frankie*, 142-143.
288 *six times*: Stephens, *Dark Remedy*, 53.
288 *up to 100,000*: Essinger, *Frankie*, 15.
288 *seventeen phocomelia cases*: Geoff Watts, "Frances Oldham Kelsey," *Lancet* 2015; 386: 1334.
290 *4.5 percent*: Paul Placek, Selma Taffel, and Mary Moien, "1986 C-Sections Rise; VBACs Inch Upward," *American Journal of Public Health* 1988; 78(5): 562-563.

290 *approximately 32 percent*: Wolf, *Cesarean Section*, 4–5.
290 *55 percent overall*: Marina Lopes, "C-Sections Are All the Rage in Brazil. So Too, Now, Are Fancy Parties to Watch Them," *Washington Post*, June 12, 2019.
290 *10 percent*: Michael Greene, "Two Hundred Years of Progress in the Practice of Midwifery," *New England Journal of Medicine* 2012; 367(18): 1732–1740.
291 *over eight million*: Editorial, "Towards the Global Coverage of a Unified Registry of IVF Outcomes," *Reproductive BioMedicine Online* 2019; 38(2), 1.
291 *About two-thirds*: Jaime Natoli, Deborah Ackerman, Suzanne McDermott, et al., "Prenatal Diagnosis of Down Syndrome: A Systematic Review of Termination Rates (1995–2011)," *Prenatal Diagnosis* 2012; 32: 142–153.
292 *Several for-profit companies*: Patrick Turley, Michelle Meyer, Nancy Wang, et al., "Problems with Using Polygenic Scores to Select Embryos," *New England Journal of Medicine* 2021; 385(1): 78–86.
292 *the first baby selected*: Carey Goldberg, "Picking Embryos with Best Health Odds Sparks New DNA Debate," *Bloomberg.com*, September 17, 2021, https://www.bloomberg.com/news/articles/2021-09-17/picking-embryos-with-best-health-odds-sparks-new-dna-debate, accessed July 21, 2022; Pete Shanks, "The First Polygenic Risk Score Baby," *Biopolitical Times*, September 30, 2021, https://www.geneticsandsociety.org/biopolitical-times/first-polygenic-risk-score-baby, accessed July 21, 2022.
292 *online Genomic Prediction panel discussion*: Genomic Prediction Clinical Laboratories, "Rank Ordering Embryos for Transfer: Patient and Clinician Perspectives on PGT-P," April 10, 2021, https://infoproc.blogspot.com/2021/04/first-baby-born-from-polygenically.html, accessed July 21, 2022.
292 *"Part of that duty"*: Goldberg, "Picking Embryos."
292 *"compounding benefits"*: Shanks, "The First Polygenic."
292 *Many scientists question*: Turley, "Problems with Using," 78–86.
295 *"In identifying some individuals"*: Kelly Ormond, Douglas Mortlock, Derek Scholes, et al., "Human Germline Genome Editing," *American Journal of Human Genetics* 2017; 101: 167–176.
295 *"Evolution has been"*: Patrick Skerrett, "Experts Debate: Are We Playing with Fire When We Edit Human Genes?" *Stat*, November 17, 2015, https://www.statnews.com/2015/11/17/gene-editing-embryo-crispr/, accessed July 21, 2022.
295 *about seventy-five nations*: Francoise Baylis, Marcy Darnovsky, Katie Hasson, et al., "Human Germline and Heritable Genome Editing: The Global Policy Landscape," *CRISPR Journal* 2020; 3(5): 365–377.
296 *scientists at Sun Yat-sen*: Gina Kolata, "Chinese Scientists Edit Genes of Human Embyros, Raising Concerns," *New York Times*, April 23, 2015.
296 *Only four of the embryos*: Puping Liang, Yanwen Xu, Xiya Zhang, et al., "CRISPR/Cas9-Mediated Gene Editing in Human Tripronuclear Zygotes," *Protein & Cell* 2015; 6(5): 363–372; Mukherjee, *The Gene*, 478.
296 *he announced to the world*: Gina Kolata, Sui-Lee Wee, and Pam Belluck, "Chinese Scientist Claims to Use Crispr to Make First Genetically Edited Babies," *New York Times*, November 26, 2018.
297 *against West Nile virus*: William Glass, David McDermott, Jean Lim, et al., "CCR5 Deficiency Increases Risk of Symptomatic West Nile Virus Infection," *Journal of Experimental Science* 2006; 203(1): 35–40.
297 *over one hundred Chinese*: David Cyranoski and Heidi Ledford, "Genome-Edited Baby Claim Provokes International Outcry," *Nature* 2018; 563: 607–608.

297 *three years in prison*: David Cyranoski, "What CRISPR-Baby Prison Sentences Mean for Research," *Nature* 2020; 577: 154–155.
297 *Huntington's disease*: Walter Isaacson, *The Code Breaker: Jennifer Doudna, Gene Editing, and the Future of the Human Race* (New York: Simon & Schuster, 2021), 341–342. Isaacson's book does an excellent job of weighing the risks and benefits, and ethical implications, of germ line gene modification.
298 *do not feel germ line experimentation*: Ibid., 323–324, 330–332.

CONCLUSION

299 *"It helps a man"*: Osler, "Chauvinism in Medicine," 689.
302 *"Chance favors the prepared mind"*: Barry, *The Great Influenza*, 68.
302 *"When they do appear"*: Ernst Chain, "The Quest for New Biodynamic Substances," *Perspectives in Biology and Medicine* 1967; 10(2): 208.
303 *more than 30 percent*: Sally Rockey, "Comparing Success Rates, Award Rates, and Funding Rates," National Institutes of Health, Office of Extramural Research, March 5, 2014, https://nexus.od.nih.gov/all/2014/03/05/comparing-success-award-funding-rates/, accessed July 22, 2022.
303 *approximately 20 percent*: National Institutes of Health, "Extramural Research Overview for Fiscal Year 2020," May 19, 2021, https://www.niaid.nih.gov/grants-contracts/fy-2020-award-data, accessed July 22, 2022.
303 *The system also appears*: Aaron Carroll, "Why the Medical Research Grant System Could Be Costing Us Great Ideas," *New York Times*, June 18, 2018; Diego Oliveira, Yifang Ma, Teresa Woodruff, et al., "Comparison of National Institutes of Health Grant Amounts to First-Time Male and Female Principal Investigators," *Journal of the American Medical Association* 2019; 321(9): 898–900.
303 *fewer than one in six*: Gina Kolata, "So Many Research Scientists, So Few Professorships," *New York Times*, July 14, 2016, 3.
304 *median annual salary*: Rodoniki Athanasiadou, Adriana Bankston, McKenzie Carlisle, et al., "Assessing the Landscape of US Postdoctoral Salaries," *Studies in Graduate and Postdoctoral Education* 2018; 9(2): 213–242.
304 *55 percent*: Chris Woolston, "Uncertain Prospects for Postdoctoral Researchers," *Nature* 2020; 588: 181–184.
304 *Emmanuelle Charpentier*: Kolata, "So Many Research."
304 *approximately half*: Editorial, "Stop Exploitation of Foreign Postdocs in the United States," *Nature* 2018; 563: 444.
306 *more than half*: Patrick Boyle, "More Women Than Men Are Enrolled in Medical School," *Association of American Medical Colleges*, December 9, 2019, https://www.aamc.org/news-insights/more-women-men-are-enrolled-medical-school, accessed July 22, 2022.
307 *"Methuselah gene"*: Robin Smith and Max Gomez, *Cells Are the New Cure: The Cutting Edge Medical Breakthroughs That Are Transforming Our Health* (Dallas, TX: BenBella Books, 2017), 191.
307 *interested in telomeres*: Ibid., 193–197; Francesca Rossiello, Diana Jurk, Joao Passos, et al., "Telomere Dysfunction in Ageing and Age-Related Diseases," *Nature Cell Biology* 2022; 24: 135–147. On the other hand, enhancing telomerase might also play a role in promoting cancerous growth. Inhibiting telomerase in certain situations might someday prove a helpful treatment for cancer.

Selected Bibliography

Ackerman, Kenneth. *Dark Horse: The Surprise Election and Political Murder of President James A. Garfield.* New York: Carroll & Graf Publishers, 2003.
Artenstein, Andrew, ed. *Vaccines: A Biography.* New York: Springer, 2010.
Bamji, Andrew. *Faces from the Front: Harold Gillies, The Queen's Hospital, Sidcup and the Origins of Modern Plastic Surgery.* Solihull, West Midlands, UK: Helion & Company, 2017.
Barnard, Christiaan, and Curtis Bill Pepper. *One Life.* Toronto: Macmillan, 1969.
Barry, John. *The Great Influenza: The Story of the Deadliest Pandemic in History.* New York: Viking, 2004.
Bazell, Robert. *Her-2: The Making of Herceptin, a Revolutionary Treatment for Breast Cancer.* New York: Random House, 1998.
Bickel, Lennard. *Rise Up to Life: A Biography of Howard Walter Florey Who Made Penicillin and Gave It to the World.* New York: Charles Scribner's Sons, 1972.
Bliss, Michael. *The Discovery of Insulin.* Chicago: University of Chicago Press, 1982.
Brock, Robert. *Robert Koch: A Life in Medicine and Bacteriology.* Washington, D.C.: ASM Press, 1999.
Brown, Kevin. *Penicillin Man: Alexander Fleming and the Antibiotic Revolution.* Gloucestershire, UK: Sutton Publishing, 2004.
Bud, Robert. *Penicillin: Triumph and Tragedy.* New York: Oxford University Press, 2007.
Cameron, Hector. *Joseph Lister: The Friend of Man.* London: Whitefriars Press, 1949.
Carter, K. Codell, trans. *Essays of Robert Koch.* New York: Greenwood Press, 1987.
Carter, K. Codell, and Barbara Carter. *Childbed Fever: A Scientific Biography of Ignaz Semmelweis.* Westport, CT: Greenwood Press, 1994.
Carter, Richard. *Breakthrough: The Saga of Jonas Salk.* New York: Trident Press, 1966.
Cassidy, Tina. *Birth: The Surprising History of How We Are Born.* New York: Atlantic Monthly Press, 2006.
Cheney, Dick. *In My Time: A Personal and Political Memoir.* New York: Threshold Editions, 2011.
Cheney, Dick, and Jonathan Reiner. *Heart: An American Medical Odyssey.* New York: Scribner, 2013.
Clark, James. *The Murder of James A. Garfield: The President's Last Days and the Trial and Execution of His Assassin.* Jefferson, NC: McFarland & Company, 1993.
Clark, Ronald W. *The Life of Ernst Chain: Penicillin and Beyond.* New York: St. Martin's, 1985.
Claxton, K. T. *Wilhelm Roentgen.* London: Heron Books, 1970.

Conant, Jennet. *The Great Secret: The Classified World War II Disaster That Launched the War on Cancer.* New York: W. W. Norton & Co., 2020.

Cooke, Robert. *Dr. Folkman's War: Angiogenesis and the Struggle to Defeat Cancer.* New York: Random House, 2001.

Cooper, David. *Open Heart: The Radical Surgeons Who Revolutionized Medicine.* New York: Kaplan, 2010.

Cooper, Geoffrey, Rayla Greenberg-Temin, and Bill Sugden, eds. *The DNA Provirus: Howard Temin's Scientific Legacy.* Washington, D.C.: ASM Press, 1995.

Cooper, Thea, and Arthur Ainsberg. *Breakthrough: Elizabeth Hughes, the Discovery of Insulin, and the Making of a Medical Miracle.* New York: St. Martin's, 2010.

Cox, Caroline. *The Fight to Survive: A Young Girl, Diabetes, and the Discovery of Insulin.* New York: Kaplan, 2009.

Curie, Eve. *Madame Curie.* Translated by Vincent Sheean. New York: Da Capo Press, 2001.

Debré, Patrice. *Louis Pasteur.* Translated by Elborg Forster. Baltimore: Johns Hopkins University Press, 1998.

DeVita, Vincent, and Elizabeth DeVita-Raeburn. *The Death of Cancer.* New York: Farrar, Straus and Giroux, 2015.

Doenecke, Justus. *The Presidencies of James A. Garfield & Chester A. Arthur.* Lawrence, KS: University Press of Kansas, 1981.

Dubos, René. *Louis Pasteur: Free Lance of Science.* Boston: Little, Brown, 1950.

Epstein, Randi Hutter. *Get Me Out: A History of Childbirth from the Garden of Eden to the Sperm Bank.* New York: W. W. Norton & Company, 2010.

Essinger, James, and Sandra Koutzenko, *Frankie: How One Woman Prevented a Pharmaceutical Disaster.* North Palm Beach, FL: Blue Sparrow Books, 2018.

Faguet, Guy. *The War on Cancer: An Anatomy of Failure. A Blueprint for the Future.* New York: Springer, 2005.

Farmer, Laurence. *Master Surgeon: A Biography of Joseph Lister.* New York: Harper & Brothers, 1962.

Farrell, Jeanette. *Invisible Enemies: Stories of Infectious Disease.* New York: Farrar, Straus and Giroux, 1998.

Fisher, Richard. *Joseph Lister.* New York: Stein and Day, 1977.

Fitzharris, Lindsey. *The Butchering Art: Joseph Lister's Quest to Transform the Grisly World of Victorian Medicine.* New York: Farrar, Straus and Giroux, 2017.

Fong, Kevin. *Extreme Medicine: How Exploration Transformed Medicine in the Twentieth Century.* New York: Penguin, 2012.

Forssmann, Werner. *Experiments on Myself: Memoirs of a Surgeon in Germany.* Translated by Hilary Davies. New York: St. Martin's, 1974.

Forrester, James. *The Heart Healers: The Misfits, Mavericks, and Rebels Who Created the Greatest Medical Breakthrough of Our Lives.* New York: St. Martin's, 2015.

Gillies, Harold. *Plastic Surgery of the Face.* London: Hodder and Stoughton, 1920.

Gillies, Harold, and D. Ralph Millard. *The Principles and Art of Plastic Surgery.* Boston: Little, Brown, 1957.

Gleave, Tom (originally published as "R.A.F. Casualty"). *I Had a Row with a German.* London: Macmillan, 1941.

Godlee, Rickman John. *Lord Lister.* London: Macmillan, 1917.

Selected Bibliography

Goetz, Thomas. *The Remedy: Robert Koch, Arthur Conan Doyle, and the Quest to Cure Tuberculosis.* New York: Gotham Books, 2014.

Gordon, Olivia. *The First Breath: How Modern Medicine Saves the Most Fragile Lives.* London: Bluebird, 2019.

Graeber, Charles. *The Breakthrough: Immunotherapy and the Race to Cure Cancer.* New York: Twelve, 2018.

Hall, Stephen. *A Commotion in the Blood: Life, Death, and the Immune System.* New York: Henry Holt & Co., 1997.

Harari, Yuval Noah. *Homo Deus: A Brief History of Tomorrow.* New York: Harper Collins, 2017.

Hare, Ronald. *The Birth of Penicillin, and the Disarming of Microbes.* London: George Allen & Unwin, 1970.

Harris, Seale. *Banting's Miracle: The Story of the Discoverer of Insulin.* Philadelphia: Lippincott, 1946.

———. *Woman's Surgeon: The Life Story of J. Marion Sims.* New York: Macmillan, 1950.

Hesketh, Robin. *Betrayed by Nature: The War on Cancer.* New York: Palgrave Macmillan, 2012.

Hibbert, Christopher. *George IV: Regent and King.* New York: Harper & Row, 1973.

Hodges, Paul. *The Life and Times of Emil H. Grubbe.* Chicago: University of Chicago Press, 1964.

Hollingham, Richard. *Blood and Guts: A History of Surgery.* New York: Thomas Dunne Books, 2008.

Imber, Gerald. *Genius on the Edge: The Bizarre Double Life of Dr. William Stewart Halsted.* New York: Kaplan, 2010.

Infield, Glenn. *Disaster at Bari.* New York: Macmillan, 1971.

Isaacson, Walter. *The Code Breaker: Jennifer Doudna, Gene Editing, and the Future of the Human Race.* New York: Simon & Schuster, 2021.

Jacobs, Charlotte. *Jonas Salk: A Life.* New York: Oxford University Press, 2015.

Kenny, Charles. *The Plague Cycle: The Unending War Between Humanity and Infectious Disease.* New York: Scribner, 2021.

Kinch, Michael. *The End of the Beginning: Cancer, Immunity, and the Future of a Cure.* New York: Pegasus Books, 2019.

Kluger, Jeffrey. *Splendid Solution: Jonas Salk and the Conquest of Polio.* New York: G. P. Putnam's Sons, 2004.

Laszlo, John. *The Cure of Childhood Leukemia: Into the Age of Miracles.* New Brunswick, NJ: Rutgers University Press, 1995.

Lax, Eric. *The Mold in Dr. Florey's Coat: The Story of the Penicillin Miracle.* New York: Henry Holt & Co., 2004.

Leaf, Clifton. *The Truth in Small Doses: Why We're Losing the War on Cancer—and How to Win It.* New York: Simon & Schuster, 2013.

Leavitt, Judith Walzer. *Brought to Bed: Childbearing in America, 1750-1950.* New York: Oxford University Press, 1986.

Lechevalier, Hubert, and Morris Solotorovsky. *Three Centuries of Microbiology.* New York: McGraw Hill, 1965.

Lerner, Barron. *The Breast Cancer Wars: Hope, Fear, and the Pursuit of a Cure in Twentieth-Century America.* New York: Oxford University Press, 2001.

Loudon, Irvine. *Death in Childbirth: An International Study of Maternal Care and Maternal Mortality: 1800-1950.* New York: Oxford University Press, 1992.

———. *The Tragedy of Childbed Fever.* New York: Oxford University Press, 2000.

MacFarlane, Gwyn. *Alexander Fleming: The Man and the Myth.* Cambridge, MA: Harvard University Press, 1984.

———. *Howard Florey: The Making of a Great Scientist.* London: Oxford University Press, 1979.

Mayhew, Emily. *The Reconstruction of Warriors: Achibald McIndoe, the Royal Air Force and the Guinea Pig Club.* London: Greenhill Books, 2004.

Meikle, Murray. *Reconstructing Faces: The Art and Wartime Surgery of Gillies, Pickerell, McIndoe & Mowlem.* Dunedin, NZ: Otago University Press, 2013.

Metchnikoff, Elie, ed. *The Founders of Modern Medicine.* Freeport, NY: Books for Libraries Press, 1939.

Millard, Candice. *Destiny of the Republic: A Tale of Madness, Medicine, and the Murder of a President.* New York: Doubleday, 2011.

Miller, G. Wayne. *King of Hearts: The True Story of a Maverick Who Pioneered Open Heart Surgery.* New York: Times Books, 2000.

Monagan, David, and David Williams. *Journey into the Heart: A Tale of Pioneering Doctors and Their Race to Transform Cardiovascular Medicine.* New York: Gotham Books, 2007.

Mosley, Leonard. *Faces from the Fire: The Biography of Sir Archibald McIndoe.* Englewood Cliffs, NJ: Prentice Hall, 1962.

Mukherjee, Siddhartha. *The Emperor of All Maladies: A Biography of Cancer.* New York: Scribner, 2010.

———. *The Gene: An Intimate History.* New York: Scribner, 2016.

Nathan, David. *The Cancer Treatment Revolution: How Smart Drugs and Other Therapies Are Renewing Our Hope and Changing the Face of Medicine.* Hoboken, NJ: John Wiley & Sons, 2007.

Nuland, Sherwin. *The Doctor's Plague: Germs, Childbed Fever, and the Strange Story of Ignác Semmelweis.* New York: W. W. Norton & Company, 2003.

Offit, Paul. *The Cutter Incident: How America's First Polio Vaccine Led to the Growing Vaccine Crisis.* New Haven, CT: Yale University Press, 2005.

Oshinsky, David. *Polio: An American Story.* New York: Oxford University Press, 2005.

Paget, Stephen. *The Surgery of the Chest.* Bristol, UK: John Wright & Co., 1896.

Parissien, Steven. *George IV: Inspiration of the Regency.* New York: St. Martin's, 2001.

Patterson, James. *The Dread Disease: Cancer and Modern American Culture.* Cambridge, MA: Harvard University Press, 1987.

Peskin, Allan. *Garfield.* Kent, OH: Kent State University Press, 1978.

Pickover, Clifford. *The Medical Book: From Witch Doctors to Robot Surgeons, 250 Milestones in the History of Medicine.* New York: Sterling Publishing, 2012.

Pound, Reginald. *Gillies: Surgeon Extraordinary.* London: Michael Joseph, Ltd., 1964.

Reagan, Ronald. *An American Life.* New York: Simon & Schuster, 1990.

Reid, Robert. *Marie Curie.* New York: Saturday Review Press/E. P. Dutton & Co., 1974.

Richtel, Matt. *An Elegant Defense: The Extraordinary New Science of the Immune System.* New York: William Morrow, 2019.

Robbins, Louise. *Louis Pasteur and the Hidden World of Microbes.* New York: Oxford University Press, 2001.

Rosen, William. *Miracle Cure: The Creation of Antibiotics and the Birth of Modern Medicine.* New York: Viking, 2017.

Selected Bibliography

Rosenberg, Steven, and John Barry. *The Transformed Cell: Unlocking the Mysteries of Cancer.* New York: G. P. Putnam's Sons, 1992.

Rutkow, Ira. *James A. Garfield.* New York: Henry Holt & Co., 2006.

Seavey, Nina, Jane Smith, and Paul Wagner. *A Paralyzing Fear: The Triumph Over Polio in America.* New York: TV Books, 1998.

Semmelweis, Ignaz. *The Etiology, Concept, and Prophylaxis of Childbed Fever.* Translated and edited by K. Codell Carter. Madison, WI: University of Wisconsin Press, 1983.

Sims, J. Marion. *The Story of My Life.* New York: D. Appleton and Company, 1884.

Smith, Hedrick. *The Power Game: How Washington Works.* New York: Random House, 1988.

Smith, Jane. *Patenting the Sun: Polio and the Salk Vaccine.* New York: William Morrow, 1990.

Smith, Robin, and Max Gomez. *Cells Are the New Cure: The Cutting-Edge Medical Breakthroughs That Are Transforming Our Health.* Dallas, TX: BenBella Books, 2017.

Stephens, Trent, and Rock Brynner. *Dark Remedy: The Impact of Thalidomide and Its Revival as a Vital Medicine.* Cambridge, MA: Perseus Publishing, 2001.

Stott, Anne. *The Lost Queen: The Life and Tragedy of the Prince Regent's Daughter.* Yorkshire, UK: Pen & Sword Books, 2020.

Tobin, James. *The Man He Became: How FDR Defied Polio to Win the Presidency.* New York: Simon & Schuster, 2013.

Troan, John. *Passport to Adventure.* Pittsburgh, PA: Neworks Press, 2000.

Truax, Rhoda. *Joseph Lister: Father of Modern Surgery.* New York: Bobbs-Merrill Company, 1944.

Wertz, Richard, and Dorothy Wertz. *Lying-In: A History of Childbirth in America.* New York: Free Press, 1977.

Williams, Peter, and Ted Harrison. *McIndoe's Army: The Injured Airmen Who Faced the World.* London: Pelham Books, 1979.

Wilson, Daniel. *Living with Polio: The Epidemic and Its Survivors.* Chicago: University of Chicago Press, 2005.

Winchester, Simon. *The Man Who Loved China: The Fantastic Story of the Eccentric Scientist Who Unlocked the Mysteries of the Middle Kingdom.* New York: Harper, 2008.

Wisnia, Saul. *The Jimmy Fund of Dana-Farber Cancer Institute.* Charleston, SC: Arcadia Publishing, 2002.

Wolf, Jacqueline. *Cesarean Section: An American History of Risk, Technology, and Consequence.* Baltimore: Johns Hopkins University Press, 2018.

Wrench, Guy Theodore. *Lord Lister: His Life and Work.* New York: Frederick Stokes Company, 1913.

Image Credits

12 iStock.com/ttsz
16 From Werner Forßmann Herzkatheter-Röntgenaufnahme, *Über die Sondierung des rechten Herzens* (Berliner Klinische Wochenschrift vom 5. November 1929), via Wikimedia Commons
26 Used with permission from Intermountain Healthcare. © Intermountain Healthcare. All rights reserved
55 Henry Mahon, Thomas Fisher Rare Book Library, University of Toronto, CC BY 2.0, via Wikimedia Commons
58 Courtesy of the Thomas Fisher Rare Book Library, University of Toronto
68 Courtesy of the Thomas Fisher Rare Book Library, University of Toronto
95 Paul Nadar, public domain, via Wikimedia Commons
97 ZEISS Microscopy from Germany, CC BY-SA 2.0, via Wikimedia Commons
101 Courtesy of the Alexander Fleming Laboratory Museum, Imperial College Healthcare NHS Trust
112 Photo by Brigadier Sidney Smith, Wellcome Trust, CC BY 4.0, via Wikimedia Commons (left); Profgeorgev CC0, via Wikimedia Commons (right)
118 Courtesy of the Alexander Fleming Laboratory Museum, Imperial College Healthcare NHS Trust
128 Courtesy of the Food and Drug Administration, public domain, via Wikimedia Commons
131 DO11.10, GNU FDL, via Wikimedia Commons
154 Jonas Salk Papers, Special Collections & Archives, University of California, San Diego
158 Courtesy of the Centers for Disease Control and Prevention, public domain, via Wikimedia Commons
181 Courtesy of the National Cancer Institute, public domain, via Wikimedia Commons
186 Courtesy of the Dana-Farber Cancer Institute
206 Courtesy of the National Library of Medicine, public domain, via Wikimedia Commons
226 Public domain, via Wikimedia Commons
235 From the Archives of the Royal College of Surgeons of England
243 Courtesy of Queen Victoria Hospital NHS Foundation Trust/East Grinstead Museum
261 Public domain, via Wellcome Collection
262 Courtesy of the Science Museum Group Collection Online, CC BY-NC-SA 4.0

Image Credits

271 Copperplate engraving by Jenő Doby from István Benedek, *Ignaz Phillip Semmelweis 1818-1865* (Gyomaendrőd, Hungary: Corvina Kiadó, 1983), public domain, via Wikimedia Commons
289 Courtesy of the National Library of Medicine, public domain, via Wikimedia Commons
371 Author photo by Todd Lajoie

Index

Page numbers in *italics* refer to photographs and illustrations

A
abortion, 260, 281, 283, 291
Abraham, Edward, 121
accoucheurs (male midwives), 255, 260–261, *261*, 263–265. *See also* midwives
Achong, Bert, 198
acinar cells, 51, 53–54, 59–60
actinomycin D, 187–188
Agnew, David Hayes, 227–228
agriculture, use of antibiotics in, 122–124
AIDS vaccine research, 159
Aldrin, Buzz, 3
Alexander, Albert, 75–77, 109–110, 114
Alexander, Stewart, 177–178
alkaloids, 187
alkylating agents, 178
Allen, Frederick, 47–48, 62–63, 67
Allison, Jim, 209
American Cancer Society, 206
American College of Chest Physicians, 34
American Heart Association, 23–24, 36
American Medical Association (AMA), 157–158, 275
American Society of Human Genetics, 295
amethopterin (methotrexate), 186, 188–191
aminopterin, 180, 186
amniocentesis, 291
amputations, 217–220, 222–223, 225, 251
androgen deprivation therapy, 193
anesthesia
  childbirth, 277–280, 283
  chloroform, 224, 277–280
  credit for invention of, 219, 306
  epidural, 283
  ether, 218–219, 277, 279
  nitrous oxide, 218–219
  religious opposition to, 278
anesthesia à la reine, 280
angiogenesis, 194, 196, 202, 209
angioplasty, 8, 10, 22–25, 39
angiosomes, 245–246
animalcules, 77–78. *See also* bacteria
anterior horn, 130–131
anthrax, 78, 82–84, 86–87, 89–93, 306
anti-angiogenesis drugs, 194
antibiotics, 99–100, 121–124. *See also* specific antibiotics
antibodies, 98, 141. *See also* immune system
anti-contagion theories, 78–79, 220
antigens, 209
anti-oncogenes (tumor suppressor genes), 200
antiseptics
  carbolic acid, 222–224
  failure to follow antiseptic practices, 216, 220, 224, 272
  Lister's work with, 221–225, 227, 230
  and mortality rates, 225
  and penicillin, 104
  surgeons' opposition to, 223–226, 230
antitoxin, 98
aorta
  and balloon angioplasty, 22–23
  and coronary angiography, 17–19
  heart anatomy, 12, *12*, 13
  and heart transplants, 40
  and left ventricular assist device (LVAD), 9

Index

and mitral stenosis,
  31–32. *See also* mitral
  stenosis
and open heart surgery,
  25–26, 39
aplastic anemia, 175
Aretaeus of Cappadocia, 50
Armstrong, Neil, 3
arrhythmia, 14, 16, 20, 24
atherosclerotic plaque, 19–20
Arthur, Chester A., 215, 230
artificial intelligence, 307
assisted reproductive
  technology, 291. *See also*
  infertility
atrial septal defects, 38, 44
Avastin (bevacizumab), 194
axicabtagene ciloleucel
  (Yescarta), 211

B

Bachmann, Adolf, 23–24
Bacillus anthracis, 83. *See also*
  anthrax
Bacillus influenzae, 105
bacteria. *See also* specific
  bacteria
  alternative treatment
    options, 124
  animalcules, 77–78
  antibiotic resistant,
    121–124
  as cause of cancer,
    198–199, 202
  classifying, 78
  contagion and anti-
    contagion theories,
    78–79
  discovery of, 77
  DNA in, 78, 121, 187
  genetically engineered,
    124
  germ theory of disease,
    78, 80, 84, 97, 225
  Gram stain, 78
  gram-negative, 78, 121
  gram-positive, 78, 121
  helpful, 78
  infectious diseases. *See*
    infectious diseases
  pathogenic, 78
  prevalence of on Earth, 78

shapes of, 78
toxins produced by, 98
bacterial infection. *See*
  antibiotics; bacteria
bacteriophages, 124
Bailey, Charles, 32–34
Baker, Sara Josephine, 282
balloon angioplasty, 8, 10,
  22–25, 39
Balmis, Francisco Javier de,
  162
Baltimore, David, 199
Banting, Frederick, 52–55, 55,
  56–66, 68–71, 306
Banting and Best Department
  of Medical Research, 70
barbers, surgeries by, 172, 217
Bari, Italy, 175–178, 192
Barnard, Christiaan, 39,
  41–42
Barnathan, Elliot, 164–165
Barr, Yvonne, 198
batteries
  future alternatives to, 44
  heart battery (sinoatrial
    node), 12, 27, 45
  for ICD (implantable
    cardioverter-
    defibrillator), 44
  for left ventricular assist
    device (LVAD), 9–10
  pacemakers, 44
Battle of Britain, 239–240
Battle of Jutland, 236
Battle of the Somme, 233
Bavolek, Cecelia, 38
Baxter, Jedediah Hyde, 227
Beatrice, Princess, 280
Beatson, George, 193
bed rest, as treatment for
  myocardial infarction, 6,
  10, 19
Beijerinck, Martinus, 139
Bell, Alexander Graham,
  84, 228
Berlin Physiological Society,
  89
Best, Charles, 54–55, 55,
  56–65, 69–71, 306
beta thalassemia, 296
beta-lactamase (penicillinase),
  121

Betsey, Anarcha, and Lucy
  (enslaved black women),
  275–276
bevacizumab (Avastin), 194
Billroth, Theodor, 27
BioNTech, 165–166, 305
birth control, 258, 283
Bishop, Michael, 200
Black, Donald Campbell, 224
bleomycin, 192
Bliss, D. Willard, 216,
  226–230
blood, oxygenation of. *See*
  oxygenation of blood
blood clots
  and atherosclerotic
    plaque, 19–20
  and balloon angioplasty,
    22. *See also* balloon
    angioplasty
  and catheterization, 22
  Cheney, Richard, 7, 9
  dissolving, 20
  and heart attacks, 13,
    19–20
  and heart lacerations, 29
  and heart-lung machine,
    38. *See also* heart-lung
    bypass machine
  pulmonary embolism, 37,
    264, 281
  thrombolytic agents, 9,
    20, 38
  uterine, 264
blood thinners, 9, 20, 38
blood transfusions, 283
blood vessels
  angiogenesis in tumors,
    194, 196, 202, 209
  connecting in plastic
    surgery, 244–246, 253
  heart anatomy, 11–13. *See
    also* heart disease
bloodletting, 217, 280–281
bortezomib (Velcade), 201
Boston Children's Hospital,
  145, 178, 182, 194
Brady, James, 248, 250
brain cancer, 194
brain death, 40–41
BRCA1, 200–201
BRCA2, 200–201

breast cancer, 173, 191, 193, 200–202, 212, 286
Brigham and Women's Hospital, 246, 248
Brougham, Henry, 264
Brown, Lesley, 290
bubonic plague, 78, 89, 167
Bull, William, 203–204
Bureau of Agricultural Chemistry and Engineering (Peoria, Illinois), 114–115
Burkitt, Denis, 197–198
Burkitt's lymphoma, 198
burns, 238–243
Bush, George H. W., 7
Bush, George W., 8
bypass surgery. *See* coronary artery bypass surgery

C
Cambridge Military Hospital, 233
Campbell-Renton, Margaret, 105
Canadian National Research Council, 71
cancer. *See also* specific types of cancer
    angiogenesis, 194, 196, 202, 209
    anti-angiogenesis drugs, 194
    bacteria as cause of, 198–199, 202
    bacteria as treatment for, 203–207
    Cancer Genome Atlas, 201
    carcinogens, 195–196, 200, 202
    causes of, 175, 196, 198–199, 202, 285–286
    chemotherapy. *See* chemotherapy
    childhood leukemia, treatment of at Boston Children's Hospital. *See* Farber, Sidney
    circulating tumor DNA (CT-DNA), 211
    complex nature of, 201–202
    and diethylstilbestrol (DES), 285–286
    fibroblast growth factor, 194
    gene mutations, 199–200, 202
    hormones, influence of, 193, 196, 202, 285
    immunotherapy, 207–211, 213
    and infectious disease, 170–171, 197–198
    inherited genetics, 202
    intratumoral heterogeneity, 202
    and messenger RNA (mRNA), 166
    molecular level discoveries, 199–202, 212–213
    mortality rates, 170, 171, 178, 191–192, 212
    oncogenes, 199–202
    personalized treatment, 201, 210–211
    polygenic, 200, 202
    prevalence of, 170–171
    prophylactic measures, 195, 201, 212–213
    proto-oncogenes, 200
    radiation as cause of, 175, 284
    radiation as early treatment for, 173–175
    research, funding for, 191–192, 212. *See also* war on cancer
    screening tests, 195–196, 201, 211, 213, 285
    surgery as early treatment for, 172–173, 175, 178, 203–204
    survival rates, 191, 212
    tumor suppressor genes (anti-oncogenes), 200
    variability of, 171–172
    viruses as cause of, 196–198, 202
    war on, 172, 191–192, 212, 302–303
Cancer Genome Atlas, 201
carbohydrates, 47, 49, 52, 54
carbolic acid, 222–224
carcinogens, 195–196, 200, 202
cardiac catheterization, 6, 8, 10, 13–16, 16, 17, 20–25, 39
Carnot, Sadi, 245
Caroline, Princess of Wales, 137
Carrel, Alexis, 245–246
CAR-T cell therapy, 210–211
Carter, Jimmy, 210
catheterization. *See* cardiac catheterization
cathode rays, 173–174
CCR5 gene, 296–297
CD4+ T cells, 141
CD8+ T cells, 141
CD28 protein, 209
ceftazidime-avibactam, 123
cervical cancer, 195–196, 190, 206
cesarean section (C-section), 273, 281, 290
cetuximab (Erbitux), 201
CGM (continuous glucose monitoring), 73–74
Chain, Ernst, 104–112, *112*, 113, 120–121, 125
Chamberland, Charles, 85–86
Chamberlen, Hugh, 263
Chamberlen, Peter (the elder), 261–263
Chamberlen, Peter (the younger), 261–263
Charles I, 263
Charlotte, Princess of Wales, 254–257, 264–265
Charpentier, Emmanuelle, 293, 304, 306
chemotherapy
    6-mercaptopurine, 187–191
    actinomycin D, 187–188
    alkaloids, 187
    alkylating agents, 178
    amethopterin (methotrexate), 186, 188–189, 191
    aminopterin, 180, 186
    bleomycin, 192

Index

chemotherapeutic agents, search for, 187–188, 192
childhood leukemia, 179–181, 187, 189–191
etoposide, 192
folic acid antagonists (anti-folates), 180–181, 183, 186, 301
maximum tolerated dose (MTD), 188
methotrexate (amethopterin), 186, 188–191
multidrug resistance, 191
Mustargen (nitrogen mustard), 178
paclitaxel (Taxol), 192
prednisone, 189
side effects of, 180, 187–191, 201
successes, 192
VAMP (vincristine, amethopterin, 6-mercaptopurine, and prednisone), 189–191
vinblastine, 187
vincristine, 187–189
Cheney, Richard (Dick), 6–10, 44
chicken cholera, 85–86
childbirth. *See also* pregnancy
accoucheurs (male midwives), 255, 260–261, *261*, 263–265
anesthesia, 277–280, 283
birth weight, increase in, 259
cesarean section (C-section), 273, 281, 290
Charlotte, Princess of Wales, 254–257, 264–265
early theories of conception and pregnancy, 260
and female anatomy, 259–260
female genitalia, taboos on viewing, 260–261, 263, 274

fetal mortality rates, 258, 289
fetal ultrasound, 283
and fistulas, 273–276
forceps, 261–262, *262*, 263–265, 282
and head size of baby, 259–260
home births, 280, 282–283. *See also* accoucheurs; midwives
hospital births, 280, 282–283
infection of uterus, 266–271
intervention, risks of, 282
male doctors, 280–282
maternal mortality rates, 258, 268–270, 272, 281–282, 289
and men, presence at, 260–261, *261*
midwives, 260–261, 263, 280–283
pain, fear of, 279
potential consequences of, 258
and pregnancies, number of, 258
puerperal fever (childbed fever), 266–270, 272, 277, 281–282
stalled labor, 262, 265, 283
childhood leukemia. *See* leukemia
Children's Cancer Research Fund, 182. *See also* Jimmy Fund
Children's Hospital (Boston), 145, 170, 178, 182, 185, 194
Children's Hospital (Cincinnati), 146
chimeric antigen receptor (CAR), 210
China, gene modification experiments, 296–297
Chinese Academy of Medical Sciences, 297
chloroform, 224, 277–280
cholera, 78–80, 89, 93, 129
cholesterol, 19, 43, 124, 307

choriocarcinoma, 186
chorionic villus sampling, 291
Churchill, Edward, 37
chymotrypsin, 51
Cincinnati Children's Hospital, 146
circulating tumor DNA (CT-DNA), 211
Clauson, Phyllis, 211
Clayton, Elaine, 19
Cleveland Clinic, 17, 39
climate change, 308
Clowes, George, 65
cocaine, 172
Cohn, Ferdinand, 83–84, 90
Cohnheim, Julius, 84
Coley, William, 203–205, 206, 207–208
Coley's toxin, 205–206
Collins, Francis, 295
Collip, James Bertram, 61–65, 69–71, 106, 306
colon cancer, 194, 212
colorectal cancer, 191
congestive heart failure, 25
Coningham, Arthur, 176
consumption, 87. *See also* tuberculosis
contagion theories, 78–80
continuous glucose monitoring (CGM), 73–74
Coolidge, Calvin, 133
coronary angiography, 17–19. *See also* coronary angioplasty
coronary angioplasty, 8, 10, 22–25, 39
coronary arteries
anatomy, 13
angiography, 17–19
bypass surgery. *See* coronary artery bypass surgery
catheterization. *See* cardiac catheterization
Cheney, Richard, 6–8
coronary angioplasty, 8, 10, 22–25, 39
diagonal, 8
endothelium, 19
future of treatment, 43–44
imaging of. *See* cardiac catheterization

Index

coronary arteries *(continued)*
  left anterior descending artery (LAD), 6, 8, 13, 23
  left circumflex artery, 13
  mortality rate for coronary artery disease, 43
  plaque, 19–20
  pulmonary artery, 11, *12*, 37, 40
  right coronary artery (RCA), 6–7, 13
  visualization of, 17–19
coronary artery bypass surgery (CABG)
  and balloon angioplasty, 23–24
  Cheney, Richard, 7–8, 10
  described, 26, *26*, 27
  and heart-lung machine, 38–39
Corynebacterium diphtheriae, 98
Covid-19 (coronavirus disease 2019), 159–162, 165–168, 305, 307
Cowan, William Maxwell, 120
cowpox, 137–138. *See also* smallpox
Cox, James, 132
Crick, Francis, 192
Cripps, Stafford, 241
CRISPR/Cas9, 293–297, 304, 307
Croft, Richard, 264–265
C-section. *See* cesarean section (C-section)
CTLA-4 protein, 209
Cummings, Karen, 211–212
Curie, Marie, 174–175, 299, 305, 396
Curie, Pierre, 174–175
Cutter Laboratories, 155–156, 159
cyclosporine, 40
cystic fibrosis, 291, 294
cytokines, 140–141, 208
cytomegalovirus, 165

D

D. T. Watson Home for Crippled Children, 148
Dale, Henry, 119
Dana, Charles, 186
Dana-Farber Cancer Institute, 186, 211
D'Angelo, James, 207
Darvall, Denise, 41
Dashiell, Elizabeth, 203
defibrillation, 8, 18, 20, 44
Delahanty, Thomas, 248, 250
Dempsey, Jack, 249
DeVita, Vincent, 190, 194
DHFR (dihydrofolate reductase), 191
diabetes insipidus, 50
diabetes mellitus
  complications of, 50, 74
  effects of, 74
  glucose monitoring, 73–74
  Hughes, Elizabeth, 46–49, 66–67, 71
  insulin. *See* insulin
  and kidneys, 49, 73
  life expectancy in early 1900s, 47
  medications for, 72–73
  as metabolic disease, 49–50
  and obesity, 50, 72, 74
  origin of term, 50
  and pancreas. *See* pancreas
  prediabetes, 72
  prevalence of, 71–72
  stem cell research, 74
  symptoms of, 46, 48–50
  treatment during early 1900s, 46–48
  type 1, 49, 71–72
  type 2, 49–50, 71–72, 74
  urine, glucose in, 47, 49–50
  vision issues as complication of, 50, 74
Diana, Princess of Wales, 264
diastole, 28, 31
diethylstilbestrol (DES), 284–286
digestive enzymes, 51–54, 59–60
dihydrofolate reductase (DHFR), 191
diphtheria, 78, 98, 116
Ditzen, Gerda, 14–16
DNA. *See also* RNA
  in bacteria, 78, 121, 187
  and CAR-T cell therapy, 210
  circulating tumor DNA (CT-DNA), 211
  and coding for proteins, 164, 166
  and CRISPR/Cas9 modification of, 293–297, 304, 307
  and diethylstilbestrol (DES), 286
  double-helix structure, discovery of, 192
  and folic acid, 179
  gene modifications in germ line cells, 294–298
  and methotrexate, 191
  recombinant technology and synthetic insulin, 73
  telomeres, 307
  and viruses, 139, 199
Domagk, Gerhard, 104
Donald, Ian, 283
Dotter, Charles, 20–22
"Dottering," 21–22
Doudna, Jennifer, 293, 306
Down syndrome, 291
DPP-4 inhibitors, 73
Dreyer, Georges, 105
Duchenne muscular dystrophy, 294
Duke of Brunswick, 265
Duke of Edinburgh, 243
Duke of Kent, Edward, 265

E

East Grinstead, England, 239–240, 243–244
Ebola virus, 166

Index

Edison, Thomas, 84
Edwards, Ralph, 183-185, 212
Edwards, Robert, 290
Ehrlich, Paul, 88-89, 97-99, 104, 125
Eisenhower, Dwight D., 19
EKG (electrocardiogram), 7, 18-19
electron microscopes, 132, 198
Eli Lilly & Company, 65, 187
Elion, Gertrude, 187
embolisms
  air, 14, 36
  blood clots, 37, 264, 281
Enders, John, 145, 148
Enterococcus faecium, 123
enzymes
  and antibiotic resistant bacteria, 123
  cardiac, 7
  Cas9, 293. *See also* CRISPR/Cas9
  digestive, 51-54, 59-60. *See also* acinar cells
  lysozyme, 100-101, 103
  and methotrexate, 191
  and mRNA, 164-165
  penicillinase (beta-lactamase), 121
  reverse transcriptase, 199
  telomerase, 307
  urokinase, 164
epidural anesthesia, 283
Episcopal Hospital, 33
Epstein, Michael Anthony, 198
Epstein-Barr virus, 198
Erbitux (cetuximab), 201
ergometrine, 283
erlotinib (Tarceva), 201
erysipelas, 203-205, 220, 223
erythropoietin, 165
estrogen, 193, 284-286
ether, 106, 218-219, 277, 279
ethical issues
  fetal genetic screening, 292-293
  germ line experiments, 295-298

and medical advances, 308
preimplantation genetic diagnosis (PGD), 291-294
etoposide, 192
Eugénie, Empress, 275
Ewing's sarcoma, 191
exercise and heart disease, 7, 43

F

face transplant, 246-248
facial injuries, 231-235, *235*, 236, 239-243, 246-248
Farber, Sidney, 178-181, *181*, 182-191, 211, 300-301
fats and metabolism, 49
Favaloro, René, 39
FDA (Food and Drug Administration)
  Avastin (bevacizumab), approval of, 194
  and Coley's toxin, 206
  and diethylstilbestrol (DES), 284-285
  Humulin (synthetic insulin), approval of, 73
  Keytruda (pembrolizumab), approval of, 210
  Kymriah (tisagenlecleucel), approval of, 211
  Mustargen, approval of, 178
  Opdivo (nivolumab), approval of, 210
  and thalidomide, 286-288
  Yervoy (ipilimumab), approval of, 209-210
  Yescarta (axicabtagene ciloleucel), 211
fetal mortality rates, 258, 289
fetal surgery, 290
fetal ultrasound, 283
fibroblast growth factor, 194
fistulas, 273-276

Fleming, Alexander, 100-105, 116-118, *118*, 119-121, 125, 306
Fletcher, Charles, 77, 109
Flexner, Simon, 131, 142
Florey, Ethel, 109
Florey, Howard, 104-112, *112*, 113-120, 125, 306
flying ambulances, 251
folic acid (folate), 179-181, 300-301
folic acid antagonists (antifolates), 180-181, 183, 186, 301
Folkman, Judah, 194
Food and Drug Administration (FDA). *See* FDA
Ford, Gerald, 7
Forssmann, Werner, 13-17
Foster, Jodie, 250
fractures, compound, 222-223
Frances O. Kelsey Award for Excellence and Courage in Protecting Public Health, 288
Francis, Thomas, Jr., 152
Franco-Prussian War (1870-1871), 85
Frei, Emil (Tom), 188-190
Freireich, Emil J. (Jay), 188-190
full face transplant, 246-248
funding for research, 113, 135, 142, 163, 191, 302-303, 305
fundraising
  for childhood leukemia, 182-186. *See also* Jimmy Fund
  for polio research, 135, 182

G

Garfield, James A., 214-217, 225-230, 248, 250
gas exchange and blood oxygenation, 11, 34-39
gene modification, 293-298, 308

Index

gene therapy, 307
Genentech, 73, 201
Generating Antibiotic Incentives Now (GAIN) Act, 124
genetics
  and bacteria, 122, 124
  and CRISPR technology. See CRISPR/Cas9
  and fetal screening, 291
  gene mutations, 199–200, 202
  genetic code. See DNA
  and germ line cells. See germ line cells
  polygenic risk score, 292–293
  and type 2 diabetes, 72
Genomic Prediction, 292
George III, 265
George Washington University Hospital, 249
Georgia Warm Springs Foundation, 135
germ line cells, 294–298, 308
germ theory of disease, 78, 80, 84, 97, 225
Gibbon, John, 37–38, 299
Gillies, Harold, 232–236, 239–240, 248
Gleave, Tom, 237–239, 241–242
Gleevec (imatinib), 201
Glidden, Gregory, 36
GLP-1 agonists, 73
glucagon, 51, 74
glucose metabolism, 49, 51–52. See also insulin
glucose monitoring, 73–74
Goeckeler, George, 32
gonorrhea, 116
government, role of in research, 305
Gram, Hans Christian, 78
Gram stain, 78
gram-negative bacteria, 78, 121
gram-positive bacteria, 78, 121
Grant, Ulysses S., 214
Greenlees, James, 222–223
Gregg, Alan, 113–114
Groote Schuur Hospital, 39
Gross, Samuel, 225
Grubbé, Émil, 174

Grüntzig, Andreas, 21–22
Grüntzig, Michaela, 22
Guinea Pig Club, 242–243, 243
Guiteau, Charles, 215, 230
gunshot wounds, 216, 231, 253. See also Garfield, James A.; Reagan, Ronald
Gustafson, Einar, 169–170, 183–185, 186, 211–212. See also Jimmy Fund
gynecology. See also obstetrics
  father of gynecology, 276. See also Sims, J. Marion
  fistulas, 273–276
  Mothers of Gynecology (Betsey, Anarcha, and Lucy), 275–276
  speculum, invention of, 274

H

H1N1 virus, 166
H5N1 avian influenza virus, 167
H7N7 virus, 167
Hahn, William, 202
Hahnemann Hospital, 32
Halsted, William, 172–173, 230
Hamilton, Frank, 225, 227
Hancock, Winfield, 214
Harari, Yuval Noah, 2
Harding, Warren, 133
Hare, Ronald, 103
Harken, Dwight, 29–32, 34
Hata, Sahachiro, 99
Hayes, Rutherford B., 214
He, Jiankui, 296–297
head and neck cancer, 201, 204–205
healthcare infrastructure, 168
heart
  anatomy, 11–12, 12, 13
  importance of in human psyche, 45
  laceration of, 27–30
  metabolism, 13, 25, 35
  and oxygenation of blood, 11–13
  pulmonary artery, 11, 12, 37, 40

  right atrium, 11–12, 14, 16, 40
  right ventricle, 11, 28, 30, 36, 40
  role of, 11
  septal defects, 35–36, 38
heart attack. See myocardial infarction
heart disease
  advances in treatment for, 43
  atherosclerotic plaque, 19–20
  and balloon angioplasty, 22–25. See also balloon angioplasty
  and cardiac catheterization. See cardiac catheterization
  as cause of death, 11
  Cheney, Richard, 6–10
  congenital heart defects, 35–36, 38
  and coronary angiography, 17–19
  early 1900s, 10–11
  failed treatments, 39
  future of treatments for, 43–44
  heart transplants, 9–10, 39–43
  mitral stenosis, 30–34
  mortality rates, 11, 20, 43
  open heart surgery, 25–29, 39
  and oxygenation of blood during surgery, 34–39
  prevention, 43
  and regenerative medicine, 43–44
  stem cell research, 43–44
  ventricular septal defect, 35–36
heart failure, 9
heart laceration, 27–30
heart transplants, 9–10, 39–43
heart-lung bypass machine, 38–39, 42, 44, 299. See also coronary artery bypass surgery (CABG)
heart-lung transplants, 43

Index

Heatley, Norman, 105–106, 108–116, 118–120, 125
Helicobacter pylori, 198–199
helper T cells, 141
Henry VIII, 264
heparin, 38. *See also* blood thinners
hepatitis B, 198
hepatitis C, 198
HER2, 201
Herbst, Arthur, 285
Herceptin (trastuzumab), 201
Herrick, James, 13
high cholesterol. *See* cholesterol
Hinckley, John, 248, 250
Hitchings, George, 187
HIV, 166, 296–297
Hodgkin, Dorothy Crowfoot, 73, 121
Hodgkin's lymphoma, 191, 212
Holter monitor, 8
hormones
  and cancer, 193, 196, 202
  and embryological development, 285
  endocrine secretion, 51
  estrogen. *See* estrogen
  estrogen, synthetic (DES), 284–286
  glucagon, 51
  incretins, 73
  insulin. *See* insulin
  and mRNA, 165–166
  somatostatin, 51
  testosterone, 193
Huanan Seafood Market, 160
Huggins, Charles, 193, 198
Hughes, Antoinette, 46–47, 66–67
Hughes, Charles Evan, 46–48, 66
Hughes, Elizabeth, 46–49, 66–67, 71
Human Epidermal Growth Factor Receptor 2, 201
human genome, 2, 171, 199–201, 210–211, 293–298
Human Genome Project, 171
human papillomavirus (HPV), 196, 198
Humulin, 73

Hunt, Mary ("Moldy Mary"), 114
Huntington's disease, 291, 294, 297
hyperglycemia, 49
hypertension and heart disease, 19

I

ICD (implantable cardioverter-defibrillator), 8–9, 44
IL-2, 208
imatinib (Gleevec), 201
immortality, 3
immune system, 140–141, 208–209. *See also* immunotherapy
immunology, 140. *See also* viruses
immunosuppressive drugs, 40, 247
immunotherapy, 207–211, 213
Imperial Chemical Industries, 115
implantable cardioverter-defibrillator (ICD), 8–9, 44
in vitro fertilization (IVF), 290–292
infectious diseases. *See also* specific diseases
  anti-contagion theories, 78–79
  bacteria as cause of. *See* bacteria
  and cancer, 170–171, 197–198
  contagion theories, 78–80
  examples of, 78–79
  identifying cause of versus treatment, 97
  and life span in U.S., 11
  mortality rates, 77, 100, 167
  prevention measures, 167
  response to, recommendations for, 167–168
  viruses as cause of. *See* viruses
infectious mononucleosis (mono), 198

infertility, 286, 290–292. *See also* IVF (*in vitro* fertilization)
influenza, 46, 122, 142–144, 161, 167
Inova Fairfax Hospital, 10
insulin
  discovery of, 52–61
  Eli Lilly & Company production of, 65
  Genentech production of synthetic insulin, 73
  and glucose metabolism, 49, 51–52
  production of in pancreas, 49, 51, 61, 66, 73. *See also* pancreas
  pumps, 73–74
  refining for use by diabetics, 61–66
  resistance, 72
  and rivalries among researchers, 306
  role of, 49
  and stem cell research, 74
  synthetic, 73
  and type 2 diabetics, 72
interferon alpha, 194
intracytoplasmic sperm injection, 290
intratumoral heterogeneity, 202
invisible bullet, 78. *See also* bacteria
ipilimumab (Yervoy), 210
Iraq, 252
iron lung treatment, 127–128, *128*, 135, 142
ischemia, 8, 20, 22, 244, 273
islets of Langerhans, 49, 51, 53, 59, 74
Ivanovsky, Dmitri, 139
IVF (*in vitro* fertilization), 290–292

J

Jackson, Charles T., 219, 306
James I, 263
James VI, 278
Jefferson, Thomas, 138, 276
Jefferson Medical College, 38

Jeffery, Ed, 64
Jenner, Edward, 137–138
Jensen, Elwood, 193
Jimmy. *See* Gustafson, Einar
Jimmy Fund, 183–186, 211–212
Johnson & Johnson, 305
Joslin, Elliott, 62, 67
Judkins, Melvin, 20
June, Carl, 211

K
Karikó, Katalin, 163–166, 304, 306
Kelsey, Frances, 286–288, 289, 305–306
Kemball, Bishop, and Co., 115
Kennedy, John F., *289*
ketoacidosis, 72
Keytruda (pembrolizumab), 210
kidney cancer, 194
kidneys, 49, 73, 188
killer T cells, 141
Kitasato, Shibasaburo, 98
Klebsiella pneumoniae, 123
Koch, Robert, 82–93, 96–97, 97, 125, 159, 205, 306
Koch's postulates, 88–89
Kollock, Phineas Miller, 273
Kymriah (tisagenlecleucel), 211

L
La Touche, C. J., 102
LAD (left anterior descending artery), 6, 8, 13, 23
Landsteiner, Karl, 131, 142
Langerhans, Paul, 51
Larrey, Dominique Jean, 251
Lax, Eric, 120
Leavitt, Judith, 282
Lederle, 115
left anterior descending artery (LAD), 6, 8, 13, 23
left atrium, 12, *12*, 30–32, 40
left circumflex artery, 13
left ventricle, 9, 12, *12*, 30–31, 36, 40
left ventricular assist device (LVAD), 9–10

Leopold (prince), 255–256, 265, 280
leprosy, 78, 288
leukemia
  acute lymphoblastic, 211
  CAR-T treatment, 211
  chemotherapeutic drugs, 187–191
  chemotherapy for, 179–181, 187, 189–191
  chronic myelogenous, 201
  curability, 191, 212
  folic acid and anti-folate treatments, 179–181, 300–301
  fundraising for research, 182–186
  molecular level drugs, 201
  mortality rates, 178, 191
  radiation as cause of, 284
  survival rates, 191
  symptoms of, 179
  treatment of at Children's Hospital in Boston, 170, 178–181. *See also* Farber, Sidney
  variability in forms of, 171
  and white blood cells, 179
Lewis, Paul, 131
Li, Wenliang, 159–161
life expectancy, 2, 5, 11, 47. *See also* Methuselah gene; mortality rates
Lillehei, Walt, 35–36
lipase, 51
Lister, Joseph, 219–225, *226*, 230, 280
Liston, Robert, 218–219
liver cancer, 198
Loeffler, Friedrich, 88
Long, Crawford Williamson, 218–219, 306
Loudon, Irvine, 266
Lovett, Robert, 133
Lower, Richard, 40
lung cancer, 191, 193–194, 201, 210
lungs, 11–12
LVAD (left ventricular assist device), 9–10
lymphocytes (B cells and T cells), 141

lymphoma, 170, 178, 183, 197–198, 211. *See also* Hodgkin's lymphoma
lysozyme, 100–101, 103

M
MacAlyane, Eufame, 278
Macleod, John, 54, 56–58, *58*, 59–60, 62–63, 65, 69–70, 306
macrophages, 141, 211
Maitland, Charles, 136
malaria, 79
mammography, 201
March of Dimes, 135, 182
Marmite, 179
Marshall, Barry, 198–199
Massachusetts General Hospital, 225
mast cells, 141
mastectomy, 173
maternal mortality rates, 258, 268–270, 272, 289
Mauriceau, François, 263
Max Planck Institute for Infection Biology, 304
maximum tolerated dose (MTD), 188
Mayer, Adolf, 139
McCarthy, Tim, 248, 250
McIndoe, Archibald, 239–244, 248
medical advances, 1–5, 283, 299–304, 306–308. *See also* technological advances
Medtronic, 74
medulla oblongata, 131
melanoma, 208, 210
Mellanby, Edward, 119
meningitis, 116
Merck, 115
mesmerism, 219
messenger RNA (mRNA), 139, 163–166, 304, 307
metabolism, 13, 25, 35, 47, 49–54, 72, 123
metal nanoparticles, 124
metformin, 72
methicillin, 123
methotrexate (amethopterin), 186, 188–189, 191

Index

Methuselah gene (cholesteryl ester transfer protein gene), 307
miasma ("bad air"), 2, 79-80, 220-221, 266, 268
microbiology, 81, 84-85
microscopes
  electron, 132, 198
  operating microscope, 246
  optical, 132, 139
  van Leeuwenhoek's construction of, 77
midwives, 255, 260-261, *261*, 263-265, 280-283
Minkowski, Oskar, 51
mitochondria, 12-13
mitral insufficiency, 32-33
mitral stenosis, 30-34
mitral valve, 12, *12*, 30-32
mobile army surgical hospital (MASH), 252
Moderna, 165-166, 305
mold, 101-103. *See also* penicillin
"mold juice," 103, 105-106. *See also* penicillin
monoclonal antibodies, 194, 201
monocytes, 141
Montagu, Mary Wortley, 136-137
Morestin, Hippolyte, 232
mortality rates
  and antiseptics, 225
  cancer, 170-171, 178, 191-192, 212
  coronary artery disease, 43
  fetal, 258, 289
  heart disease, 11, 20, 43
  infectious disease, 77, 100, 167
  leukemia, 178, 191
  maternal, 258, 268-270, 272, 281-282, 289
  myocardial infarction, 20
  poliomyelitis, 130, 135
  smallpox, 136
Morton, William, 218-219, 277, 306
Mothers of Gynecology (Betsey, Anarcha, and Lucy), 275-276

Moyer, Andrew, 115
multidrug resistance
  antibiotics, 121-124
  chemotherapeutic drugs, 191
multiple myeloma, 201, 288
Murray, Joseph, 287
Murrow, Edward R., 154
Mustard, William, 35
mustard gas, 176-178
Mustargen (nitrogen mustard), 178
myc, 200
Mycobacterium tuberculosis, 88-90
Myler, Richard, 23
myocardial infarction, 6-8, 10, 13, 17, 19-20, 24-25. *See also* coronary arteries

N
Napoleon, 138, 251
National Cancer Act, 191
National Cancer Institute, 188, 190, 196, 207
National Foundation for Infantile Paralysis, 135, 146, 149-151, 153, 155, 182, 186
National Institute of Health (NIH), 155, 295, 303-304
nationalism, 85
Nelms, Sarah, 137
neonatal medicine, 289
nephropathy, 74
neu, 200
neutrophils, 141
Nightingale, Florence, 80
NIH (National Institute of Health), 155, 295, 303-304
nitrous oxide, 218-219
nivolumab (Opdivo), 210
Nixon, Richard, 172, 191
Nobel Prize recipients
  Baltimore, David, 199
  Banting, Frederick, 69
  Bishop, Michael, 200
  Carrel, Alexis, 246
  Charpentier, Emmanuelle, 293, 304
  Doudna, Jennifer, 293
  Dulbecco, Renato, 199

Edwards, Robert, 290
Ehrlich, Paul, 99
Enders, John, 145
Fleming, Alexander, 120
Florey, Howard, 120
Forssmann, Werner, 17
Huggins, Charles, 198
Koch, Robert, 96
Landsteiner, Karl, 131
Macleod, John, 69
Marshall, Barry, 198
Robbins, Frederick, 145
Rous, Peyton, 198
Temin, Howard, 199
Varmus, Harold, 200
Waksman, Selman, 121
Warren, Robin, 198
Watson, James, 192
Weller, Thomas, 145
Noble, Clark, 187
Noble, Robert, 187

O
obesity, 50, 72, 74, 308
obstetrics, 280-284, 289-291. *See also* childbirth
O'Connor, Basil, 135, 146-148, 151-152
oncogenes, 199-202
oncologists, 171, 179
Opdivo (nivolumab), 210
open heart surgery, 25-29, 39
organ donation, 43, 74, 247. *See also* transplants
organ transplants. *See* transplants
Osler, William, 1, 299
ovarian cancer, 200-201
Oxford University, 77, 104-105, 109, 112-113, 115-121
oxygenation of blood
  and bloodletting, 281
  coronary bypass grafts, 26
  heart, role of, 11-13
  during heart surgery, 34-39
  heart-lung machine, 25. *See also* heart-lung bypass machine
oxytocin, 283

P

p53, 200
pacemaker
   implantable cardioverter-defibrillator (ICD), 8
   sinoatrial node, 12
   wireless, 44
paclitaxel (Taxol), 192
Paget, Stephen, 27
pancreas
   acinar cells, 51, 53–54, 59–60
   appearance of, 50
   digestive enzymes, 51–54, 59–60
   endocrine secretion of hormones, 51
   functions of, 51
   GLP-1 receptors, 73
   and glucose metabolism, 49, 51–52
   insulin production, 49, 51, 61, 66, 73. *See also* insulin
   islets of Langerhans, 49, 51, 53, 59
   and pathophysiology of diabetes, 51–52
   stones (pancreatic lithiasis), 53
   transplants, 74
pancreatic cancer, 201–202
pandemics
   Covid-19. *See* Covid-19
   frequency of, 167
   and international cooperation, need for, 306–307
   polio. *See* poliomyelitis
   Spanish flu, 46, 161
Pap smears, 195–196
Papanicolaou, George, 195–196, 299
Parke, Davis & Company, 205
Parr, Catherine, 264
Parr, Jerry, 249
Pasteur, Louis, 80–82, 85–87, 90–95, 95, 96, 125, 159, 221, 306
Pasteurella multocida, 85
pasteurization, 81
PD-1, 210

pembrolizumab (Keytruda), 210
Pendleton Civil Service Reform Act of 1883, 230
penicillin
   Chain, Ernst. *See* Chain, Ernst
   chemical composition, 121
   discovery of, 101–104
   diseases treated with, 116
   first human patients to be treated with, 109–110
   Fleming, Alexander. *See* Fleming, Alexander
   Florey, Howard. *See* Florey, Howard
   Heatley, Norman. *See* Heatley, Norman
   initial lack of interest in for therapeutic use, 103–104, 107–108
   "mold juice," 103, 105–106
   patent for, 112–113, 115
   penicillium mold, *101*
   problems with manufacturing, 110
   production of in U.S., 114
   resistance to, 121–123
   source of mold in Fleming's lab, theories on, 102
   synthetic, 121
   undertreatment, effect of, 121
   and World War II, impact of, 106, 108–109, 111, 115–116
penicillinase (beta-lactamase), 121
Penicillium chrysogenum, 114–115
Penicillium notatum, 102–103, 111, 114
Penicillium rubrum, 102
pericardial tamponade, 28
Pfizer, 115, 166, 305
phage therapy, 124, 293
phagocytes, 141
pharmaceutical companies
   diethylstilbestrol (DES), promotion of, 284–286

   new antibiotics, development of, 123–124
   penicillin, lack of interest in, 108, 110
   penicillin, production of penicillin, 115
Philadelphia General Hospital, 33
Philip, Prince (Duke of Edinburgh), 243
Phipps, James, 137–138
phocomelia (seal limb), 288. *See also* thalidomide
Physiatric Institute, 47–48, 62
Pickerill, Henry, 234
pitchblende, 175
plastic surgery, 234, 236, 239, 244, 246–248
platelets, 20, 178–179
pneumonia, 116
poliomyelitis. *See also* Sabin, Albert; Salk, Jonas
   bacteria ruled out as cause of, 131
   bulbar polio, 130–131
   emergence of in 1900s, improved sanitation as reason for, 129–130
   epidemics, 132, 142
   fear of, 128–129, 132, 135, 142, 159
   funding for research on, 135
   iron lung treatment, 127–128, *128*, 135, 142
   mortality rate, 130, 135
   natural immunity developed prior to 1900s, 130
   origin of term, 130
   paralysis from, 126–127, 129–132, 147, 155–156, 158–159
   prevalence of during 1900s, 142
   prevention of, early measures for, 129
   Roosevelt, Franklin Delano, 132–135
   Schwartz, Arvid, 126–128
   in Soviet Union, 157

Index

spinal cord involvement, 130–131, *131*
symptoms of, 126, 129–130
vaccine for, development of. *See* Sabin, Albert; Salk, Jonas
virus as cause of, 131–132
Polk State School for the Retarded and Feeble-Minded, 148
polonium, 175
polygenic risk score, 292–293
Pomahač, Bohdan, 246–247
Popper, Erwin, 131
postdoctoral scientists and researchers, 303–305
Pott, Percivall, 195
prednisone, 189–191
pregnancy. *See also* childbirth
abortion, 260, 281, 283, 291
amniocentesis, 291
birth control, 258, 283
diethylstilbestrol (DES), 284–286
early theories of conception and, 260
fetal genetic screening, 291–292
infertility, 286, 290–292
number of pregnancies, 258
preimplantation genetic diagnosis (PGD), 291–294
and thalidomide, 286–288
X-rays during, 284
preimplantation genetic diagnosis (PGD), 291–294
prontosil, 104
prostate cancer, 193, 212
proteins. *See also* specific proteins
and bacteria, 121
and cancer treatments, 187, 194, 210–211
Cas9 and CRISPR/Cas9 gene modification, 293. *See also* CRISPR/Cas9
CD28, 209

cholesteryl ester transfer protein gene (Methuselah gene), 307
CTLA-4, 209
and function of cells, 164
in human diet, 49
Human Epidermal Growth Factor Receptor 2, 201
and immune system, 140–141
and immunotherapy, 207, 209–210
and mRNA, 164–166
and oncogenes, 199–201
PD-1, 210
and stem cells used to grow artificial organs, 252–253
T cell receptors, 209
vascular endothelial growth factor (VEGF), 194, 209
and viruses, 139–141, 166, 199
proto-oncogenes, 200
Pryce, Merlin, 101–102
pseudouridine, 165
pulmonary artery, 11, *12*, 37, 40
pulmonary embolism, 37, 264, 281

Q
Queen Victoria Hospital (QVH), 239–241, 243
Queen's Hospital, 235

R
rabies vaccine, 93–95, 306
radiation, 173–175, 200, 210, 284
radium, 175, 195, 299
ras, 200
RCA (right coronary artery), 6–7, 13
Reagan, Nancy, 249
Reagan, Ronald, 248–251
regenerative medicine, 2, 43–44, 253, 307
Rehn, Ludwig, 27–30
retinitis pigmentosa, 294
retinoblastoma, 200

retinopathy, 74
reverse transcriptase, 199
Reyburn, Robert, 215–216
rheumatic fever, 17, 31, 33, 116. *See also* mitral stenosis
Rhoads, Cornelius, 178
Richards, Alfred Newton, 115
Richardson-Merrell, 287
right atrium, 11–12, *12*, 14, 16, 40
right coronary artery (RCA), 6–7, 13
right ventricle, 11, *12*, 28, 30, 36, 40
RNA
messenger RNA (mRNA), 139, 163–166, 304, 307
reverse transcriptase, 199
and viruses, 139, 199
Robbins, Frederick, 145
Rockefeller, John D., Jr., 203
Rockefeller Foundation, 111, 114
Rockefeller Institute, 140, 197, 203
Romeis, Peter, 16
Röntgen, Wilhelm, 173–174, 228
Roosevelt, Franklin Delano, 132–135
Rosenberg, Steven, 207–208
Ross, Allan, 7
Rous, Peyton, 196–198
Rous sarcoma virus (RSV), 197, 199
Roux, Émile, 92
Ryder, Teddy, 68

S
Sabin, Albert, 145–147, 149, 151, 155–158, *158*, 159, 306
Salk, Donna, 153
Salk, Jonas, 145–154, *154*, 155–159, 163, 306
Salk, Peter, 153, 158
Salk Institute for Biological Sciences, 159
salvarsan, 99
Sandler, Robert, 180

Index

sanitation, 129-130, 167. *See also* sewage systems
sarcomas, 203-205
SARS-CoV-2. *See* Covid-19
Schatz, Albert, 121
Schlumpf, Maria, 22
Schmidt, Pamela, 36
Schneider, Richard, 14
Schwartz, Arvid, 126-128
scientific knowledge, advances in, 1-3, 302-303. *See also* medical advances
screening tests
 for cancer, 195-196, 201, 211, 213, 285
 prenatal, 291
scrotal cancer, 195
seal limb (phocomelia), 288. *See also* thalidomide
Second Pearl Harbor, 175-177
Semmelweis, Ignaz, 266-271, 271, 272, 277
Semmelweis reflex, 272
septal defects, 35-36, 38, 44
serotonin, 20
sewage systems, 79-80, 130, 222, 270
sexually transmitted diseases, 116, 196. *See also* specific diseases
Seymour, Jane, 264
SGLT-2 inhibitors, 73
Shaw, Laura, 20-21
Shepard, Alan, 3
Shumway, Norman, 39-43
sickle cell disease, 294
Sidney Farber Cancer Center, 186
Simpson, James Young, 220, 277-279
Sims, J. Marion, 272-276
sinoatrial node, 12, 27
6-mercaptopurine, 187-191
skin flaps, 233-234, 236, 241
Slamon, Dennis, 201
slaves, experimentation on, 272, 275-276
smallpox, 79, 135-138, 162-163, 276
Smellie, William, 263
Smith, Hugh, 140
smoking, 6, 19, 43, 195, 212
Snow, John, 79-80, 279-280

somatic cells, 294, 297
somatostatin, 51
SONAR (SOund Navigation And Ranging), 283
Sones, Mason, 17
Soviet Union, 157
Spanish flu, 46
speculum, invention of, 274
spinal cord, 130-131
spontaneous generation, 81
Squibb, 115
src, 199-200
St. Mary's Hospital, 100, 117, 119
staphylococcus, 101-103, 121, 123
Staphylococcus aureus, 121, 123
Starr, Harold Morley, 237
Stein, Fred, 203-204
stem cell research, 1, 43-44, 74, 252, 307
STEM education, 303
stenosis, 19, 21, 23, 30-34
stents, in coronary angioplasty, 8, 25
Steptoe, Patrick, 290
Stockton, Walter, 32
stomach cancer, 198-199, 207
Streptococcus pyogenes, 203-204
streptomycin, 121, 187
Subbarao, Yellapragada, 180
sulfa drugs, 76, 104, 108-109, 111, 116, 240, 272
sulfapyridine, 76
sulfonamides. *See* sulfa drugs
sulfonylureas, 72
Sun Yat-sen University, 296
superior vena cava, 12
surgeries. *See also* traumatic injuries
 amputations, 217-220, 222-223, 225, 251
 anesthesia. *See* anesthesia
 antiseptics, use of. *See* antiseptics
 by barbers, 172, 217
 cancer treatment, 172-173, 175, 178, 203-204
 cesarean section (C-section), 273, 281, 290

 coronary artery bypass, 7-8, 10, 23-24, 26, 26, 27, 38-39
 fetal, 290
 fistulas, 275-276
 gunshot wounds, 216, 231, 253. *See also* Garfield, James A.; Reagan, Ronald
 heart lacerations, 27-30
 left ventricular assist device (LVAD) implantation, 9
 mitral stenosis, 30-34
 open heart, 25-29, 39
 operating theaters in mid-1800s, 217, 220
 oxygenation of blood during, 34-39
 transplants. *See* transplants
Syme, James, 220
syphilis, 78, 99, 116
systole, 28, 31

T
T cells, 141, 208-210
tamoxifen, 193, 201
Tandem, 74
Tarceva (erlotinib), 201
Taxol (paclitaxel), 192
Taylor, Ian, 244-245
Taylor, Linda, 208
technological advances, 3, 84, 171, 251-253, 283, 291-293. *See also* CRISPR/Cas9; medical advances
telomerase, 307
telomeres, 307
Temin, Howard, 199
testicular cancer, 191, 212
tetanus, 116
thalidomide, 286-288
Theiler, Max, 140
Thompson, Leonard, 63-65
thrombolytic agents, 9, 20, 38
thromboxane, 20
Thuillier, Louis, 90-91, 93
tisagenlecleucel (Kymriah), 211
tissue engineering, 253

Index

TNK-tPA (tissue plasminogen activator), 20
tobacco, 195. *See also* smoking
Tonks, Henry, 235
Townsend, Smith, 215
transplants
   artificial organs, 252–253
   full face transplant, 246–248
   heart transplants, 9–10, 39–43
   heart-lung transplants, 43
   islet cells, 74
   pancreas, 74
   rejection, 40, 43, 247
   and stem cells, 307
trastuzumab (Herceptin), 201
traumatic injuries
   advances in treatment of, 252–253
   amputations, 217–220, 222–223, 225, 251
   anesthesia. *See* anesthesia
   antiseptics. *See* antiseptics
   blood vessels, method for connecting, 244–248
   burns, 238–243
   compound fractures, 222–223
   damage control surgery, 252
   erysipelas infections, 203–205, 220, 223
   ethics, 251
   face transplant, 246–248
   facial injuries, 231–235, 235, 236, 239–243, 246–248
   Garfield, James A., 214–217, 225–230, 248, 250
   gunshot wounds, 216, 231, 253. *See also* Garfield, James A.; Reagan, Ronald
   hospitals in mid-1800s, reputation of, 220
   immediate treatment of, 251
   infections, 203–205, 220, 223, 227–229
   operating theaters in mid-1800s, 217
   plastic surgery. *See* plastic surgery
   psychological damage, 242
   pus, 221, 227–229
   surgical practices in 1800s, 217–218
   technology, advances in, 252–253
   transport of victims, 251–252
   triage, 231, 251
   unsterile conditions in mid-1800s, 217–218, 220, 224
   and war, 231–244, 251–252. *See also* World War I; World War II
Treponema pallidum, 99
triage, 231, 251
triangulation technique, 245–246
Trisomy 13, 291
Trisomy 18, 292
Trisomy 21, 291
Troan, John, 149–150, 152
Truth or Consequences (radio program), 183–185, 212
trypsin, 51
tubed pedicle flap, 236, 240–241, 243, 243
tuberculin, 96–97
tuberculosis, 78, 87–89, 93
tumor suppressor genes (anti-oncogenes), 200
Tunney, Gene, 249
type 1 diabetes. *See* diabetes mellitus
type 2 diabetes. *See* diabetes mellitus
typhoid, 129

U

Ullrich, Axel, 201
ultrasound, fetal, 283
United Network for Organ Sharing (UNOS), 43
University of Minnesota Hospital, 35–36, 126
University of Toronto, 54, 57, 61, 65, 70
uranium, 174–175
urine, and diabetes, 47, 49–50
urokinase receptors, 164

V

vaccines
   AIDS vaccine research, 159
   anthrax, 86–87, 90–91
   antigenicity, 141
   attenuated, 86–87, 141, 145
   for cancer, 307
   Covid-19, 162, 165–166, 305, 307
   influenza, 142–144, 163
   killed virus, 141, 145
   and mRNA. *See* messenger RNA (mRNA)
   polio. *See* Sabin, Albert; Salk, Jonas
   rabies, 93–95, 306
   reliance on living biological organisms, 163
   worldwide rates of, improving, 167
vaginal cancer, 285
Valadier, Charles, 232
VAMP (vincristine, amethopterin, 6-mercaptopurine, and prednisone), 189–191
van Leeuwenhoek, Antonie, 77
van Roonhuysen, Roger, 263
vancomycin, 123
varicella (chicken pox), 145
Variety Club of New England, 182
variola, 136. *See also* smallpox
Varmus, Harold, 200
vascular endothelial growth factor (VEGF), 194, 209
VEGF (vascular endothelial growth factor), 194, 209
Velcade (bortezomib), 201
venae cavae, 12, 25, 40
ventricular fibrillation, 8–9, 20, 31

ventricular septal defect, 35–36, 44
ventricular tachycardia, 8
Vibrio cholerae, 93
Vicarage, Willie, 235–236
Victoria (queen), 224–225, 265–266, 280
Victoria, Princess of Saxe-Coburg-Saalfeld, 265
Vietnam War, 252
vincristine, 189–191
viral infections. *See also* viruses
    Covid-19. *See* Covid-19
    poliomyelitis. *See* poliomyelitis
    smallpox. *See* smallpox
viruses. *See also* specific viruses
    antigens, 141
    as cause of cancer, 196–198, 202
    as cause of polio, 131–132. *See also* poliomyelitis
    derivation of term, 138
    described, 138–139
    DNA, 139
    and electron microscope, 132, 139
    goal of, 139
    human papillomavirus (HPV), 196, 198
    and immune system. *See* immune system
    messenger RNA, 139
    and proto-oncogenes, 200
    and RNA, 139, 199
    size of, 139
    tobacco mosaic disease discoveries, 139
    and vaccines, 140

von Behring, Emil, 97–98
von Mering, Josef, 51
von Pettenkofer, Max, 80

W

Waksman, Selman, 121, 187
Walter Reed Army Medical Center, 252
waltzing, 241
war on cancer, 172, 191–192, 212, 302–303
Warm Springs, Georgia, 134–135
Warner, Constance, 33–34
Warren, John Collins, 219
Warren, Robin, 198
Washkansky, Louis, 41–42
Watson, James, 192
Watson, Thomas, 38
weather, role of in growing penicillium mold, 102–103
Weaver, Harry, 149–150
Weaver, Warren, 111
weight loss, 43, 46–47, 49, 74
Weinberg, Robert, 202
Weissman, Drew, 165–166
Weizmann Institute of Science, 158
Weller, Thomas, 145
Wells, Horace, 219, 306
West Nile virus, 297
Western Reserve Medical School, 54
white death, 87. *See also* tuberculosis
Whitehead, Emily, 211
widow-maker. *See* left anterior descending artery (LAD)
Wiens, Dallas, 246–248
Willis, Thomas, 50
Wills, Lucy, 17

Wilmington Memorial Hospital, 33
Wilms' tumor (kidney cancer), 188
Winchell, Walter, 152
Woman's Hospital, 275
women scientists, 305–306. *See also* individual women
World War I, 178, 231–236, 283
World War II, 106, 108, 111, 115–116, 175–178, 237–244
Wright, Almoth, 117
Wuhan Central Hospital, 159–160

X

X-rays, 172–174, 284

Y

yellow fever, 79, 140
Yervoy (ipilimumab), 210
Yescarta (axicabtagene ciloleucel), 211

Z

Zika virus, 165–166
Zola, Signor, 205

About the Author

Andrew Lam, M.D., is a practicing retina surgeon in Western Massachusetts and an assistant professor of ophthalmology at the University of Massachusetts Medical School. Before becoming a physician, he earned a history degree from Yale University. He is the author of three award-winning books: *Saving Sight*, a memoir of his surgical training and history of ophthalmology's greatest innovations, and *Two Sons of China* and *Repentance*—both novels of the Second World War. His writing has appeared in numerous publications, including the *New York Times* and the *Washington Post*, and he has served as a contributing commentator for PBS NewsHour, New England Public Radio, and many other media outlets. He resides in Longmeadow, Massachusetts, with his wife and four children. Learn more at www.AndrewLamMD.com.